THE CHICAGO PUBLIC LIBRARY

FORM 19

Polymer–Solid Interfaces

Organizing Committee

J J Pireaux
P Bertrand
J L Brédas

International Advisory Committee

J E Castle (Guildford)
A Cros (Marseille)
M Grunze (Heidelberg)
J Israelachvili (Santa Barbara)
S P Kowalczyk (Yorktown Heights)
G Marletta (Catania)
E Ochiello (Novara)
P A Pincus (Santa Barbara)
J A Rabe (Mainz)
N V Richardson (Liverpool)
E Sacher (Montreal)
W R Salaneck (Linköping)

Sponsored by:

European Economic Community (Value Programme, DG XIII)
Fonds National de la Recherche Scientifique
Loterie Nationale
Communauté Française de Belgique
BASF Aktiengesellschaft
Colgate-Palmolive R & D
Du Pont de Nemours
Fina Research
IBM Belgium
Institut Français du Pétrole
Laborlux
Perkin-Elmer
Sabena World Airlines
Shell Research CRCS
Thomson CSF

Polymer–Solid Interfaces

Proceedings of the First International Conference,
Namur, Belgium, 2–6 September 1991

Edited by J J Pireaux, P Bertrand and J L Brédas

Institute of Physics Publishing
Bristol and Philadelphia

British Library Cataloguing-in-Publication Data. A catalogue record for this book is available from the British Library.

ISBN 0-7503-0192-9

Library of Congress Cataloging-in-Publication Data are available

Published by IOP Publishing Ltd, a company wholly owned by the Institute of Physics, London

IOP Publishing Ltd
Techno House, Redcliffe Way, Bristol BS1 6NX, UK

US Editorial Office: IOP Publishing Inc., The Public Ledger Building, Suite 1035, Independence Square, Philadelphia, PA 19106

Printed in Great Britain by Galliard (Printers) Ltd, Great Yarmouth, Norfolk

Contents

v

Section II: Metals on Polymers

Section III: Polymers: Bulk/Surface/Interface Properties

Preface

For a long time (polyethylene was discovered at ICI in 1933, polyethylene terephthalate at du Pont de Nemours in the mid-1970's) polymers have simply been used as new raw materials, the exploitation of their bulk properties leading to cheap substitutes for other compounds. Recently, high-technology and much more expensive polymers have been developed, tailored for specific applications. The production volume of some of these polymers is small, but is counterbalanced by a very high added value. Nowadays applications are often based on precisely designed thermal, mechanical, chemical, dielectric, or interfacial properties that are obtained by incorporating the polymer in hybrid structures. These are multiphase or multicompound materials including a polymer with a metal, semiconductor, ceramic, glass or other polymer(s).

Applications of polymer–solid interfaces are now legion in our everyday lives. Consumers benefit widely from new food packaging (eg. milk or juice tetra brick), decorative tissues, lacquers and paints (in the home as well as, for example, the automotive industry), new telecom components based on microelectronic circuits, recreative and business devices using magnetic (audio and video) tapes, floppy and hard discs, etc..

A deeper knowledge of polymer–solid interface characteristics would allow the production of still better suited materials. For that purpose, a multidisciplinary approach is mandatory. It is indeed necessary to gather and correlate experimental results from diverse applied horizons (some of them are cited above), theoretical modelling of interfaces involving a polymer, and fundamental studies of simplified model materials.

We therefore took the opportunity to organise an International Conference on Polymer–Solid Interfaces, to mix scientists from industry and academia, presenting and discussing experimental and theoretical results. Our goal was to trigger a cross-pollination of ideas and allow the participants to sort out general trends. The conference topics (highlighted by ten invited lectures) were deliberately broadly based as opposed to a strictly dedicated meeting. A comparison of knowledge obtained from different polymer–solid interfaces was expected to result in a better understanding of the physics and chemistry at interfaces, including adhesion and ageing aspects.

For the first time a new type of contribution was solicited, within the context of a scientific contest of review papers. On the basis of extended abstracts, the International Scientific Advisory and Organizing Committees selected among twelve propositions, three self-nominated reviews. Since these three presentations turned out to be of a very high quality their full-length texts introduce sections I, II and IV of these conference proceedings.

From some one hundred and twenty participants, sixty scientific communications were presented, twenty orally and forty as posters. The majority of these presentations are included in this volume.

As expected,these scientific contributions are mostly oriented towards the understanding of the complex physico–chemical processes occurring at polymer–solid interfaces. Indeed, many questions concerning, for example, the determination of the reaction site, types of bonds, their spatial orientation and their relation with adhesion are still to be resolved.

Along these lines, different systems are considered, evolving from model interfaces to real systems of technological importance. The model interfaces are not only an idealization of the real ones but can also lead to the design of new materials with unexpected properties. This opens the way for the development of smart materials.

In these proceedings two types of model systems will be considered: Polymer deposited on different solids (Section I) and metal deposited on polymers (Section II). They start from very idealized systems to more complex technological questions.

For 'polymers deposited on solids', systems of increasing complexity are presented. Firstly, small organic molecules adsorbed on well defined monocrystalline surface planes, then the deposition of macromolecules at a covering lower than one monolayer. Here, entropic restrictions lead to highly oriented chains exhibiting lamella and domain structure; their strong interaction with the substrate may be visualized by STM images. As for organic 'monolayers', a few contributions are devoted to Langmuir–Blodgett oriented films. These are also of practical interest, for example as lubricant layers improving wear properties of magnetic recording media. Organic 'multilayers' can also be produced. Based on these 'molecularly engineered' materials new applications are developed: wave guides, sensors, storage media, rectifiers, nanoscale composites, etc.. Finally, the role of 'polymer thin films' is very important in microelectronics to solve the problems of packaging and planarization of the devices, and also in other fields, for example safety glass.

The section 'metals on polymers' is also organized according to an increasing complexity scale. First, studies of the adsorption of metal atoms on polymer surfaces are presented. Their interaction with the macromolecular chains gives rise to a new chemistry as one has to consider the formation of organo–metallic complexes with significant orientation and charge transfer effects. In this field the major contribution of the theoretical approaches has to be recognized. Then, thin metal layers deposited on polymers are studied. Examples will show that the nature of the metal and the substrate surface functionalities govern the interface formation. In the case of a weak interaction, the metal atoms can diffuse into the polymer and metal oxyde clusters can be produced. Here, the interfacial physicochemistry is mainly related to the adhesion.

Section III collects more general contributions on polymer properties. Interaction between experiment and theory will prove to be very fruitful. To characterize the surfaces and the interfaces at a molecular level, different experimental tools are now available: (AR)XPS, UPS-SR ARUPS, HREELS, NEXAFS, Static SIMS, AFM/STM, small angle x-ray diffraction, etc.. On the theoretical and computational side different approaches are proposed: Hartree–Fock (ab initio and semi-empirical), Density Functional theory, Molecular Dynamics, etc.. Selected results are printed in this section.

In order to produce the adhesion at the interface 'polymer surface modifications' are often required (Section IV). These modifications may be produced by different treatments: wet chemistry, ion beam, laser ablation and, mainly, plasma. It is then significant to characterize the new surface functionalities and the degradation associated with these treatments. The stability of the newly formed chemical groups has also to be examined in detail.

J J Pireaux
P Bertrand
J L Brédas

List of Participants

M Abraham
IMM Institut für Mikrotechnik
Mainz
Germany

S Affrossman
University of Strathclyde
Glasgow
UK

V Andre
BASF AG
Ludwigshafen
Germany

Y Aravot
Volcani Centre
Bet-Dagan
Israel

F Arefi
ENSCP
Paris
France

J Baguet
Meuse Optique Contact
Bar-le-Duc
France

P Bertrand
Université Catholique de
*   Louvain*
Louvain-la-Neuve
Belgium

C Boes
University of Münster
Münster
Germany

A M Botelho Do Rego
Centro de Quimica Fisica
*   Molecular Da UTL*
Lisbon
Portugal

J-L Brédas
Université de Mons-Hainaut
Mons
Belgium

A Calderone
Université de Mons-Hainaut
Mons
Belgium

W Caseri
Institut für Polymere
Zurich
Switzerland

R Caudano
FUNDP
Namur
Belgium

A K Chakraborty
University of California
Berkeley
USA

M Chtaib
Laborlux
Esch-sur-Alzette
Luxembourg

D T Clarck
ICI
Wilton
UK

A M Cros
Faculté des Sciences de Luminy
Marseille
France

P Dannetun
IFM
Linköping
Sweden

D J David
Monsanto
Springfield
USA

H De Deurwaerder
CoRI
Limelette
Belgium

P G De Gennes
ESPCI/Collège de France
Paris
France

Y D P De Puydt
Université Catholique de
*   Louvain*
Louvain-la-Neuve
Belgium

J Delhalle
FUNDP
Namur
Belgium

J L Dewez
Université Catholique de
*   Louvain*
Louvain-la-Neuve
Belgium

K A Dill
University of California
San Francisco
USA

A D Doren
Université Catholique de
*   Louvain*
Louvain-la-Neuve
Belgium

J L Droulas
Université Claude
*   Bernard Lyon 1*
Villeurbanne
France

O El Khattabi
Université Libre de Bruxelles
Brussels
Belgium

D F Evans
University of Minnesota
Minneapolis
USA

F Fally
FUNDP
Namur
Belgium

F Faupel
Institut für Metallphysik der
   Universität Göttingen
Göttingen
Germany

C Frederiksson
Université de Mons-Hainaut
Mons
Belgium

J F Friedrich
Erkner
Germany

F Garbassi
Istituto Guido Donegani
Novara
Italy

J A Gardella
State University of New York
Buffalo
New York

Y Geerts
ULB
Brussels
Belgium

U Gelius
Uppsala University
Uppsala
Sweden

G Geuskens
Université Libre de Bruxelles
Brussels
Belgium

W Gopel
Universität Tübingen
Tubingen
Germany

O Gottsleben
Akzo Research Laboratories
Obernburg
Germany

Ch Gregoire
FUNDP
Namur
Belgium

A Hagemeyer
BASF
Ludwigshafen
Germany

O L J F Heuschling
Université Catholique de
   Louvain
Louvain-la-Neuve
Belgium

P Hollemaert
FUNDP
Namur
Belgium

D J Holmes
Courtaulds Coatings Group
   Research
Tyne and Wear
UK

Z Hruska
Solvay & Cie SA
Brussels
Belgium

E H Humbeeck
Université Catholique de
   Louvain
Louvain-la-Neuve
Belgium

F Iacona
Dipartimento di Scienze
   Chimiche Dell'Universiti
Catania
Italy

R Jerome
Université de Liège
Liege
Belgium

J Jupille
CNRS Saint-Gobain
Aubervilliers
France

K P Karrer
Centre de Recherches des
   Carriéres
Saint-Fons
France

M Kohler
Solvay SA
Brussels
Belgium

S P Kowalczyk
IBM
Yorktown Heights
USA

L Lacaze
Universités Paris 7 et Paris 6
Paris
France

P Y Lai
Universität Mainz
Mainz
Germany

C Lapersonne
Universités Paris 7 et Paris 6
Paris
France

R Lazzaroni
Université de Mons-Hainaut
Mons
Belgium

Q T Le
FUNDP
Namur
Belgium

G Lecayon
DRECAM/SRSIM
Gif-sur-Yvette
France

C Lefebvre
FUNDP
Namur
Belgium

G Legeay
IRAP
Le Mans
France

D Leonard
Université Catholique de
*   Louvain*
Louvain-la-Neuve
Belgium

T Lippert
University of Bayreuth
Bayreuth
Germany

E M Liston
Gasonics International Plasma
*   Corporation*
Hayward
USA

P Louette
FUNDP
Namur
Belgium

G Marletta
Universita di Catania
Catania
Italy

D Mathieu
DRECAM/SRSIM
Gif-sur-Yvette
France

W P McKenna
Eastman Kodak
Rochester
USA

F Menu
Solvay & Cie SA
Brussels
Belgium

K L Mittal
IBM
Thornwood
USA

Y N Nakayama
Uppsala University
Uppsala
Sweden

A N B Naves De Brito
Uppsala University
Uppsala
Sweden

F P Netzer
Universität Innsbruck
Innsbruck
Austria

T P Nguyen
Laboratoire de Physique
*   Cristalline - IMN*
Nantes
France

S Nowak
University of Fribourg
Fribourg
Switzerland

E Orti
University of Valencia
Burjassot
Spain

M Piens
CoRI
Limelette
Belgium

J-J Pireaux
FUNDP
Namur
Belgium

C Quet
Groupe Elf Aquitaine
Artix
France

L Quillet
Universités Paris 7 et Paris 6
Paris
France

J P Rabe
Max-Plank Institut für
*   Polymerforschung*
Mainz
Germany

M G Ramsey
University of Innsbrück
Innsbrück
Austria

L J M Reginster
Shell
Louvain-la-Neuve
Belgium

M Rei Vilar
LASIR (CNRS)
Thiais
France

N V Richardson
Surface Science Research
*   Centre*
Liverpool
UK

S Ries
Thomson CSF
Orsay
France

J Riga
FUNDP
Namur
Belgium

J Ritsko
IBM
Yorktown Heights
USA

P G Rouxhet
Université Catholique de
*   Louvain*
Louvain-la-Neuve
Belgium

L Sabbatini
University Bari
Bari
Italy

E Sacher
Ecole Polytechnique
Montreal
Canada

W R Salaneck
IFM-Linköping University
Linköping
Sweden

P P Schmidt
Office of Naval Research
Arlington
USA

M Schott
Universités Paris 7 et Paris 6
Paris
France

K Seki
Nagoya University
Nagoya
Japan

A Selmani
Ecole Polytechnique
Montreal
Canada

A M Servais
DRECAM/SRSIM
Gif-sur-Yvette
France

J S Shaffer
University of California
Berkeley
USA

J F Silvain
Laboratoire CNRS Maurice
Letort
Villers-lez-Nancy
France

H Sotobayashi
Fritz-Haber-Institut der MPG
Berlin
Germany

S Stafström
Linköping University
Linköping
Sweden

M Stamm
Max-Planck Institut fur
Polymerforschung
Mainz
Germany

S Stanton
Alcoa Technical Center
USA

G B Street
IBM
San Jose
USA

O F P Thoelen
Mons
Belgium

N Thorne
Centre de Recherches de
Voreppe SA
Voreppe
France

O N Tretinnikov
BSSR Academy of Sciences
Minsk
Belorussia

N Ueno
Chiba University
Chiba
Japan

P C M Van Woerkom
Akzo Corporate Research
Arnhem
The Netherlands

T G Vargo
State University of New York
Buffalo
USA

J Verbist
FUNDP
Namur
Belgium

G Wegner
Max-Planck Institut für
Polymerforschung
Mainz
Germany

L T Weng
Université Catholique de
Louvain
Louvain-la-Neuve
Belgium

R H West
UMIST
Manchester
UK

R C White
Colombia University
New York
USA

F Druschke
IBM
Sindelfingen
Germany

I Wilson
ALCAN International
Banbury
UK

SECTION I

POLYMERS ON SOLIDS

Progress and future directions in the theory of strongly interacting polymer–solid interfaces

Arup K. Chakraborty

Center for Advanced Materials

Lawrence Berkeley Laboratory

and

Department of Chemical Engineering

University of California

Berkeley, CA 94720.

Abstract

Polymer / solid interfaces need to be synthesized for many technological applications. A special class of polymer / solid interfaces are those wherein there are strong and specific interactions between the polymer segments and the substrate. A prominent example of such systems are polymer / metal interfaces. In recent years, the technological importance of these interfacial systems and the challenging scientific questions that must be addressed in order to develop an understanding of the nature of these interfaces has motivated significant research efforts in this area. In this review, we discuss the progress that has been made in applying quantum and statistical mechanical approaches toward the elucidation of the relationship between the chemical constitution of the polymer and the substrate and interfacial chain structure and dynamics.

Prepared for an invited review lecture at the International Conference on Polymer - Solid Interfaces, Namur, Belgium, September 2 - 6, 1991.

Introduction

The synthesis of interfaces between organic polymers and solid substrates is of importance in a variety of applications in the microelectronics, aerospace, automotive and food packaging industries. Specific examples include thin film insulators in multilayer structures, interconnects, packaging of microelectronic devices, fiber - filled polymer composites, stabilization of colloidal dispersions, and polymer lined metal containers for protective food packaging. In addition, these systems are of relevance to biomedical and biotechnological applications such as biocompatibility of artificial organs and affinity chromatography based protein separations. In recent years, the technological import of polymer - solid interfaces has led to significant research efforts being directed toward the understanding of these systems. Formidable scientific challenges must be confronted and overcome in order to elucidate the nature of these interfacial systems of extreme technological importance. One crucial issue pertains to the structure and dynamics of polymer chains near a solid substrate. In particular, a legitimate goal that research in this area should strive to achieve is to develop an understanding of how the chemical constitution of the organic polymer and the solid substrate relate to the structure and dynamics of interfacial chain molecules. This is a worthwhile direction to pursue because many relevant macroscopic properties such as adhesion tension and diffusion barrier properties are directly related to the structure of the interfacial region and how it evolves with time, temperature, and other important variables. One is immediately cautioned, however, that the connections between microscopic structure and dynamics and macroscopic interfacial properties are not at all obvious, and establishing these connections presents an important challenge that is just beginning to be addressed. In this review, we will be concerned only with interfacial chain structure and dynamics; only a few cursory comments regarding how connections may be established between this and macroscopic properties will be made.

Polymer - solid interfaces may broadly be classified into two categories. One class is comprised of systems wherein the segment - surface interactions are weak and dispersive. An example of such a system is the polyethylene - graphite interface. A second category of polymer - solid interfaces that has been the subject of much recent attention is comprised of systems wherein there are strong and specific interactions between segments of the polymer chains and the solid substrate. A prominent example of such interfaces are those between organic polymers and metallic substrates. As a result of many theoretical and experimental studies [e.g., 1-5], much is now known about the structure of interfacial chains adjacent to weakly interacting surfaces. However, progress in understanding chain dynamics at these interfaces has been far more modest [6, 7]. Strongly interacting polymer - solid interfaces have been the subject of much more recent investigation, and an understanding of the nature of these systems is just beginning to emerge. However, most experimental [e.g., 8-12] and theoretical [13-18] studies have been directed toward investigating the structure of the interface. While it has been demonstrated [18] that interfacial chain dynamics is crucial for determining the structure of strongly interacting polymer - solid interfaces, an understanding of the same is still in its infancy [19]. In this review, we will focus on the structure and dynamics of strongly interacting polymer - solid interfaces, as exemplified by the interfaces between polymers and metallic substrates. Furthermore, we will be specifically concerned with the progress that has been made in developing an understanding of these systems based on theoretical and computational studies. The review does not comprehensively discuss all previous efforts in detail; rather, the focus will be on issues that this author deems to be most critical.

This review is organized as follows. We begin by outlining the major issues that need to be addressed in order to develop an understanding of the structure and dynamics of the interfacial systems of concern. This will be followed by a discussion of the approach that is appropriate for addressing these issues. Attention will then be focussed on the progress that has been made in pursuing the theoretical and computational aspects of this approach. Finally, major unresolved issues and this author's opinion regarding future directions of research that should help clarify some of these issues will be outlined.

Critical Issues and Research Strategy

We identify the key questions that must be answered in order to elucidate interfacial chain structure and dynamics at strongly interacting polymer - solid interfaces in the context of a specific example; viz., polymer - metal interfaces. However, the issues are in general the same for all systems characterized by segment - surface interactions that are strong and specific (in a sense to be made clear shortly).

As we have noted earlier, polymer - metal interfaces are technologically important in several applications in the microelectronics, aerospace, and food packaging industries. These systems are distinct from several other polymeric interfaces that have been studied more extensively, in the sense that, herein there is the possibility that certain functional groups of the organic polymer may chemisorb on the metallic substrate. As compared to physisorption, chemisorptive interactions are not only stronger, but are also specific. In other words, only certain functional groups of the organic polymer interact very favorably with the surface. As an example, consider the interactions of the polyimide, PMDA - ODA (Fig. 1), with various metals.

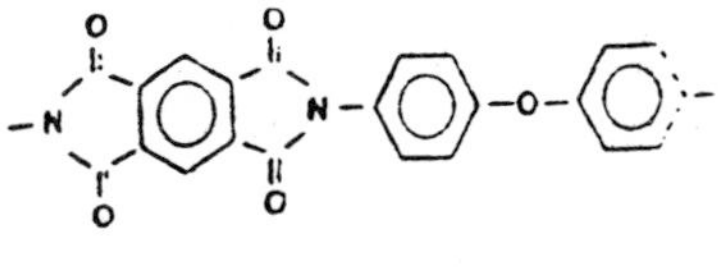

Fig. 1

It has been shown that with aluminum the primary interaction is with the carbonyl group [11], while with chromium there are favorable interactions with the aromatic ring and the carbonyl group [13]. We shall return to these details later, but here we just note that the substrate metal interacts favorably with specific functional groups of the organic polymer. Thus, there are strong and specifically directed enthalpic driving forces that favor preferential adsorption of certain functional groups. The interplay between these strong and specific enthalpic driving forces and the entropic constraints associated with confining chain molecules near a solid surface determines the structure and dynamics of the interfacial chains.

In order to relate interfacial chain structure and dynamics to chemical constitution, the following key issues must, therefore, be addressed. Firstly, the potentially reactive functional groups of the organic polymer must be identified. In other words, the nature of the interactions between a few metal atoms that make up the substrate and segments of the organic polymer must be clarified. Secondly, since in many applications the interfacial chains are in contact with a metal surface, the segment - surface interactions must be elucidated. Finally, the conformations adopted by the interfacial chains when they are subject to strong and specific segment - surface interactions must be understood. It will be made clear that for strongly interacting polymer - solid interfaces the conformations adopted by the adsorbed polymer molecules is closely related to the interfacial chain dynamics associated with the formation of the interface.

The most suitable approach for addressing the three key issues outlined above must neccessarily involve a continuous exchange of information between theory and experiment. In Fig. 2 such an approach is depicted schematically.

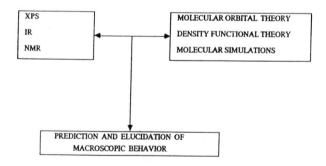

Fig. 2

Spectroscopic techniques can be used to experimentally probe the nature of the interfaces under consideration. X-ray photoelectron spectroscopy (XPS) and infrared (IR) spectroscopy can be used to characterize the nature of the segment - level interactions. Nuclear magnetic resonance (NMR) spectroscopy can help elucidate interfacial chain conformations and dynamics. XPS has been used extensively [e.g., 8, 11, 15, 20] to study the interactions of overlayers of metal atoms on polymer surfaces. These studies will be discussed in the context of their relationship with certain theoretical investigations later. Infrared reflection absorption spectroscopy (IRRAS) and attenuated total reflection (ATR) have been used to study the interactions of polymer segments interacting with native metal oxide surfaces [21] and special alloy materials [7]. The use of NMR spectroscopy to characterize interfacial chain conformations and dynamics is still in its infancy, largely because information pertinent to these issues is just beginning to emerge.

The experimental studies alluded to above, must be complemented by theoretical and computational investigations. In fact, only a continuous exchange of information between theoretical and experimental efforts will allow the development of structure - property relationships. The theoretical and computational approach that could help develop relationships between chemical constitution and macroscopic properties (such as adhesion) must involve quantum and statistical mechanical methods. The systems that we are concerned with involve strong and specific interactions. As such, the first problem that we encounter is that the interaction potentials for segment - substrate interactions are not simple, and no simple forms (e.g., Lennard - Jones potentials) that describe them are known *a priori*. In other words, the energy hypersurfaces that characterize the interactions between polymer segments and the substrate must first be determined in order to theoretically or computationally study interfacial chain architecture and dynamics. These energy hypersurfaces that characterize the chemisorptive interactions exhibit a rich structure, and must be determined as a function of the relevant internal degrees of freedom of the segment, and the orientation and distance of the segment from the substrate. Calculation of these energy hypersurfaces is possible only via quantum mechanical methods. The results of the quantum mechanical calculations provide segment - surface interaction potentials that reflect the strength and specificity of the interactions under consideration. Interfacial chain structure and dynamics subject to these interaction potentials can then be studied using statistical mechanical theories and/or molecular simulations. Thus, the theoretical and computational approach that would enable us to relate chemical constitution to macroscopic properties must neccessarily be a hierarchical one, that is both validated by experiment and serves as a guide for future experimentation. This is depicted schematically in Fig. 3.

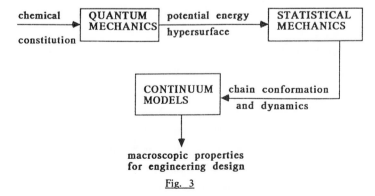

Fig. 3

Firstly, given the chemical constitution of the substrate and the organic polymer, quantum mechanical theories, such as Hartree - Fock (HF) and Density Functional (DFT) theories, can be used to determine the segment - substrate interaction potentials. We will discuss the merits and demerits of the various quantum mechanical methods that can be used, and the progress that has been made in applying these techniques toward the elucidation of segment - surface interactions. The interaction potentials that are obtained via quantum mechanical calculations serve as one of the important inputs to the second level, the statistical mechanical machinery that must be employed to study interfacial chain structure and dynamics. We will discuss the progress that has been made in this regard by focussing on stochastic methods and molecular dynamics simulations that have been used to address these issues. In order to relate chemical constitution to macroscopic properties studies at a third level are required. As shown in Fig. 3, information regarding chain architecture and dynamics that results from the statistical mechanical calculations should be used in continuum or phenomenological models that relate the structure of the interface to macroscopic properties. Very little work has been done at this highest level of theory; however, progress in this endeavor is critical for the development of relationships between macroscopic properties of relevance to engineering design and chemical constitution of the materials. In this review we will be concerned exclusively with the progress that has been made in applying quantum and statistical mechanical methods toward the elucidation of interfacial chain structure and dynamics.

Interactions of Polymer Surfaces with Metal Atoms

We begin by considering the issue of segment level interactions between the organic polymer and metal atoms. Historically, this was the first issue of relevance to strongly interacting polymer - solid interfaces that was studied via spectroscopic experiments. Following the efforts of Clark and co-workers [22] who pioneered the use of XPS to characterize clean polymer surfaces, Burkstrand [23, 24] initiated studies of polymer - metal interfaces formed by evaporating small dosages of metal atoms onto polymer surfaces. This was followed by a systematic and concerted effort in several laboratories (most prominently at IBM, Yorktown Heights) that aimed to characterize the interactions of metal overlayers on polymer surfaces [e.g., 25-28]. Many of these studies focussed on the polyimide, PMDA-ODA, which promises important applications as thin film insulators and for microelectronics packaging due to its good thermal stability and dielectric properties. XPS studies of small dosages of metal overlayers on PMDA - ODA surfaces has recently been reviewed by di Nardo [20], and the reader is referred to this article for a detailed exposition of experimental procedures and the difficulties and limitations associated with this technique.

X-ray photoelectron spectra provide a measure of the distribution of binding energies of the core electrons in a sample. Of course, the binding energies of the core electrons are a function of the molecular environment. Thus, studying the changes in binding energy of the core electrons of the organic and the metal can provide information regarding the nature of their interactions. This has been, in essence, the strategy employed by the experimentalists who have worked on studying the interactions of low coverages of metal overlayers evaporated onto polymer surfaces. These experiments provide information regarding the interactions of a small cluster of metal atoms interacting with the polymer segments. Note that this is distinct from interactions with metal surfaces, an issue that will be of considerable concern later in this article. Computational investigations have been carried out to help elucidate the considerable body of experimental work using XPS. These studies have been carried out within the framework of Hartree-Fock molecular orbital theory, which is an approach toward solving the many - electron quantum mechanical problem. As we have noted earlier, a theoretical understanding of the nature of the chemical interactions under consideration can only be obtained through the application of quantum mechanical methods. Prior to describing what has been learnt regarding the interactions of metal atoms with organic polymer segments from a combination of XPS and Hartree - Fock studies, we outline the basic elements of Hartree-Fock molecular orbital theory. We do so for two reasons. Firstly, many body quantum mechanical theories have only recently been used for the investigation of polymer interfaces, and secondly, a discussion of the limitations of these computations will motivate the application of density functional theory toward a study of the interactions of polymer segments with metal surfaces. Our discussion will be very brief since Hartree - Fock methods are discussed extensively in several excellent texts [29, 30], and the application of these methods toward investigating the interactions of a few metal atoms with oligomers of the organic polymer is fairly straightforward.

The state of a system can be described by a probability amplitude or wave function, $\Psi(r, t)$. In other words, $\Psi(r, t)$ is interpreted such that $|\Psi(r, t)|^2$ dr is the probability of finding the system in the region between r and $r + dr$ at time t. The spatial and temporal evolution of the wave function is described by the Schrødinger equation which may be written as

$$[- (h^2/8\pi^2 m) \nabla^2 + V(r, t)] \Psi(r, t) = - (h/2\pi i) \{\partial \Psi(r, t) / \partial t\} \qquad (1)$$

where ∇^2 is the Laplacian operator, $V(r, t)$ is the potential experienced by the particles, h is Planck's constant, m is the mass of the particle, and $i = \sqrt{-1}$. The time - dependent Schrødinger equation is quite complicated, and the search for solutions to it appears to be a formidable task. Fortunately, for many systems of chemical interest, the potential, $V(r, t)$, does not depend upon time. It can be shown easily [29] that for these cases the solutions to the Schrødinger equation are separable; i.e., the wave function may be written as a product of a spatial part and a time - dependent part. For such systems, one needs to solve only for the spatial dependence, $\psi(r)$. The time - independent Schrødinger equation may be written as

$$H \psi(r) = E \psi(r) \qquad (2)$$

where H is the Hamiltonian, and E is the energy of the system. The Hamiltonian, H, for a molecular system with n electrons and k nuclei can be expressed as follows

$$H = -(h^2/8\pi^2 m) \Sigma_j \nabla_j^2 - \Sigma_l \Sigma_j (Z_l e^2 / r_{lj}) + \Sigma_i \Sigma_j (e^2 / r_{ij}) \qquad (3)$$

where m is the mass of the electron, Z_l is the atomic number of nucleus l, e is the charge of the electron, r_{lj} is the distance between electron j and nucleus l, and r_{ij} is the interelectronic separation. Several approximations (such as the Born - Oppenheimer approximation) have been made in expressing the

Hamiltonian to be as shown in Eq. 3. The first term in Eq. 3 represents the kinetic energy of the electrons; the second term reflects the interaction between the electrons and the nuclei, and the third term represents the interelectronic repulsions. The interelectronic repulsion terms are of a form such that the Schrødinger equation cannot be solved exactly for a many - electron system. Thus, approximate methods are required to solve the Schrødinger equation for problems of interest. Hartree - Fock (HF) theory provides one such approach that leads to approximate solutions to the many - electron problem.

HF theory is an approach wherein the interelectronic repulsions are treated in a mean field sense; however, the correlations arising from the requirement that the wave function be antisymmetric to particle exchange (Pauli Principle) are incorporated into the theory. This is done by assigning each electron to one - electron functions called spin orbitals. Each spin orbital is the product of a spatial part and a spin part. The wave function for the many - electron system is expressed as a determinant (Slater determinant) of spin orbitals. The resulting one - electron equations in canonical form are

$$F \, \Phi_i \, (q) = \varepsilon_i \, \Phi_i \, (q) \qquad (4)$$

$$F \, \Phi_i = [\, -(1/2) \, \nabla^2 + v \, (q) + \int \Sigma_{j=1..N} \, \{\Phi_j{}^* \, (q') \, \Phi_j \, (q') / |q\text{-}q'|\} \, dq'] \, \Phi_i \, (q)$$

$$- \int \Sigma_{j=1..N} \, \{ \, \Phi_i(q) \, \Phi_j{}^* \, (q') \, \Phi_j \, (q) / |q\text{-}q'|\} \, dq' \qquad (5)$$

where $v \, (q)$ is the potential imposed by the nuclei and other external sources, $\Phi_i(q)$ is the spin orbital for the ith electron, and q represents both spin and spatial coordinates. The last term in the Fock operator, F, represents the non-local exchange operator that incorporates the effects of exchange or Pauli correlations. It should be clear that Hartree-Fock theory as stated in Eqs. 4 and 5 does not account for coulomb correlations (i.e., the interelectronic repulsions are still treated in a mean - field sense). This point will become important in our discussions of segment - surface interactions.

The HF equations are solved variationally, with the one - electron spatial orbitals being expanded in terms of a truncated complete set of functions; i.e.

$$\Phi_i = \Sigma_k \, a_{ki} \chi_k \qquad (6)$$

where the basis functions, χ_k, are usually chosen to be atomic orbitals of the constituent atoms. Using the variational theorem with Eq. 6 describing the spatial orbitals leads to the Roothan - Hall equations which can be solved self consistently to obtain the eigenvectors and eigenvalues. Once the molecular orbitals are thus obtained, several one - electron properties, electron density distributions and total energies can be calculated. It is to be noted that the computational scheme outlined above requires that several interelectronic repulsion integrals be calculated and stored. The computation time for Hartree - Fock calculations scale with the number of basis functions as K^4, where K is the number of basis functions employed in the calculation. This makes calculations for a large number of electrons prohibitive. We shall return to this point in more detail when discussing appropriate schemes for studying segment - surface interactions. At this point we merely note that *ab initio* Hartree - Fock calculations are very demanding, and so for the study of the interactions of metal atoms with polymer segments, some workers have carried out semiempirical molecular orbital calculations [30]. In semiempirical molecular orbital methods several of the integrals that arise in HF theory are either approximated or ignored. The resulting sacrifice in accuracy is accounted for by introducing parameters into the treatment that are optimized to match experimental data. Several semiempirical schemes are available; an example is the modified neglect of differential overlap (MNDO) method proposed by Dewar and Thiel [31, 32].

From the viewpoint of being complementary to photoelectron spectroscopies, information regarding changes in the electron density distributions and the eigenvalues of the various occupied core levels upon interaction with metal atoms is of primary interest. This is so because changes in electron density distributions are direct consequences of bonding between the metal and the organic, and within certain approximations, the eigenvalues can be used to calculate photoelectron spectra. The major approximation in question is Koopman's theorem [33]. This states that the negative of the eigenvalue for a certain molecular orbital is the ionization potential for the electron corresponding to that energy level. Clearly, this ignores the relaxation of the molecular orbital manifold upon ionization. As we shall see, in many cases, the relaxation energy does not change much over small ranges of binding energy [34]. Thus, Koopman's theorem may be used successfully to predict the relative positions of peaks in the photoelectron emission spectra. In cases where this is not the case, so called ΔSCF calculations [29, 30] have to be performed, wherein the manifold of states is calculated for both the neutral species and the ion.

We begin reviewing the progress that has been made in applying Hartree - Fock molecular orbital (HFMO) theory toward the elucidation of the interactions between metal atoms and polymer segments by focussing first on the system that has been studied the most, viz., the interactions of the polyimide, PMDA-ODA, with metal atoms. Computational studies of these systems have largely been due to Rossi and co-workers [13-15]. These studies have been carried out with a view toward elucidating experimental XPS data, and consequently, a rather clear picture of the interactions between metal atoms and PMDA-ODA segments has emerged. In the HF studies in question, model compounds of the polyimide have been studied. This is so because performing accurate HF calculations for long polymer chains is not feasible. Furthermore, since the aim of these investigations was to elucidate the local interactions between metal atoms and organic polymer segments, studying long polymer chains was not essential. The model compounds were chosen such that the potentially reactive functional groups of the organic polymer were not only part of the model compound, but were also in a molecular environment similar to that in the polymer. For example, Fig. 4 shows the model compound that was used by Rossi et al. [13] to study the interactions of PMDA-ODA with chromium.

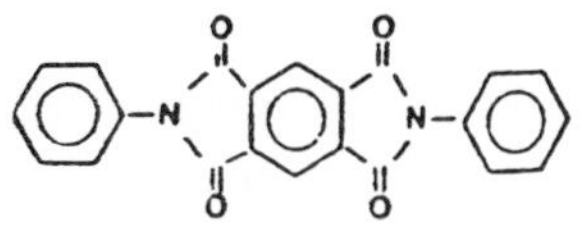

Fig. 4

The model compound chosen by these authors is (N, N' - diphenylpyromellitimide), which is an excellent model for the segment of the polymer strand found between two adjacent ether oxygens.

Rossi and co-workers have carried out *ab initio* Hartree - Fock calculations using minimal basis sets [29]. As such, the energetics of interaction calculated by these authors are not expected to be particularly accurate. However, this was not the main purpose of their work, which rather aimed to elucidate qualitative features of the photoemission data. In studying the interactions of polyimide model compounds with chromium, these authors did use extended basis sets for chromium. The investigations of this group of workers include a study of the electronic properties of polyamic acid (a precursor for forming the PMDA-ODA polymer) [15] and fluorine containing polymers [14]. However, here we focus on their study of the interactions of chromium with PMDA-ODA model compounds.

Rossi et al. [13] first report the electronic properties of the model compound (Fig. 4). Their results show that the molecular orbitals are delocalized not only over the central benzene portion of the molecule, but also over the imide fragments. Using Koopman's theorem and replacing the calculated spectral lines with gaussian functions (to reflect the spectrometer optics and sample inhomogeneities) Rossi et al. calculated the XPS spectrum for the model compound. Fig. 5 is taken from their paper, and shows a comparison of the calculated XPS spectrum for the model compound and the experimental spectrum for the polyimide, PMDA-ODA. The dotted line represents the experimental spectrum, and the solid line is the calculated spectrum.

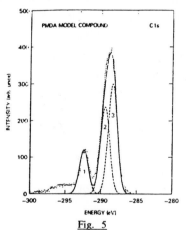

Fig. 5

The two spectra compare very well, and therefore, it is fair to say that the local electronic properties of the polyimide are well reproduced by the model compound chosen by Rossi et al. [13].

In order to study the nature of the interactions of chromium with the model compound Rossi and co-workers placed the chromium atom atop the central benzene ring of the model compound. The results of these calculations show that there is charge transfer from the chromium atom to the four carbon atoms (C-C=O) of the aromatic ring that are connected to the carbonyl groups, and the oxygen atoms of the carbonyl groups. This implies that interaction with chromium atoms should lower the binding energy of the C (1s) electrons corresponding to the C-C=O atoms. Indeed, this is what is observed in the experimental spectrum. Experiments [27] show that the C (1s) peaks corresponding to C-C=O and the carbonyl groups are attenuated and broadened upon interaction with chromium. In fact, Rossi et al. have shown that upon assuming that only 50% of the segments interact with chromium atoms, the calculated and experimental spectra agree with each other very well.

In summary, the results of the calculations reported by Rossi and coworkers [13] demonstrate that HFMO theory can be used successfully to identify the potentially reactive functional groups of the organic polymer, and elucidate the details of the nature of the interactions with metal atoms. To further demonstrate the use of such an approach toward addressing the question of how polymer segments interact with metal atoms we now discuss one more example of a similar study.

We will concern ourselves now with the interactions of acrylic polymers (e.g., poly (methyl methacrylate)) with aluminum atoms. These systems have important applications as photoresist materials, and have been studied using HFMO theory by Chakraborty et al. [16]. Their work, just as the work of Rossi and coworkers, aimed to address the issue of the interactions of metal atoms with polymer segments. These authors also investigated the interactions of model compounds with metal atoms. The major difference between the study reported by Chakraborty et al. [16] and that due to Rossi and coworkers is that the former study did not make *a priori* assumptions regarding the location of the interacting metal atoms with respect to the organic model compound. Such assumptions could not be made for the acrylic polymer / aluminum atom system because the geometries adopted during complexation were not obvious. Furthermore, as was made clear by these authors, the organic model compounds undergo changes in conformation and geometry as a result of interactions with aluminum atoms. Thus, Chakraborty et al. [16] minimized the energy of the model compound / aluminum complexes with respect to important geometric degrees of freedom in order to study the nature of the interactions in these systems. *Ab initio* HF calculations with energy minimization is extremely compute intensive if reasonably large basis sets are used. As such, Chakraborty et al. [16] performed semiempirical molecular orbital calculations (using the MNDO Hamiltonian) to obtain the minimum energy structures. Single point *ab initio* calculations were then performed to study these minimum energy structures.

Chakraborty et al. [16] first calculated the electronic properties of the model compound for poly (methyl methacrylate) (PMMA). Fig. 6 shows the model compounds chosen by these authors to represent the monomer and dimer of PMMA.

Fig. 6

The calculated electronic properties were validated against experiment by comparing calculated and experimental XPS spectra. Fig 7 shows the calculated experimental spectra for the C (1s) and O (1s) emissions, and Fig. 8 depicts the corresponding experimental spectra [21].

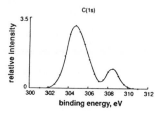

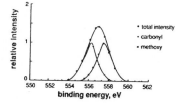

Fig. 7

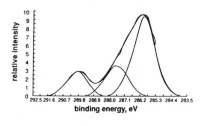

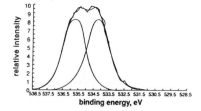

Fig. 8

As is clear, the calculated and experimental spectra agree very well. As such, the model compounds chosen by these authors may be considered to be good models for the polymer under consideration. In fact, as suggested by Chakraborty et al. [16], these model compounds may be considered reasonable for the study of carbonyl bearing polymer segments with aluminum atoms.

In contrast to the study by Rossi et al. [13], Chakraborty and coworkers studied the interactions of a s-p bonded metal (aluminum). While on one hand the absence of d - level electrons simplifies the computational requirements, these workers found rather complicated interactions which stem from the fact that aluminum being a group III metal is electron deficient. This implies simply that since aluminum has fewer valence electrons than levels, it tends to form bridge bonded complexes with carbonyl bearing segments. In view of this, Chakraborty et al. [16] studied dimers of PMMA model compounds interacting with two aluminum atoms. Studying the interactions of monomers of the model compound with one aluminum atom (akin to the study of Rossi et al. [13] with chromium) would exclude the possibility of bridge bonded interactions.

Based on their studies of the interactions of dimers of PMMA with two aluminum atoms, Chakraborty et al. [16] found that aluminum atoms interact primarily with the carbonyl functionality of PMMA. However, there are weaker interactions with the methoxy functionality. This trend was also found to be the case for interactions with poly (acrylic acid), henceforth labelled PAA. These authors performed studies as a function of the distance of the aluminum atoms from the model compound. In other words, they studied the evolution of the interaction along a reaction coordinate. The results of these calculations lead to the following picture for the interactions of PMMA with aluminum atoms. As the aluminum atoms begin to interact with the organic there is charge transfer to the carbonyl carbons from the aluminum atoms and charge transfer from and to the carbonyl oxygens. This has been rationalized [16] by carefully examining the nature of the interactions between the frontier orbitals of the interacting species. In the early stages of the interaction, there are bridge bonded interactions which are such that the double bond of the carbonyl group is weakened but does not completely lose its double bond character. As the aluminum atoms approach closer, the double bond is broken and strong aluminum - oxygen - carbon bonds are formed. These computational results have been compared with experiment by Chakraborty et al. [16], leading to the elucidation of the nature of the interactions between carbonyl group containing polymers and aluminum atoms. Herein we outline the conclusions of their study.

There are five papers that have been published in recent years [11, 12, 25, 35, 36] wherein experiments pertaining to the interactions of aluminum with carbonyl bearing polymers have been discussed. Two of these studies [35, 36] have examined the interactions of aluminum with polyacrylic acid, while two others [11, 25] focussed on aluminum interacting with PMDA-ODA. While the computational investigations of Chakraborty et al. [16] have been performed for PMMA model compounds, these authors have provided reasons as to why their results may be compared to experiments conducted with PAA and PMDA-ODA. The major reason that allows such a comparison to be made is that the primary interaction in all of these cases involves the carbonyl group. The discussion provided by Chakraborty et al. [16] regarding the comparison of their computational results with experiment leads to an elucidation of many features of the interaction between carbonyl containing polymers and aluminum, and may be summarized as follows. Allara's [35] observation that upon exposing PAA films to aluminum vapor the infrared stretching frequency for the carbonyl group (1730 cm-1 in the free polymer) is attenuated and a new band appears at lower frequency (between 1500 and 1600cm-1), is consistent with the calculated results which show that the C=O bond loses double bond character and Al-O-C bonds are formed. DeKoven and Hagans [36] have used XPS to study the interactions of PAA with aluminum atoms. While their experiments do indicate that there are interactions with the carbonyl group, they also find that there are significant interactions with the hydrocarbon backbone of the polymer. Strong interactions with the hydrocarbon backbone are not found by Chakraborty et al. [16], and such interactions seem unreasonable. It is possible that this may be an artifact of the XPS experiments performed by DeKoven and Hagans. The most likely cause of such an artifactual result is degradation of the polymer caused by radiation damage. Atanasoska at al. [11] have carried out very careful coverage dependent studies of the interactions of aluminum with PMDA-ODA. They find that the primary interaction is with the carbonyl group. They also find remarkable changes with coverage. The coverage dependence is noteworthy because there is a correspondence between the coverage dependent experimental study and the computational study [16] performed as a function of distance of the aluminum atoms from the organic. In other words, the results obtained by Chakraborty et al. [16] regarding the evolution of the interaction as the two aluminum atoms approach the dimer correspond closely to the experimental study as a function of coverage. The issue of whether there is a legitimate reason for this correspondence, or whether this is merely coincidental remains unresolved. Pireaux and coworkers' seminal contributions [12] using High Resolution Electron Energy Loss Spectroscopy (HREELS) to study the interactions under consideration appeared after the computational study we have been discussing, and so a comparison of theory with these experiments has not appeared in the literature. However, the computational results do agree with the results of HREELS experiments in broad terms. We note, however, that some of the results do not agree in detail.

Based on our discussion of the computational work that has been done regarding the application of Hartree-Fock methods toward the study of the interactions of metal atoms with model compounds for the organic polymer, it should be clear that this is now a mature effort. In other words, for many cases HF calculations of this kind can be routinely performed to study the local interactions between polymer segments and a few metal atoms. Furthermore, these studies help elucidate experimental data and identify the functional groups that interact strongly with substrate atoms. We now turn to the more challenging question of how polymer segments interact with metal surfaces, and the interfacial chain structure and dynamics that arise as a consequence of such interactions. One final note before we leave the subject of interactions with metal atoms. It has been shown both experimentally [21] and via a discussion in ref. 12 that the potentially reactive functional groups may be quite different for polymer segments interacting with pure metal atoms and that for interactions with native metal oxide surfaces. For example, for PMMA interacting with aluminum the carbonyl group interacts more strongly than the methoxy group. However, for interactions with native aluminum oxide surfaces the methoxy group interacts more strongly leading to the formation of carboxylate anions on the surface [21]. Progress in developing a theoretical understanding of the nature of polymer - native metal oxide surfaces is virtually nonexistent. However, many of the broad physical considerations are the same as that for polymer - metal interfaces.

Interactions of Organic Oligomers with Metal Surfaces

In many applications, the issue of importance is the interfacial chain structure and dynamics adjacent to metal surfaces, as opposed to the behavior of polymer chains with an overlayer of metal atoms. Of course, it is obvious that the electronic properties of metal surfaces are not the same as that of a few metal atoms [37], and thus the energetics of the interactions between polymer segments and metal atoms may be different from that obtained from studies such as the ones we have been considering [13-16]. Obvious as this point is, only recently has this issue received experimental and theoretical attention [10, 17]. Grunze and coworkers were perhaps the first to explicitly address this point experimentally [10]. Our focus here will be on recent theoretical efforts [17, 38] to understand the nature of segment - surface interactions and the consequent interfacial chain structure and dynamics at strongly interacting interfaces. We begin these discussions by considering the issue of how one may calculate the interaction energies between polymer segments and metal surfaces.

As we have noted, HF theory is quite capable of elucidating the nature of interactions between polymer segments and metal atoms. However, it is the opinion of this author that it is not the appropriate framework to study interactions with metal surfaces. This is largely due to two reasons. Firstly, as we have noted, HF theory does not account for coulomb correlations. Of course, HF configuration interaction methods [29] which use multideterminantal wave functions do account for coulomb correlations. However, these methods are far too compute intensive to be used for the systems under consideration. The energetic consequences of neglecting coulomb correlations are more pronounced for low electron densities. In fact, the coulomb correlation energy is comparable to the exchange energy at low densities [39]. Thus, for systems containing a large number of low density electrons the energetics calculated using mean field theories will be significantly in error. For small organic molecules or metal atoms the energy is dominated by that of the high density core electrons. Thus, energetics calculated using HF theory are quite accurate for these systems. In contrast, metal surfaces have a large number of low density electrons that constitute the valence band. Thus, calculating the energetics of metal surfaces and their interactions with other species using mean field theories leads to significant errors. Thus, HF theory without configuration interaction (CI) is not appropriate for calculating the interactions of organic oligomers with metal surfaces. HF/CI methods would prove to be accurate, but these methods are too compute intensive. In fact, calculating single determinantal wave functions for metal surfaces is prohibitive. This is so because a HF calculation is at least a 3N dimensional variational calculation (where N is the number of electrons), or said another way, time and memory requirements scale with the fourth power of the number of basis functions used. This implies that as the number of electrons (and consequently number of basis functions) increases, computer time and memory requirements increase dramatically. It should be clear then that HF calculations are prohibitive and inaccurate for examining the interactions of organic oligomers with metal surfaces, and HF/CI calculations are computationally intractable. Thus, in order to address this issue we must turn to a different approach to solve the many - electron problem. The approach that is most suitable for our purpose is the density functional theory (DFT) of the inhomogeneous electron gas. The reasons for this assertion will become clear in the following discussion wherein we outline the elements of DFT.

Over the last decade, one theoretical framework that has been used extensively to study systems with a large number of interacting particles is DFT. Both quantal and classical systems have been studied using this approach. However, the origins of DFT lie in the quantum mechanical literature [40]. The basic idea that underlies DFT is to replace the complicated N-electron wave function, $\Psi(r_1, r_2, ... r_N)$, and the associated Schrødinger equation with a much simpler function, the electron density distribution, $n(r)$, and its associated computational scheme. A theory wherein density is the fundamental variable would be truly remarkable. This is so because the electron density distribution is always a three-dimensional function. There is a long history of such theories, which until 1964 had the status of models. However, two profound theorems due to Hohenberg and Kohn [40] established the legitimacy of using the density distribution as a fundamental variable. The first theorem proves that the external potential, $v_{ext}(r)$, is (to within an additive constant) a unique functional of $n(r)$; since, in turn, $v_{ext}(r)$ and $n(r)$ determine the hamiltonian, the full many - particle ground state is a unique functional of the electron density distribution. This establishes the density as a fundamental variable. The second Hohenberg - Kohn theorem proves an important variational relationship; viz., the correct ground state density yields the minimum energy of the system. Thus, the density distribution of a system of interacting particles can be obtained by minimizing the energy functional with respect to the density distribution while maintaining a constant number of particles. Such a minimization leads to the following Euler - Lagrange equation,

$$\delta E[n(r)] / \delta n(r) = \mu \qquad (7)$$

where $E[n(r)]$ is the energy functional, and μ is a lagrange multiplier that can be identified to be the chemical potential [41].

In order to implement DFT for performing practical calculations, we must first represent the energy functional in explicit terms. The energy functional may be written as

$$E[n(r)] = \int v_{ext}(r) n(r) dr + T[n(r)] + \{1/2\} \iint \{\{n(r) n(r')\}/|r - r'|\} dr dr' + E_{xc}[n(r)]$$

$$(8)$$

where $T[n(r)]$ is the kinetic energy functional, and $E_{xc}[n(r)]$ is the exchange - correlation functional that represents all the many - body terms that are ignored in the mean field approximation for the interelectronic repulsions. If $T[n(r)]$ and $E_{xc}[n(r)]$ were known explicitly, then the density distribution for a system of interacting electrons could be determined by solving the following Euler - Lagrange equation

$$v_{ext}(r) + \delta T[n(r)]/\delta n(r) + \int \{n(r')\}/|r - r'|\} dr' + \delta E_{xc}[n(r)]/\delta n(r) = \mu \quad (9)$$

In order to solve Eq. 9 for any system, we need general forms for $T[n(r)]$ and $E_{xc}[n(r)]$. Such general expressions do not exist for systems with inhomogeneous electron density distributions, and so approximations must be made. These approximate methods and their relative merits are now discussed.

We consider first the exchange - correlation energy functional. Physically, it represents corrections to the energy determined via a purely mean field treatment of the interelectronic repulsions. The most commonly used approximation for the exchange - correlation energy is the local density approximation (LDA). This approximation is based on the following assumption. Consider an inhomogeneous electron density distribution, $n(r)$. In the LDA one assumes that the exchange - correlation energy per particle at any location is the same as that for a homogeneous electron gas with the same density. This is to say that for the purpose of calculating the exchange - correlation energy (and the associated potential) LDA assumes that an inhomogeneous density distribution may be considered to be locally homogeneous. Of course, this immediately brings up the issue of how one defines the region over which one applies this local approximation. The simplest possible choice is to assume that the contribution to the exchange - correlation energy from every point in space is that for a homogeneous electron gas with that point density. This physical idea can be expressed mathematically as

$$E_{xc}[n(r)] = \int \varepsilon_{xc}(n) n(r) dr \quad (10)$$

where $\varepsilon_{xc}(n)$ is the exchange - correlation energy per particle of a homogeneous electron gas with density n.

In order to calculate the exchange - correlation energy using Eq. 10 we need to know $\varepsilon_{xc}(n)$. The exchange - correlation energy for a homogeneous electron gas may be divided into two pieces; that due to exchange and that due to coulomb correlations. Thus,

$$\varepsilon_{xc} = \varepsilon_x + \varepsilon_c \quad (11)$$

ε_x for a homogeneous electron gas is known from the Dirac formula [40, 41]. Using this expression and the LDA leads to the following expression for the exchange energy.

$$E_x = C_x \int \{n(r)\}^{4/3} dr \quad (12)$$

where C_x is a known constant.

The next issue pertains to finding ε_c for a homogeneous electron gas. Historically, many authors [e.g., 42] have worked on developing approximate analytical expressions for this quantity in the limits of high and low electron densities. We do not discuss these developments here, but rather, we point out the methodology that has been used in the calculations that will be the subject of our attention. This method is also the most accurate.

In 1980, Ceperley and Alder [43] performed Monte - Carlo simulations to obtain the total energy of a homogeneous electron gas at various densities. By substracting out the kinetic, exchange, and mean field repulsion energy terms from these results the coulomb correlation energy per particle can be obtained as a function of density. This result is exact in the sense that it is obtained from calculations wherein all the interactions were explicitly included. The results obtained by Ceperley and Alder have been parametrized by various authors. One such accurate and easy to use version is that due to Perdew and Zunger [44]. This is the version employed to study the interactions of organic polymer segments with metal surfaces. We note that the LDA is quite accurate, and various studies on different types of materials have shown that the calculated results are as reliable as high quality *ab initio* HF calculations [41, 45].

We now turn attention to the kinetic energy functional. In work pertaining to polymer / metal interfaces the kinetic energy functional has been represented in two different ways. We outline both approaches here, and point out the advantages and disadvantages of using both methods.

The simplest way to represent the kinetic energy functional is the Thomas - Fermi expression [46], which predates the formal development of DFT. The Thomas - Fermi expression is based on many approximations. In particular, it is constructed by using a LDA for the kinetic energy, and the kinetic energy expression for a homogeneous system of non-interacting electrons. As such, results obtained using this kinetic energy functional are not very accurate, and may be considered to be correct only in a semiquantitative sense. The Thomas - Fermi kinetic energy functional has been used to study polymer - metal interactions [17], and may be written as follows

$$T_{TF} [n (r)] = C_{TF} \int \{n (r)\}^{5/3} dr \qquad (13)$$

where C_{TF} is a known constant. If one uses the Thomas - Fermi equation for the kinetic energy functional, direct solutions of the Euler - Lagrange equations can be obtained for the electron density distributions. This simplicity is, however, obtained by sacrificing quantitative accuracy. For more accurate calculations the Thomas - Fermi approximation is inadequate, and the kinetic energy must be determined more precisely. This can be done by the method devised by Kohn and Sham [].

Kohn and Sham [47] devised an ingenious way to calculate the kinetic energy very accurately. We will not develop these equations here since descriptions of the same are available in recent texts [41, 46] and the original paper. We simply note that using this method the kinetic energy can be calculated very accurately (but not exactly) at the expense of greater computational expense. The computational expense is enhanced primarily because orbitals are reintroduced into the Kohn - Sham formulation. The celebrated Kohn - Sham equations can be written as follows in atomic units

$$[-(1/2)\nabla^2 + v_{eff} (r)] \Phi_i (r) = \varepsilon_i \ \Phi_i (r) \qquad (14a)$$

$$v_{eff} (r) = v_{ext} (r) + \int \{n (r') / |r - r'|\} \ dr' + \delta E_{xc} [n (r)] / \delta n (r) \qquad (14b)$$

$$n (r) = \Sigma_{i=1..N} |\Phi_i (r)|^2 \qquad (14c)$$

where N is the number of electrons, $v_{eff} (r)$ is the effective potential, Φ_i are the Kohn - Sham orbitals, and ε_i are the eigenvalues. In most practical applications the Kohn - Sham equations are solved self consistently with the LDA for E_{xc}. The effective potential, which incorporates both Pauli and coulomb correlations, is a local potential. Thus, the computational expense of performing Kohn - Sham calculations is roughly the same as that required for calculations using the Hartree hamiltonian.

However, it is obviously more expensive than solving Euler - Lagrange equations. Current efforts [38] in further studying the nature of the segment level interactions at polymer - metal interfaces are based on performing quantitatively accurate Kohn - Sham calculations.

In the opinion of this author, the Kohn - Sham formulation is the appropriate theoretical framework to probe the segment level interactions at polymer - metal interfaces. In order to perform these calculations, however, a model for the surface must be chosen. Two models of metal surfaces have been used in this connection. So far, these models have been applied toward studying segment - surface interactions with only one metal (aluminum). Aluminum is a simple s-p bonded metal. A simple model for such a surface is the jellium model [37]. In this model, the detailed potential experienced by the electrons due to the ions located at the lattice positions is replaced by a uniform positive background that terminates abruptly at the interface. The electrons move in the field of this smoothed out potential. In Fig. 9, the positive background and the electron densities for two different bulk densities (or Wigner - Seitz radius, r_s) are shown.

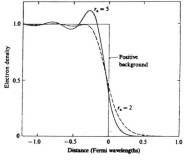

Fig. 9

Shaffer et al. [17] have used the jellium model of an aluminum surface to study its interactions with oligomers of PMMA. The jellium model reproduces many properties of simple metals very well [48], and thus it can be used profitably to examine the problem at hand. However, it is still an approximate model for s-p bonded metals, and furthermore, it is utterly inappropriate for the study of transition metals with d - band electrons. For more accurate calculations of aluminum and transition metal surfaces, the surface has been modeled [38] as a finite cluster of metal atoms. The size of the cluster is chosen to match the cohesive energy and Fermi level of the metal. For aluminum roughly 40 atoms are found to be adequate for such a description. We now discuss the progress that has been made in using such models and DFT toward the study of specific segment - surface interactions. This discussion will be followed by a description of the interfacial chain structure and dynamics that results from such segment level interactions.

Efforts to study the interactions with surfaces are scant [17, 38]. Herein, we will not present the details of the analysis presented by these authors, but rather, we will outline the major conclusions from these efforts and the questions that have arisen as a consequence of these investigations.

Shaffer and coworkers [17, 38] have recently carried out investigations of the segment level interactions between PMMA and aluminum surfaces. These authors have used the jellium model for the aluminum surface, and have used Greens function techniques to calculate the energetics of the interaction as a function of internal and external degrees of freedom. Shaffer et al. [17] have used the Thomas - Fermi kinetic energy functional, the Dirac formula for exchange and the Perdew - Zunger expression for coulomb correlations. The results of these calculations are to be viewed as being semiquantitative. In ongoing studies, Shaffer and Chakraborty [38] have solved the Kohn - Sham equations for the interactions between organic segments and large clusters of aluminum atoms. The calculated results are in qualitative agreement with the calculations using the jellium model. Herein, we discuss the major physical conclusions of the efforts due to Shaffer and coworkers [17, 38] in analyzing the nature of the interactions between PMMA oligomers and aluminum surfaces.

Fig. 10 shows some of the degrees of freedom that have been explored in calculating the energetics of interaction between PMMA monomers and the aluminum surface. The orientation and location of the monomer unit is specified by the distance of a reference atom from the surface and two Eulerian angles. Various internal degrees of freedom, such as the torsion angles shown in Fig. 10, may also be important in determining the interaction energy.

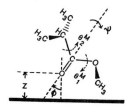

Fig. 10

Fig. 11 shows the energy of interaction for PMMA monomers interacting with an aluminum surface as a function of the distance of the carbonyl oxygen from the surface. The different curves correspond to different values of the orientation angles, ϕ and ψ.

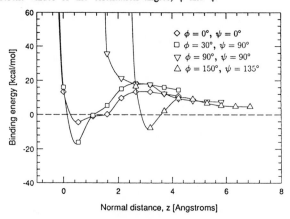

Fig. 11

The energetics for the interactions of PMMA monomers with an aluminum surface depend strongly on the orientation in which the monomer approaches the surface. This can be seen clearly by considering (for example) the curves in Fig. 11 that correspond to values of ϕ equal to 0° and 30°. The interaction is stronger for ϕ equal to 30°. The physical reasoning for this is that the primary interactions involve the Π_u^* orbital and the filled, non-bonding lone pair orbitals of the organic. When the carbonyl bond is tilted toward the surface, the orientation of the Π_u^* orbital is such that it leads to a more favorable interaction with the surface. In contrast to the situations wherein the monomer attaches itself to the surface, the curve corresponding to ϕ equal to 90° represents an orientation in which the alkyl groups approach closest to the surface. In this case, the dominant feature is a strong repulsion at close distances. Further evidence of the strong orientation dependence is provided by the other curves shown in Fig. 11. One important point regarding these results is that there are two strongly interacting functional groups per segment in PMMA, the methoxy group and the carbonyl group. This leads to an even richer structure of the energy hypersurface for the interactions under consideration, as has been demonstrated in Fig. 11 which is taken from the paper by Shaffer et al. [17]. These authors have explored the energy hypersurface in some detail, and have provided an empirical force field that can be used for simulating interfacial chain structure subject to such interactions. We note one important point regarding the results depicted in Fig. 11. The interactions under consideration are strong, and several minima are deeper than the thermal energy, kT, at room temperature. This will be seen to have important implications for interfacial chain structure and dynamics.

Shaffer et al. have also calculated portions of the energy hypersurface for the interactions of *meso* dimers of PMMA with aluminum surfaces. Fig. 12 shows a representative portion of their calculated results.

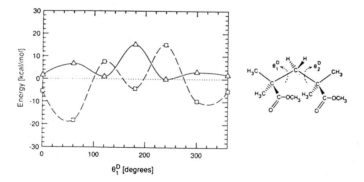

Fig. 12

The calculated results are plotted in Fig. 12 in a manner that requires some explanation. Compared to the monomer, several new internal degrees of freedom are introduced when considering the dimer. Two of these are the torsion angles shown in the inset, θ_1^D and θ_2^D. In Fig. 12, the points represented by squares correspond to the strongest interaction energy for a set of values of the two torsion angles and the orientation angles; in other words, each point corresponds to the minimum of the binding energy well on curves similar to those discussed for the monomer. For cases in which binding does not occur, the positive interaction energy is taken at a point 1.5 Å from the surface. The curve drawn through the squares in Fig. 12 thus represents the variation of the strongest binding energy (or the magnitude of the repulsive interaction for non binding situations) with θ_1^D for a fixed value of θ_2^D.

Several interesting points are to be noted from Fig. 12. First, it is clear that the binding energy is highly dependent on both the internal configuration of the PMMA dimer and its overall orientation of approach. There are distinct configuration dependent minima, as well as configurations that do not bind to the surface. Another interesting feature can be seen by examining the points depicted by triangles in Fig. 8. These points represent the conformational energy of the free PMMA dimer as a function of the backbone torsion angles. The conformation energies were calculated using the intermolecular potentials reported by Vacatello and Flory [49] in their work pertaining to the conformational statistics of PMMA. Comparing the curve drawn through the squares with the binding energy curve leads to an important observation. Many of the conformations of the dimer that lead to strong binding with the surface correspond to cases wherein the conformational energy of the free dimer is high. Conversely, some conformations that are favorable for the free dimer bind only weakly or not at all with the surface. The competition between the conformational energies and binding energies will have an important effect on the near - surface structure of PMMA - aluminum interfaces. The data reported in Fig. 8 show that sequences of bound segments will be stabilized on strongly interacting surfaces in rotational states that are strained for the free polymer. The changes induced in the conformational statistics of PMMA molecules interacting with aluminum surfaces illustrates an important difference between the chemisorption of small molecules and the adsorption of polymer chains interacting with a surface via specific interactions. For small molecules, there is no issue of competition between rotational conformation energies and binding energies associated with specific functional groups. As has been noted by Shaffer et al., the existence of rotationally strained states on the surface has experimental support. Konstandinidis et al. [21] have performed NMR experiments probing the conformations of PMMA molecules interacting with native aluminum oxide surfaces. Shaffer et al. have provided arguments demonstrating that the data reported by Konstandinidis and coworkers does show that the rotational conformation statistics of adsorbed PMMA molecules is quite different from that observed in the bulk.

The energy hypersurfaces reported by Shaffer et al. show that the energetics of the interactions under consideration are highly dependent on the configuration and orientation of the segments as they approach the surface. In other words, the specificity and strength of the interactions leads to an energy hypersurface that is characterized by several local minima. Furthermore, several of the minima (and barriers between them) are deeper (or higher) than the thermal energy, kT, at room temperature. There are many implications of such segment - surface interactions for interfacial chain structure and dynamics. These exciting and important issues are the subject of discussion in the next section.

The progress that has been made in developing an understanding of the interactions of polymer segments with strongly interacting surfaces based on theoretical and computational studies may be summarized as follows. Recent studies [17, 38] have developed a methodology for studying these interactions, and obtained interesting results regarding the nature of the segment - surface interactions. These methods must now be employed to study a variety of different systems (such as transition metal surfaces and various polymers), and the results must be systematically validated by testing directly against experiments. In working toward this goal, progress in experimental approaches that can provide quantitative measures of the interaction energies is also neccessary. We note that theoretical studies of the segment level interactions of organic oligomers with native oxide surfaces is virtually nonexistent, and presents an important challenge for the future.

Interfacial Chain Structure and Dynamics

Let us begin by examining the implications of the energy hypersurfaces that characterize the segment - surface interactions (see preceding section) for the conformations adopted by the chain molecules in the near - surface region.

Since the segment level interactions are highly configuration dependent and the energetics involved (minima and barriers) are often larger than the thermal energy, Chakraborty et al. [18] have argued that the interfacial chains adsorb in non-equilibrium conformations. In order to clearly see the reasoning that underlies this assertion, consider a model for the PMMA - aluminum system. A simple model that captures the most important physical issues that we are concerned with is a linear chain with two pendant groups per segment that interact with the surface by binding to it with energies E_1 and E_2, respectively. Let E_1 be greater than E_2, and let both energy scales be significantly larger than kT. If a segment approaches the surface in an orientation and configuration such that it attaches itself to the surface with binding energy E_2, then the segment will be adsorbed in a local minimum that is not a global minimum. Since the energy hypersurface that characterizes the system is such that local minima are often separated from each other and the global minimum by barriers that are larger than the thermal energy at room temperature, the system is kinetically constrained from accomodating itself in a global minimum of free energy. Based on this and other reasons, Chakraborty et al. [18] have argued that long chain molecules subject to such segment - surface potentials are expected to adsorb in orientations and conformations that lead to frozen in, non-equilibrium structures. In such systems, the adsorbed polymer layer may be considered to be a quasi two - dimensional analog of a bulk polymer glass. A bulk polymer glass is a collection of non-equilibrium structures due to kinetic restrictions to motion below the glass transition temperature, T_g. For strongly interacting interfaces the adsorbed layer is also a collection of non-equilibrium structures. However, the reason for kinetic constraints to motion are quite different. In this case, the nature of interactions with the surface lead to trapped states on the surface. The analogy between a bulk polymer glass and adsorbed chains in the systems under consideration exists only in the sense that in both cases there are collections of non-equilibrium structures. Chakraborty et al. have extended this analogy to define temperatures that are akin to the glass transition temperature for strongly adsorbed polymer layers.

We note that speculations regarding the existence of glass like structures for adsorbed polymer layers has been discussed previously by Kremer [50]. However, Kremer based his ideas on the fact that upon adsorption from solutions below T_g there is an enhanced density near the surface. Thus, there may be too little solvent present to keep the polymer plasticized, and this may lead to the existence of a glassy layer adjacent to the surface. As Chakraborty et al. [18] have pointed out, the origin of the existence of non-equilibrium structures near polymer interfaces characterized by strong and specific segment level interactions is totally different, and there is no connection with the bulk glass transition temperature. In fact, the phenomena under consideration could happen above or below T_g.

In order to provide evidence for their suggestion that non-equilibrium structures constitute the near - surface structure of strongly interacting polymer - solid interfaces, Chakraborty et al. [18] have performed molecular dynamics (MD) simulations for short chains adsorbed from solution. The system they have studied is the model polymer described above. In these simulations the binding wells for the two pendant groups are parabolic with well depths of 17.2 kT and 8.6 kT. The backbone atoms and the solvent molecules interact with the wall via truncated and shifted Lennard - Jones potentials. The polymer is modelled in a manner such that constant bond lengths are maintained, there are harmonic bond bending potentials, and a 3-fold symmetric modification of the torsion potential for butane is used. The MD results reported by Chakraborty et al. show that each realization of the adsorbed chain structure leads to a different non-equilibrium conformation, thus lending support to the preceding arguments.

The formation of locked - in, non - equilibrium structures at strongly interacting polymer - solid interfaces may have many practical consequences that have been mentioned by Chakraborty et al.. An important point that should be noted is that the structure of the adsorbed layer will be highly history dependent. For example, the nature of the adsorbed layer may depend on the concentration of the solution from which the adsorption is carried out. Ponce et al. [51] have found that the measured peel strength of PMMA adsorbed on native aluminum oxide is a strong function of solution concentration. This provides some evidence that the structure of the adsorbed layer is dependent on the history of sample preparation. Another interesting consequence of the formation of non-equilibrium, locked - in structures has been speculated upon [18]. The formation of such structures with strong segment - surface interactions is expected to lead to high adhesive strength. However, it is possible that cohesive failure may occur at the diffuse interface between the glass like adsorbed layer and the polymer chains that are not directly attached to the surface. Recently, some experimental evidence for such cohesive failure has been provided [52]. However, this possibility demands much investigation in order to set it on a sound footing. While such cohesive failure may occur at weakly interacting interfaces, it is less likely.

The existence of non-equilibrium structures at the interface raises the interesting question of how the interfacial chains relax toward equilibrium as a function of time and temperature (or segment - surface interaction energy scale). Unfortunately, for the systems under consideration, MD simulations cannot address these issues because they cannot access the long time scales associated with these relaxation processes. The first step of adsorption from solution has been shown to be fast [7, 18]. It is the subsequent relaxation of the chain molecules toward equilibrium, and the dependence of the rates and nature of these relaxation processes on the temperature that cannot be addressed by brute force MD simulations. Recently, Chakraborty and Adriani [19] have developed a simple stochastic model that addresses these issues. This model is to be viewed as merely a staring point for more detailed studies of interfacial chain dynamics. However, it is a simple physical model that leads to several interesting predictions. We now outline the interesting dynamical issues that have been explored by Chakraborty and Adriani [19].

Consider the model polymer that we have described earlier; viz., a linear chain with two pendant groups per segment that bind to the surface strongly. Further, consider a single chain of the model polymer to be adsorbed in some arbitrary non-equilibrium conformation. Each pendant sticker can now be considered to be either adsorbed or desorbed. Each sticker may therefore be viewed to be a two - state system, and Chakraborty and Adriani noted that there is an isomorphism between the adsorbed chain structure and a spin 1/2 Ising model [53]. The state of each sticker group, σ_j, can take the values of +1 (spin - up) or -1 (spin - down), corresponding to a desorbed or adsorbed sticker, respectively. The collection of 2N spins constitutes Chakraborty and Adriani's model for an adsorbed chain of length N segments. The dynamics of adsorbed chain relaxation is described by the time evolution of the spin states. As such, the model developed by Chakraborty and Adriani belongs to the class of kinetic Ising models first discussed by Glauber [54]. We note that kinetic Ising models have heretofore been considered as models for bulk glassy systems by several authors [e.g., 55, 56].

The temporal evolution of the spin system that constitutes a model for the adsorbed chain is described by a Master equation [57] with transition probabilities that satisfy detailed balance. A description of the relevant physical processes that determine the dynamical phenomena under consideration is contained in the rules that are invoked in constructing the transition probability matrix for spin flips in the Master equation. Chakraborty and Adriani [19] have formulated these rules by carefully examining the physical situation. The reader is directed to the original papers for details regarding the formulation of the rules that are formulated and invoked by these authors. Herein, we merely state the rules, and focus on the results obtained by Chakraborty and Adriani.

If the sticker groups were independent then the rates of adsorption and desorption of individual stickers could be described by transition state theory (TST). However, for the situation under consideration the sticker groups are not independent. There are several impediments in the escape routes available for desorption and channels for adsorption. These impediments arise due to many reasons such as chain connectivity, steric constraints imposed by neighboring segments, rotational strain in the bonds connecting a given segment to adjacent ones, etc.. The rates of adsorption or desorption of individual stickers are thus strongly correlated with the state of the neighboring segments, and this leads to a highly cooperative dynamic behavior for adsorbed chain relaxation. In order to account for the cooperativity that characterizes adsorbed chain relaxation Chakraborty and Adriani formally write the rate at which a spin flips up against the field imposed by the surface to be

$$W_{j,\,up} = k_0\, f\,(\text{local spin state}) \exp\{-E_{j,\,well}\,/\,kT\} \quad (15)$$

where k_0 is an intrinsic time constant (or attempt frequency), $E_{j,\,well}$ is the depth of the binding well for the sticker in question, and f is a constraint function that depends on the local environment of spin states. The corresponding rate for a spin to flip down is determined by the condition of detailed balance. Note that without the function f, Eq. 15 would be the result obtained from TST.

Chakraborty and Adriani have developed the following expression for the constraint function based on purely physical considerations

$$f = \alpha\, \beta\, \gamma \quad (16)$$

The function α reflects the influence of steric constraints imposed by neighboring segments and certain aspects of restrictions due to chain connectivity. β embodies the restraints imposed by the existence of rotationally strained conformations on the surface [17, 21], and γ reflects the entropic penalties associated with shortening loops and tails upon adsorption of a given sticker. Chakraborty and Adriani [19] have provided explicit expressions for α, β, and γ. Since their rules determining the rates of spin flips were developed based on physical induction rather than rigorous deduction, Chakraborty and Adriani tested the validity of their model by comparing the results of the kinetic Ising model with those obtained from brute force MD simulations for short chains. On a coarse grained time scale they found that the dynamics were similar, and that the final states obtained at various temperatures were identical. These authors then presented results for the dynamics of adsorbed chain relaxation for long chains over time scales that are not accessible via MD simulations.

Chakraborty and Adriani have presented results for the temporal evolution of the average magnetization. The average magnetization, Ψ, is defined as follows

$$\Psi = \{ (M-M_{eq.})\,/\,(1-M_{eq.})\} \quad (17)$$

where $M_{eq.}$ is the magnetization at equilibrium (at a given temperature) and M is the average magnetization at any time. Of course, the average magnetization is directly and trivially related to the number of sorbed segments. In Fig. 13, Chakraborty and Adriani's results for the temporal evolution of the magnetization are shown for the weaker stickers of a polymer chain that is 500 segments long. For the case shown in Fig. 13, the binding energies for the strong and weak stickers are 7.68 kT and 3.84 kT, respectively. This case therefore represents the relaxation behavior at low temperatures. At the low temperatures being considered it is clear that the adsorbed chains get trapped in a non-equilibrium conformation, and thereafter remains largely frozen. Although Chakraborty and Adriani have shown results upto only 150 nanoseconds, it is evident that at this temperature the adsorbed chain will remain in a

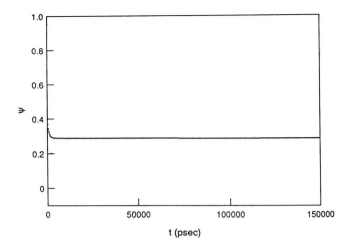

Fig. 13

non-equilibrium frozen state over experimental time scales. At this temperature, the adsorbed chains may be considered to be analogous to a bulk polymer glass.

In Fig. 14, the temporal evolution of Ψ is shown for a higher temperature. In this case, the system is seen to be slowly evolving toward equilibrium. However, the kinetics of relaxation cannot be described by one time constant. In fact, as can be seen from the line drawn through the data in the main plot and the inset, the calculations predict that a stretched exponential function, $\exp\{-(t/\tau)^{\delta}\}$, fits the data very well (with $\delta = 0.20$).

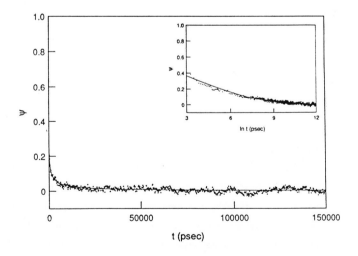

Fig. 14

Chakraborty and Adriani have reported that, as the temperature below which relaxation to equilibrium does not occur over experimental time scales is approached, over a range of temperatures, the temporal evolution of the magnetization is described by such Kohlrausch - Williams - Watts (KWW) expressions (with δ ranging from 0.2 to 0.5). Other dynamical quantities, such as the one spin time correlation function, are also found to exhibit similar behavior. Furthermore, Chakraborty and Adriani have found similar behavior for polymer chains of varying length and different architecture. However, there are some differences in the details of the observed dynamic behavior for different chain architectures.

The relaxation behavior observed by Chakraborty and Adriani (e.g., Figs. 13 and 14) is analogous to that observed for glass forming liquids. Chakraborty and Adriani have found further evidence for glass-like behavior by examining the variation of an average relaxation time with temperature. These authors defined an average nonlinear relaxation time as the area under the curve, Ψ (t). In Fig. 15 their results are shown in the form of an Arrhenius plot.

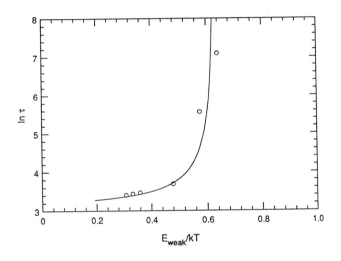

Fig. 15

For the case considered by Chakraborty and Adriani, they found that for temperatures below E / 2k the relaxation time diverges (no relaxation in experimental time scales). A temperature somewhat above E / 2k may thus be viewed to be analogous to a glass transition temperature. Fig. 15 also shows that at high temperatures the variation of the relaxation time with temperature is of the Arrhenius form, while as the temperature below which relaxation does not occur in experimental time scales is approached from above there is a sharp increase in slope and curvature. The results reported by Chakraborty and Adriani can be fit by the Vogel - Fulcher function [58], and is thus characteristic of glass forming liquids.

The physical reasons for the glassy relaxation behavior that has been reported by Chakraborty and Adriani for adsorbed polymer chains merits some discussion. Ever since Kohlrausch [59] it has been known that the dynamics of strongly interacting materials can often be described by the stretched exponential function rather than the standard Debye form. Palmer et al. [60] have shown that a class of models wherein the degrees of freedom may be considered to be hierarchically constrained leads to relaxation behavior that can be described by KWW expressions. Hierarchical constraints imply that the degrees of freedom constitute a series of levels. For a high level degree of freedom to relax or change state, some degrees of freedom at lower levels must adopt certain prescribed states. As noted by Chakraborty et al., their model for adsorbed chain relaxation belongs to this dynamical universality class. In this case, the spin being examined constitutes the highest level degree of freedom, and the neighboring spin states (lower level degrees of freedom) must adopt certain prescribed states such that impediments are removed from desorption routes or adsorption channels. For adsorbed chains, Chakraborty and Adriani have, therefore, provided a microscopic description of the physical reasons that naturally lead to hierarchically constrained dynamics.

Our current understanding of the structure and dynamics of interfacial chains adjacent to strongly interacting solid surfaces may be summarized as follows. When there are strong and specific interactions between polymer segments and the surface, the energy hypersurface that characterizes these interactions exhibits many configuration and orientation dependent local minima that are often separated by barriers that are larger than the thermal energy, kT. In such situations, the interfacial chains adsorb in non-equilibrium conformations. The structure of the interfacial chains may be viewed to be analogous to a bulk polymer glass. At low temperatures, these non-equilibrium structures do not relax to equilibrium in experimental time scales because they are trapped in non-equilibrium conformations. At higher temperatures, the dynamics of relaxation is highly cooperative and can be described by the KWW expression. Furthermore, as the temperature below which relaxation to equilibrium does not occur in experimental time scales is approached from above, average relaxation times exhibit strongly non-Arrhenius behavior characteristic of glass forming liquids. These results indicate that the dynamic behavior of adsorbed chains near strongly interacting surfaces is similar to that observed for the so called fragile liquid described by Angell [61]. We note that recently Frantz and Granick [7] have performed insightful experiments that are consistent with the glassy relaxation predicted by Chakraborty and Adriani [19]. However, direct experimental evidence for the predicted dynamics does not exist.

Current Status and Future Directions

Let us recapitulate what we have learnt regarding the nature of strongly interacting polymer - solid interfaces. As noted at the outset, there are three levels at which theoretical studies are required (Fig. 3).

The quantum mechanical studies may be divided into two categories. The first category is comprised of studies aimed to elucidate the nature of the interactions between metal atoms and polymer segments. These studies are analogs of experimental investigations wherein small dosages of metal atoms are evaporated onto polymer substrates, and are directly relevant to certain applications. As we have seen, HF theory has been very successful in identifying the strongly interacting functional groups and elucidating experimental data. The HF studies also serve as important precursors to the quantum mechanical studies aimed toward obtaining the energy hypersurfaces that characterize the segment - surface interactions. Knowledge of the nature of the interactions greatly reduces the computational task because certain regions of the hypersurface need not be probed extensively if there are no strong interactions to be expected there. In short, it is fair to say that the application of HF calculations toward the study of the interactions of metal atoms with polymer segments is a mature endeavor, and what remains is to routinely apply it toward the investigation of different systems.

The second category of quantum mechanical calculations involve the calculation of the energy hypersurfaces that characterize the nature of the segment - surface interactions. This is an emerging effort, and only two recent studies have applied DFT explicitly toward this end. While a lot has been learnt from these studies, much remains to be accomplished. Several systems need to be examined and the approach validated for quantitative accuracy by comparison with spectroscopic experiments. Furthermore, one issue that requires much attention is the systematic study of how the interactions of a segment with the surface affects the interactions of a neighboring segment with the surface. In other words, for a given system, one must establish how many segments comprise a unit such that additive potentials may be constructed for these units. Such a systematic investigation has not been undertaken for any system. For systems wherein several monomer units must be incorporated in each unit, calculating the energy hypersurface using the approach that we have discussed may be prohibitive. In such cases, perhaps the valence effective hamiltonian techniques due to Bredas and coworkers [62] may be an appropriate avenue to explore. Another uncharted area is the theoretical study of the detailed nature of the interactions between polymer segments and native metal oxide surfaces. Although this is a difficult problem, progress in this regard is neccessary because this system is of practical importance in many applications.

It has been established that the qualitative features of the energy hypersurfaces that characterize the segment level interactions for strongly interacting polymers exhibit several local minima that are often separated by high barriers [17, 18]. Thus, the chains adsorb in non-equilibrium conformations which then may evolve to equilibrium. Detailed studies of the structure of the interface and the temporal evolution of the structure with time and temperature are virtually nonexistent. There are several difficulties associated with applying brute force techniques toward the investigation of interfacial chain structure and dynamics for strongly interacting polymer / solid interfaces. As far as the structure is concerned, the difficulty is that, for non-equilibrium systems, several realizations must be obtained in order to attain reliable averages. As far as the dynamics are concerned, as we have seen, the dynamics are extremely cooperative and brute force approaches cannot probe the appropriate time scales. However, this is an important problem, and progress in developing approaches that can address these issues is needed. A beginning has been made by Chakraborty and Adriani [19] who have developed a stochastic approach to study the qualitative features of the dynamics. However, it is not possible to probe the details of interfacial chain structure and dynamics for specific systems via kinetic Ising models. Presently, it is this author's opinion that the most promising approaches toward making progress in this regard appear to be the application of stochastic approaches wherein the transition probabilities are obtained from the quantum mechanically calculated energy hypersurface and dynamically corrected transition state theory [63, 64]. Cellular automata approaches also offer interesting possibilities for the simulation of long time dynamics[65, 66]. The interesting dynamic behavior of these uniquely confined polymer chains, if elucidated in detail, could shed much light on practical issues such as how these systems should be processed during their synthesis such that they attain specific properties. In addition, the problem poses a difficult yet exciting challenge from a fundamental point of view. Of course, theoretical efforts in this direction must be complemented by experimental progress, and NMR spectroscopy is perhaps the most appropriate tool for such investigations [67].

Very little is currently known regarding the relationship between interfacial chain structure and dynamics and macroscopic properties such as adhesion and diffusion barrier properties. There are some recent efforts that have confronted this daunting problem [68, 69]. Efforts in this vein are neccessarily more coarse grained than the issues that we have been concerned with in this article. It is hoped that as our abilities to predict interfacial chain conformations and dynamics for strongly interacting polymer - solid interfaces based on a knowledge of chemical constitution of the polymer and the substrate improve, there will be greater motivation to develop approaches for predicting macroscopic properties from a knowledge of the same. Of course, as we have pointed out, some pioneering researchers have already started to shed light on this very difficult and important problem.

Acknowledgments: Partial financial support for the preparation of this article was provided by the Director, Office of Basic Energy Sciences, U.S. DOE, under Contract DE-AC03-76SF00098, and the Shell Foundation. I would like to thank various collaborators who have influenced my ideas regarding strongly interacting polymer - solid interfaces: P.M. Adriani, H.T. Davis, M.M. Denn, J.S. Shaffer, R.E. Taylor, D.N. Theodorou, M.V. Tirrell.

References

1] G. Hadziioannou, S. Patel, M. Tirrell, *J. Am. Chem. Soc.*, **108**, 2869 (1986).

2] E. Helfand, *Macromolecules*, **9**, 307 (1976).

3] D.N. Theodorou, *Macromolecules*, **21**, 1391 (1988).

4] I. Bitsanis, G. Hadziioanou, *J. Chem. Phys.*, **92**, 3827 (1990).

5] J.M.H.M. Scheutjens, G.J. Fleer, *J. Phys. Chem.*, **83**, 1619 (1979).

6] P.G. deGennes, in *New Trends in Physics and Physical Chemistry of Polymers*, Edited by L.H. Lee (Plenum, New York, 1990).

7] P. Frantz, S. Granick, *Phys. Rev. Lett.*, **66**, 899 (1991).

8] J.M. Burkstrand, *ACS Symposium Series*, **162**, 339 (1981).

9]   H. Leidheiser, Jr., P.D. Deck, *Science*, **241**, 1176 (1988).

10]   H. Grunze, R.N. Lamb, *Chem. Phys. Lett.*, **133**, 283 (1987).

11]   Lj. Atanasoska, S.G. Anderson, H.M. Meyer, III, Z. Lin, J.H. Weaver, *J. Vac. Sci. Technol.*, **A5**, 3325 (1987).

12]   J.J. Pireaux, M. Vermearsch, N. Degosserie, G. Gregoire, Y. Novis, M. Chtaib, R. Caudano, in *Adhesion and Friction*, Edited by M. Grunze, H. Kreuzer (Springer-Verlag, Berlin, 1989).

13]   A.R. Rossi, P.N. Sanda, B.D. Silverman, P.S. Ho, *Organometallics*, **6**, 580 (1987).

14]   B.D. Silverman, P.N. Sanda, J.C. Clabes, P.S. Ho, A.R. Rossi, *J. Polym. Sci. Polym. Chem. Ed.*, **24**, 3325 (1986).

15]   L.P. Buchwalter, B.D. Silverman, L. Witt, A.R. Rossi, *J. Vac. Sci. Technol.*, **A5**, 226 (1987).

16]   A.K. Chakraborty, H.T. Davis, M. Tirrell, *J. Polym. Sci., Polym. Chem. Ed*, **28**, 3185 (1990).

17]   J.S. Shaffer, A.K. Chakraborty, M. Tirrell, H.T. Davis, J.L. Martins, *J. Chem. Phys.*, in press (1991).

18]   A.K. Chakraborty, J.S. Shaffer, P.M. Adriani, *Macromolecules*, in press (1991).

19]   A.K. Chakraborty, P.M. Adriani, *submitted to Phys. Rev. Lett.*.

20]   N.J. diNardo, proceedings of the *Symposium on Metallized Plastics, Fundamental and Applied Aspects*, Electrochemical Society Meeting, Chicago, October (1988).

21]   K. Konstandinidis, B. Thakkar, A.K. Chakraborty, L.W. Potts, R. Tannenbaum, J.F. Evans, M. Tirrell, *Langmuir*, in press (1991).

22]   D.T. Clark, H.R. Thomas, D. Shuttleworth, *J. Polym. Sci. Polym. Lett. Ed.*, **16**, 465 (1978).

23]   J.M. Burkstrand, *J. Vac. Sci. Technol.*, **16**, 363 (1979).

24]   J.M. Burkstrand, *J. Appl. Phys.*, **52**, 4795 (1981).

25]   J.W. Bartha, P.O. Hahn, F. LeGoues, P.S. Ho, *J. Vac. Sci. Technol.*, **A3**, 1390 (1985).

26]   J.L. Jordan-Sweet, C.A. Kovac, M.J. Goldberg, J.F. Morar, *J. Chem. Phys.*, **89**, 2482 (1988).

27]   R. Haight, R.C. White, B.D. Silverman, P.S. Ho, *J. Vac. Sci. Technol.*, **A6**, 2188 (1988).

28]   J.L. Jordan, C.A. Kovac, J.F. Morar, R.A. Pollak, *Phys. Rev.*, **B36**, 1369 (1987).

29]   W.J. Hehre, L. Radom, P.v.R. Schleyer, J.A. Pople, *Ab Initio Molecular Orbital Theory*, Wiley, New York (1986).

30]   T. Clarke, *A Handbook of Computational Chemistry*, Wiley, New York (1985).

31]   M.J.S. Dewar, W. Thiel, *J. Am. Chem. Soc.*, **99**, 4899 (1977).

32]   M.J.S. Dewar, W. Thiel, *J. Am. Chem. Soc.*, **99**, 4907 (1977).

33]   T. Koopmans, *Physica,'s Grav.*, **1**, 104 (1933).

34]   D.T. Clark, B.J. Cromarty, A. Dilks, *J. Polym. Sci. Polym. Chem. Ed.*, **16**, 3173 (1978).

35]   D.L. Allara, *Polym. Sci. Technol.*, **12B**, 751 (1980).

36]   B.M. deKoven, P.L. Hagans, *Appl. Surf. Sci.*, **27**, 199 (1986).

37]   A. Zangwill, *Physics at Surfaces*, Cambridge University Press, Cambridge (1989).

38]   J.S. Shaffer, A.K. Chakraborty, to be published.

39] D.M. Ceperley, B.J. Alder, *Phys. Rev. Lett*, **45**, 566 (1980).

40] P. Hohenberg, W. Kohn, *Phys. Rev.*, **B136**, 864 (1964).

41] R.G. Parr, W. Yang, *Density Functional Theory of Atoms and Molecules*, Oxford University Press, Oxford (1989).

42] M. Gellman, K.A. Brueckner, *Phys. Rev.*, **106**, 364 (1957).

43] D.M. Ceperley, B.J. Alder, *Phys. Rev. Lett*, **45**, 566 (1980).

44] J.P. Perdew, A. Zunger, *Phys. Rev.*, **B23**, 5048 (1981).

45] O. Gunnarsson, J. Harris, R.O. Jones, *J. Chem. Phys.*, **67**, 3970 (1977).

46] N.H. March in *Theory of the Inhomogeneous Electron Gas*, Edited by N.H. March, S. Lundqvist (Plenum, New York, 1983).

47] W. Kohn, L.J. Sham, Phys. Rev., A140, 1133 (1965).

48] N.D. Lang in *Theory of the Inhomogeneous Electron Gas*, Edited by N.H. March, S. Lundqvist (Plenum, New York, 1983).

49] M. Vacatello, P.J. Flory, *Macromolecules*, **19**, 405 (1986).

50] K. Kremer, *J. Phys.*, **47**, 1269 (1986).

51] S. Ponce, D. Gamet, H.P. Schreiber, *J. Coatings Technol.*, **26**, 37 (1985).

52] D. Hill, M. Salmeron, M.M. Denn, *paper to be presented at the annual meeting of the Am. Inst. of Chem. Eng.*, Los Angeles, November (1991).

53] D. Chandler, *Introduction to Modern Statistical Mechanics*, Oxford University Press, Oxford (1987).

54] R.J. Glauber, *J. Math. Phys.*, **4**, 294 (1963).

55] G.H. Fredrickson, H.C. Anderson, *Phys. Rev. Lett.*, **53**, 1244 (1984); *J. Chem. Phys.*, **83**, 5822 (1985); G.H. Fredrickson, S.A. Brawer, **ibid.**, **84**, 3351 (1986).

56] J.L. Skinner, *J. Chem. Phys.*, **79**, 1955 (1983); J. Budimir, J.L. Skinner, **ibid.**, **82**, 5232 (1985).

57] N.G. van Kampen, *Stochastic Processes in Physics and Chemistry*, North Holland, Amsterdam (1981).

58] C.A. Angell, D.L. Smith, *J. Phys. Chem.*, **86**, 3845 (1982).

59] R. Kohlrausch, *Ann. Phys.* (Leipzig), **12**, 393 (1847).

60] R.G. Palmer, D.L. Stein, E.A. Abrahams, P.W. Anderson, *Phys. Rev. Lett.*, **53**, 958 (1984).

61] C.A. Angell, A. Dworkin, P. Figuire, A. Fuchs, H. Szware, *J. Chim. Phys.*, **82**, 773 (1985).

62] J.L. Bredas, R.R. Chance, R. Silbey, G. Nicolas, Ph. Durand, *J. Chem. Phys.*, **75**, 255 (1981).

63] A.F. Voter, J.D. Doll, *J. Chem. Phys.*, **82**, 80 (1985).

64] R. Czerminski, R. Elber, *Int. J. of Quant. Chem.*, **S24**, 167 (1990).

65] J.M.V.A. Koelman, *Phys. Rev. Lett*, **64**, 1915 (1990).

66] Y. Bar-Yam, Y. Rabin, C.H. Bennett, M.A. Smith, N. Margolus, T. Toffoli, *submitted to Nature.*

67] R.E. Taylor, A.K. Chakraborty, A. Pines, private comm. of work in progress (1991).

68] P.G. deGennes, *paper presented at the International Conference on Macromolecules*, Montreal, July (1990).

69] D.N. Theodorou, private comm., 1991.

Covalently coupled polymer–semiconductor interfaces: the precursor 1,4 phenylenediamine on Si(100)

Th. Kugler, U. Thibaut*, M. Abraham, G. Folkers*, and W. Göpel

Institut für Physikalische und Theoretische Chemie and
*Institut für Pharmazie, Auf der Morgenstelle 8
W-7400 Tübingen, Germany

ABSTRACT: The formation of inorganic/organic hybrid systems under ultrahighvacuum (UHV) conditions was studied for the adsorption of 1,4 phenylendiamine (PDA) and aniline on Si(100)(2x1) single crystal surfaces. Experimental X-ray photoemission spectra (XPS), ultraviolet-photoemission spectra (UPS) and high-resolution electron energy loss spectra (HREELS) are compared with theoretical results obtained in the framework of semiempirical cluster calculations using the MNDO-PM3 approach.

1. INTRODUCTION

The covalent electronic coupling between molecules with specific electronic or optical properties and semiconductor surfaces offers a promising approach to molecular engineering based upon conventional semiconductor device technology. In this context, we utilize the covalent coupling of a bifunctional organic molecule to a clean silicon surface in a reaction involving one functional group per molecule and the well-ordered two dimensional dangling bond system of the reconstructed semiconductor surface. By this technique a two dimensional organic layer of molecules is obtained, which can be modified in a subsequent step by reacting the remaining functional group of the adsorbed molecules with a second organic component. This adsorbate system can thus be used as a precursor for a thicker ordered polymer-overlayer on the semiconductor surface. To study the first step of this "molecular layer epitaxy" we have chosen the model systems 1,4 phenylenediamine (PDA) and aniline as organic components and the technologically most important Si(100)(2x1) surface as substrate.

2. EXPERIMENTAL

The experiments were performed in an ultra-high-vacuum (UHV) chamber with XPS, UPS and HREELS-facilities. For details see [W. Göpel 1989].

The samples were cut from Si(100)-wafers with native oxide, doped (40 mΩcm) from Wacker-Chemitronic, Burghausen (FRG). They were precleaned by repeated rising in EtOH$_{Uvasol}$ and H$_2$O$_{deion}$. Clean Si(100)(2x1) single crystal faces were prepared in

UHV by first heating the samples at 500°C for several hours and subsequently removing the passivating oxide layer by direct resistive heating to above 1000°C. After cooling down, PDA or aniline was admitted into the chamber via a leak-valve at a pressure of 1×10^{-5} mbar. The maximum exposures to obtain equilibria were 2400 L for PDA and 9000 L for aniline. The organic substances were freshly sublimated or destilled before use. Temperatures were measured using a Pt/Pt-Rh-thermocouple fixed to the tantalum clamps of the sample-holder.

XPS-spectra were taken using a Mg-K$_\alpha$ X-ray source. To monitor the thermal decomposition of the adsorbate, the samples were heated to successively higher temperatures for 5 min. XPS-spectra were taken after cooling down to room temperature. UPS spectra were recorded using HeI-radiation. The HREELS measurements were done in the specular mode (LH-ELS 22).

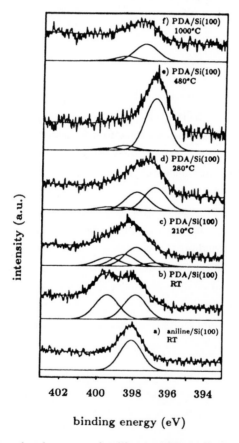

Fig.1   N1s core-level spectra of aniline and PDA after adsorption on a Si(100)(2x1)-surface and subsequent thermal treatment: a) aniline after adsorption, b-f) PDA after adsorption and annealing to 210, 280, 480 and 1000°C.

For the vibrational measurements the primary energy of the electron beam was 5 eV, for the electronic transitions 15 eV.

Semiempirical calculations were performed using MOPAC 5.0 [J.J.P. Stewart, MOPAC 5.0] in the MDNO-PM3 paramerization [J.J.P. Stewart 1989] on a CONVEX C220 two processor machine. Structures were displayed on an E&S PS 390 graphics system within the SYBYL 5.3 [TRIPOS.SYBIL 5.3, 1989] molecular modeling environment.

3. RESULTS AND DISCUSSION

The N1s-core level spectra of PDA and aniline each adsorbed on Si(100)(2x1) characterize a specific chemical shift due to chemisorption of the bifunctional organic molecule PDA. This shift involves only one of the two amino groups and leaves unaltered the second one. The peak at lower binding energy and the only peak in the spectrum of aniline are both attributed to an amino group which is directly bonded to the silicon surface after release of hydrogen atoms (Fig.1a,b). Thermal decomposition of PDA adsorbed on Si(100)(2x1) results in the gradual disapperance of the peak at higher binding energy (which is attributed to the free amino group) and a shift of the peak maximum towards lower binding energy. This indicates a dehydrogenation of the free amino group by the formation of N-Si-bonds. At even higher temperatures silicon nitride is formed (Fig.1 c-f).

The UPS results show a decrease in the work function upon chemisorption (Fig.2).

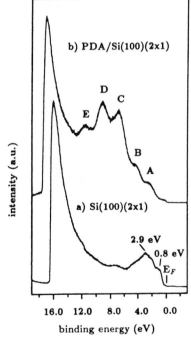

Fig.2   HeI ultraviolet photoelectron spectra (E_D = 21.2 eV) of the clean Si(100)(2x1)-surface (a) and after adsorption of PDA (b).

This effect results from both a charge-transfer from the aromatic ring system to the substrate and from an adsorbate-induced quenching of the intrinsic surface dipole moment of the asymmetric dimer reconstructed surface. A quantitive separation of both effects, however, cannot be given at the moment. The existence of the free amino groups in PDA adsorbed on Si(100)(2x1) is confirmed by the occurrence of NH_2 stretching and scissor vibrations in vibrational HREELS (Fig.3).

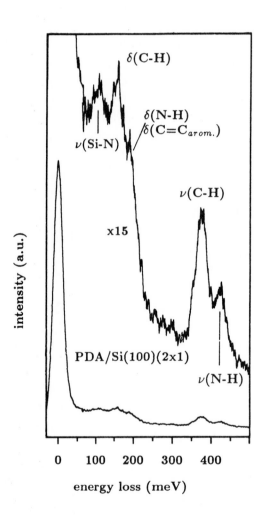

Fig.3   High-resolution electron energy loss spectrum (HREELS) of PDA adsorbed on Si(100)(2x1) with vibrational transitions (primary energy 5 eV, FWHM 28 meV).

Electronic transitions of the benzene ring probed by electronic HREELS support the assumption of the aromatic ring system persisting unaffected after chemisorption. The energetic shift of the electron transitions around 4. 1 eV can be explained in terms of a donor-acceptor-interaction between organic HOMO and substrate acceptor states. Computer aided molecular modeling of PDA adsorbed on the Si(100) semiconductor surface was done by simulating a cluster of finite size (Fig.4)

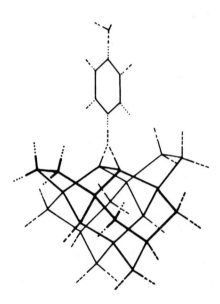

Fig.4   Stick representation of the model used for simulating the Si(100)(2x1)-surface with PDA attached on top of the free-valency silicon atoms of the surface top-layer.

This gives insight into steric and electronic factors determining the structure and reactivity of inorganic/organic hybrid systems. The calculation of the net atomic charges and the visualization of the MO contour-Ψ-plots support the inductive and mesomeric effects deduced from the experimental results. Based on the calculated eigenvalues, the UPS-spectrum of PDA adsorbed on Si(100)(2x1) is simulated by convolution with a Gaussian function. In order to simulate the surface sensitivity of UPS, states predominantly localised in the silicon substrate ($|\Psi|^2 > 90\%$) were omitted. There is a surprisingly good agreement between the calculated density of states of our suggested surface atom complex and the experimental UPS data (Fig.5).

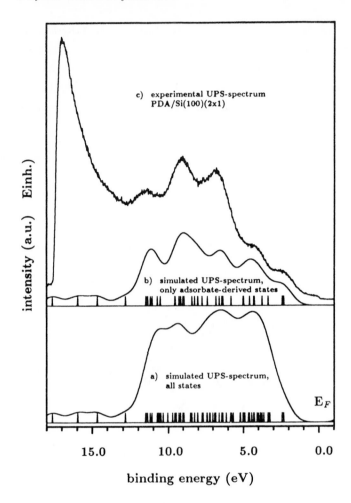

Fig.5   Complete spectrum of all eigenvalues of the adsorbate cluster and convolution by a Gaussian function with a FWHM of 1.4 eV (a); spectrum of the eigenvalues of predominantly "adsorbate-induced states". This approach simulates the surface sensitivity of UPS by excluding states with a predominant contribution to $|\Psi|^2$ localized at the silicon atom positions being > 90%. The eigenvalues were broadend by a Gaussian function with a FWHM of 1.4 eV (b); experimental UPS-spectrum of PDA adsorbed on Si(100)(2x1) (c).

Assuming the suggested binding geometry to be relevant, a tentative assignment of the UPS peaks can be done by comparison with the corresponding MO contour-Ψ-plots.

ACKNOWLEDGEMENTS

We acknowledge the technical assistance by W. Neu and the support in computer simulation by A. Dominik. This work was supported financially by the BASF, Ludwigshafen, (FRG), by the BMFT, Bonn, (FRG) (Contract No.: TK 0323/0) and by the CEC under the Framework Program of Technology in the program ESPRIT: Basic Research Action 3200 "OLDS".

4. LITERATURE

W. Göpel, Sensors and Actuators **16** (1989) 167-193
J.J.P. Stewart, MOPAC 5.0, (QCPE 455). Frank J. Seiler Research Laboratory, Colorado Springs, USA
J.J.P. Stewart, J.Comp.Chem., **10** (1989) 209
J.J.P. Stewart, J.Comp.Chem., **10** (1989) 221
TRIPOS.SYBYL 5.3. Tripos Ass., Saint Louis, USA (1989)

Preparation, polymerization and spectroscopic characterization of thin distyrylpyrazine films on silver and silicon surfaces

M. Abraham[*], S. Lach, A. Dominik, R. Lege, K. Rauer, V. Hoffmann, D. Oelkrug, and W. Göpel

Institut für Physikalische und Theoretische Chemie, Auf der Morgenstelle 8, D-7400 Tübingen

[*]Permanent Address: IMM Institut für Mikrotechnik GmbH, Ackermannweg 10, Postfach 2240, D-6500 Mainz

ABSTRACT: Thin organic polycrystalline films of DSP and poly-DSP were prepared by Knudsen evaporation and deposition from solution and melt on different substrates, i.e. Ag(111), oxidized silicon wafer (SiO_2/Si), and fused silica. The monomeric and differently polymerized films were characterized by scanning electron microscopy, photoemission (XPS, UPS)-, UV/VIS-, luminescence- and infrared spectroscopy. Electronic and vibrational structures are in good agreement with theoretical results from semiempirical MNDO calculations (MOPAC with PM3 parameter sets). Different morphologies are observed for the DSP-monomer on Ag(111) and SiO_2/Si, and the different intermolecular interactions, molecular orientations and photopolymerization structures are analyzed in detail.

1. INTRODUCTION

Preparative and spectroscopic control of the polymer/solid interface is an important approach for the growth of ordered polymer films with good adhesion and well-defined electric and optical properties. In this context, crystalline polymer films are of particular interest if they can be prepared by vacuum sublimation and subsequent photopolymerization. This makes possible their organic/inorganic heteroepitactic processing [Philips 1990]. The 2,5-distyrylpyrazine (DSP) [Braun and Wegner 1983, Nakanashi et al 1980 and references therein] represents such a compound, that can be vacuum-sublimed and photopolymerized in its monomeric crystalline state to form a highly ordered polymer with cyclobutane rings in the main chain. Its thermal polymerization on the other hand can be excluded due to the Woodward Hoffmann rules. The polymerization occurs only in the orthorhombic α-phase, whereas the monoclinic γ-phase is photostable. The photochemical process is initiated by

UV/VIS-irradiation at > 22000 cm^{-1} and leads to a less extended π-system and hence to a blue shift in the absorption.

In a recent paper [Abraham et al 1991] we reported on the controlled desorption of DSP monomers and lower oligomers by temperature-programmed mass spectrometry and on the in-situ monitoring of photochemical reactions in vacuum-deposited monomeric films by High-Resolution-Electron-Energy-Loss-Spectrometry (HREELS) and optical techniques.

The aim of the present paper is to investigate in detail the interface formation between DSP and different substrates and its influence on photopolymerization as well as three dimensional film growth.

2. EXPERIMENTAL

The monomers were purified by gradient sublimation. Their films were grown by vacuum sublimation from a Knudsen cell in ultrahigh vacuum (UHV, 10^{-9} Torr) or in high vacuum (HV, 10^{-6} Torr). For comparison, films were also prepared from solution and melting polycrystalline DSP on substrates. To ensure evaporation of only monomers, the Knudsen cell was kept at a temperature well below the critical temperature for the desorption of higher oligomers [Abraham et al 1991]. Clean Ag(111) surfaces, silica covered silicon wafers, and fused silica were used as substrates. During formation of DSP-layers on SiO$_2$/Si-wafers, the substrate was facultatively cooled from the back side with liquid air. Highly oriented α-DSP-layers were obtained on SiO$_2$/Si surfaces by freezing them out from their melt in a horizontal temperature gradient on a Kofler's hot stage.

To study and characterize the interface formation we applied Ultraviolett Photoemission and X-Ray Spectroscopies (UPS and XPS), UV/VIS-, Fluorescence Spectroscopy and Infrared Dichroic measurements as analytic tools with details on the experimental setup described earlier [Abraham et al 1991]. The three dimensional film structure was characterized subsequently by scanning electron microscopy (SEM) on (Au) metallized films.

3. RESULTS AND DISCUSSION

3.1 Photoemission Spectroscopy of Monomeric DSP

UPS with its information depth in the order of a few Ångstroms is a valuable tool to study the interface formation provided that corresponding bulk spectra are available as standards. Therefore, the typical UPS-spectrum of a bulk monomeric film was studied first. This film was thick enough to screen completely any influence from the substrate. Figure 1 shows the emission near the valence band edge.

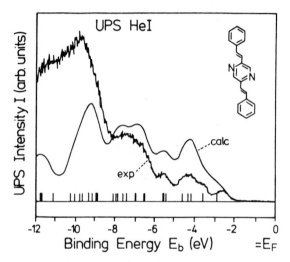

Fig. 1 Comparison of the experimental HeI photoemission spectrum of DSP with the calculated densities of valence states. The unbroadened molecular-orbital eigenvalues are also shown.

To assign the features in the spectrum to occupied molecular orbitals we performed semiempirical calculations of isolated DSP-monomers. The MNDO-procedure was used in the framework of the program package MOPAC with PM3-parameter sets. The program leads to eigenvalues and eigenfunctions for the optimized geometry at the total energy minimum. To simulate the broader UPS-spectrum the as-obtained eigenvalues (bars in Figure 1) were convoluted with broad Gaussians with 1 eV FWHM. This leads to the continuous spectrum in Figure 1. Although UPS final state or transition matrix effects have not been taken into account, the position and the number of the characteristic peaks is in good agreement with experimental results. We therefore conclude, that the emission in this energy range results from π-derived levels. Their hybridization is pronounced, so that the peaks cannot be assigned to localized MOs. The good quality of our MOPAC-calculations can also be deduced from calculations of the vibrational eigenmodes and their comparison with IR data. Theoretically, they were obtained from the force constants derived from the MOPAC optimized geometry. Figure 2 compares the experimental IR-spectrum in the CH-stretch range with as-calculated vibrational eigenmodes. The calculation makes it possible to assign the observed peaks to eigenmodes of different groups as shown in this figure.

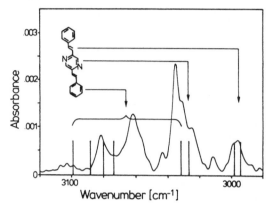

Fig. 2   Infrared absorption spectrum of μ-DSP in the C-H stretch range, and vibrational modes calculated from MOPAC-force constants.

3.2 DSP Monomers on Ag(111) and SiO_2/Si

3.2.1 Survey

The growth of DSP-monomer films on Ag(111) and on oxidized silicon wafers was studied by thickness-dependent UPS and XPS, luminescence spectroscopy, IR-dichroic measurements, and subsequently by Scanning Electron Microscopy.

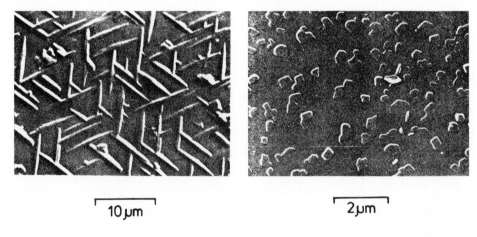

10 μm   2 μm

Fig. 3   SEM-pictures of DSP deposited on Ag(111) (left) and SiO_2/Si (right).

The SEM-pictures of DSP-films were taken after gold metallization. Results obtained for Ag(111) substrates are shown in Figure 3a in comparison with those obtained for oxidized silicon wafer substrates in Figure 3b. Films grown in UHV on clean Ag(111) surfaces with thicknesses of a several tens of nm show thermally stable needle-like structures with three growth directions of μ -DSP needles at 120° from another. Films grown on amorphous oxidized silicon surfaces show a completely different growth in small rhombic plates. The degree of crystallinity in this film increases with time even at room temperature.

The completely different microstructures of the two monomeric films determine also their ability to undergo subsequent photochemical transformations. DSP films on oxidized silicon are photopolymerizable, whereas needle-like films on Ag(111) are photostable. The degree of photopolymerization was checked by fluorescence and absorption spectroscopy. In line with our infrared measurements we conclude, that the microstructure of the needle-like films represents the well-known photostable μ - phase and the films grown on silica the well-known photopolymerizable α-phase.

3.2.2 DSP-Monomers on Ag(111)

The UPS spectra are shown in Figure 4 for a clean Ag(111) sample and after evaporation of DSP with increasing coverage.

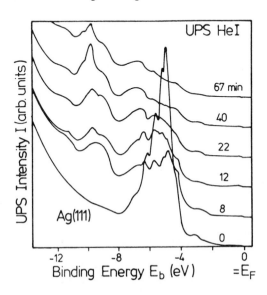

Fig. 4   HeI photoemission spectra of Ag(111) before and after DSP deposition up to 67 min corresponding to about 20 nm nominal thickness.

The substrate spectrum is characterized by its strong Ag 4d-emission. This structure changes dramatically after adsorption of even small amounts of DSP in the monolayer range. A new structure develops in the valence band range around 6 eV below the Fermi-level E_F. The shape of this spectrum is similar to that observed for the submonolayer adsorption of pyrazine on Ag(111) [Otto and Reihl 1986]. In the spectral region between 8 and 11 eV below E_F a new DSP-derived structure appears, which changes its shape up to a coverage of about 10 monolayers with the evaporation rate chosen in our experiments. The bulk UPS-spectrum of DSP is obtained after more than 40 min of evaporation. The changes in the spectrum below the bulk limit are determined by the screening of the Ag-substrate emission, and by the specific interaction of the aromatic molecules with the substrate.

To understand this interaction in more detail, the results may be compared with those on the adsorption of pyrazine and pyridine on silver. The latter has been studied extensively in the context of the SERS-effect [Otto and Reihl 1986, Otto et al 1985]. Comparing with the pyrazine data we conclude, that the Ag 4d orbitals are involved in the interaction with N-atoms of the pyrazine ring. This result is confirmed by our MO-calculations, which assign the structure around 7 eV to pyrazine-derived eigenfunctions.

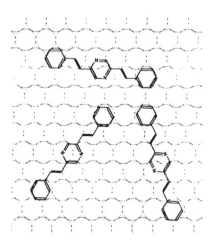

Fig. 5   Model of the adsorption geometries of DSP on Ag(111), with the N-atoms of the pyrazine rings located exactly on-top of Ag-atoms.

On Ag(111), pyridine adsorbs in a vertical configuration most probably via the lone pair of the N-atom, whereas pyrazine on the other hand adsorbs in a flat configuration. Therefore, a flat configuration of DSP on Ag(111) is a possible adsorption geometry. Three possible geometries for epitaxial growth conditions of DSP in the flat configuration are shown in Figure 5 for the Ag(111) surface, in which each N-atom of the pyrazine ring is located above an Ag-atom. The steric configuration of DSP is the same as the optimized geometry deduced from our MOPAC-calculations. Evidently, the on-top geometry is possible. It requires an interaction of the localized Ag 4d orbitals with the lone pair orbitals of the pyrazine ring.

In this adsorption geometry, three domains of the experimentally observed C_3-symmetry are possible. These domains may act as nucleation centers for the subsequent three-dimensional growth of the film. This also explains the SEM-pictures with their structures of C_3-symmetry.

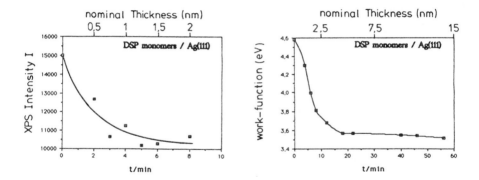

Fig. 6   Variations of the Ag $3d_{5/2}$ peak intensity (left) and the work-function (right) with the evaporation time of DSP on Ag(111). The nominal thickness is derived from the attenuation of the Ag $3d_{5/2}$ peak intensity.

The UPS-spectra on silver always show a metallic Ag Fermi-edge, even after one hour of evaporation. Evidently, even for the larger thicknesses, part of the Ag-surface inbetween DSP needles is always present within the escape depth of UPS (in the order of 1 nm). This is a clear indication for a cluster-growth mechanism. The decay of the XPS Ag 3d emission shows an exponential change with evaporation time (the first part of which is shown in Fig. 6a). This again is in line with the assumption of a cluster-growth mechanism [Venable et al 1984]. The work-function decay, however, as determined from the high-energy onset of the UPS-intensity saturates after 20 minutes (Fig. 6b), indicating that for film thicknesses in the order of 5 nm the relative amount of DSP and Ag free surface contributions (as observed with a depth resolution of UPS photoelectrons) remains constant.

3.2.3 DSP-monomers on SiO_2/Si

The different microstructures of the two DSP films on Ag and SiO_2/Si are related with their different substrate interactions. The adsorption of DSP onto oxidized silicon in particular was characterized by fluorescence and absorption spectroscopy. From the data we conclude that the interaction between the substrate and the molecules is very weak if compared with the intermolecular interaction between individual DSP molecules.

More detailed information on the intermolecular interactions in the submonolayer range was obtained from the fluorescence emission and excitation spectra of DSP on oxidized silicon wafers. At very low coverages ($\approx 10^{13}$ molecules cm^{-2}) the typical fluorescence with $_{max}$ = 24000 cm^{-1} of isolated non-aggregated monomers can be observed provided that the monomers are deposited from CH_2Cl_2 solution or under moderate vacuum.

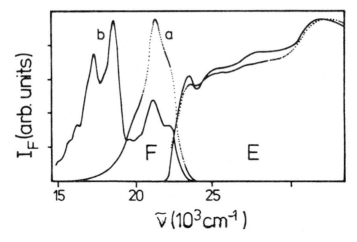

Fig. 7   Fluorescence spectra (F) of DSP deposited and measured in UHV on Si-wafers. Nominal film thicknesses are a) 0.1 nm, b) 1.1 nm. The excitation spectra (E) of both fluorescence regions a and b are almost identical.

Sublimation under UHV conditions on the contrary always results in a fluorescence band that is distinctively red shifted ($\tilde{\nu}_{max}$ = 212000 cm^1) compared to the monomer, even at very low nominal film thicknesses of about 0.1 nm = 10^{13} molecules $\cdot cm^{-2}$ (Figure 7). This band coincides with the higher energy part of the dual fluorescence in crystalline DSP. We assign it to the exciton emission of physically aggregated molecules. Aggregates are obviously much more stable on UHV-treated silicon substrates than the monomers and form under these conditions because of the high molecular mobility of monomers on UHV-clean surfaces. With increasing film thickness, the lower energy part of the dual fluorescence of crystalline DSP ($\tilde{\nu}_{max}$ = 18700 cm^{-1}) is also observed on silicon (Figure 7). This band originates from deep traps in the film and reveals the same first excitation band (exciton maximum at 23500 cm^{-1}) as the higher part of emission.

To characterize DSP-films on silicon further, we performed infrared dichroic transmission measurements with linearly polarized radiation. The polarization direction could be rotated around the axis perpendicular to the film plane (z), while the sample could be tilted around an axis in the film plane (y). This arrangement allows the determination of all spatial components of transition moments.

Statistically oriented samples with spherical distribution of transition moments must show the same absorbance for all polarization directions. All dichroic ratios, i.e. ratios of absorbances for two different directions of polarization, must be equal to unity. Samples with anisotropic distribution of transition moments must show dichroic ratios differing from unity. Therefore, the determination of absorbance with respect to all three space directions yields information about type and extent of molecular orientations in the films.

The DSP-monomers deposited by sublimation on SiO$_2$/Si wafers without vertical temperature gradient show no deviation from the spherical distribution.

DSP-layers deposited on SiO$_2$/Si-wafers cooled from the back side are still isotropic in the x/y-plane parallel to the layer. The absorbance, however, of polarization directions with components perpendicular to the surface, differ from those in the plane. This indicates a rotationally symmetric orientation with respect to the z-axis.

The α-DSP polycrystalline layers frozen from the melt with a horizontal temperature gradient (y-direction) show orientation with respect to the three space directions. The dichroic behaviour with regard to a rotation of the plane of polarization around the z-axis by the angle β relative to the x-axis could be classified to only three different groups. For the analysis, thirteen different bands were evaluated quantitatively which were proven to be sufficient isolated and narrow. In the first group of bands ("x-type bands") the dichroic ratios $R_{\beta x} = A_\beta/A_x$ increase monotonously with $\cos^2\beta$ (Figure 8a). The dichroic ratios $R_{\beta y} = A_\beta/A_y$ of the bands of the second group ("y-type") decrease monotonously (Figure 8b). The dichroic ratio $R_{\beta x}$ of the third group of bands nearly remains unity (Figure 8c).

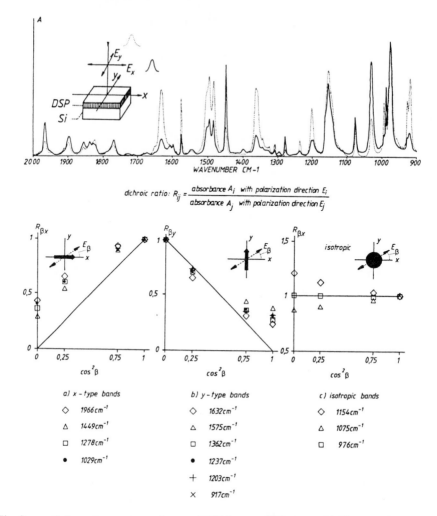

Fig. 8   Infrared spectra of a α-DSP-layer (700 nm thick) on silicon with polarization direction E_x (full line) and E_y (dotted line); dichroic ratios of x-type (a), y-type (b) and isotropic bands (c).

The point group of the isolated molecule, as well as the lattice site group, representing its static coupling in the model of an "oriented gas", is C_i. In this point group all infrared active vibrations belong to the species A_u. The corresponding transition moments could be oriented in any space direction but are fixed relative to the coordinates of the molecule for each single normal vibration. Therefore a large number of different dichroic behaviours should be expected, with maximum relative absorbances for any polarization direction. Only for additional influence from dynamic coupling the observed behaviour in Figure 8 can be understood. Dynamic interactions between the four equivalent molecules in the elementary cell of α-DSP evidently lead to a splitting of the infrared active modes of species A_u of the point group C_i of the isolated molecule into the 4 components A_u, B_{1u}, B_{2u} and B_{3u} of the factor group D_{2h}. The A_u-component is infrared inactive. The B_{1u}, B_{2u} and B_{3u}-components correspond to vibrations with transition moments in a-, b- or c-direction, respectively, of the orthorhombic crystals. Therefore only three different dichroic behaviours should be expected, with their maximum absorbance only for polarization directions along the crystallographic axes.

For factor group splittings into the three components B_{1u}, B_{2u} and B_{3u} larger than the bandwidths, the dichroic ratios $R_{\beta x} = A_\beta / A_x$ and $R_{\beta y} = A_\beta / A_y$ are expected to vary from 0 to 1 as it is indicated by full lines in Figures 8a and 8b. Factor group splitting was observed in the infrared spectrum of polycrystalline α-DSP only for a few strong bands. In comparison with the infrared spectrum of DSP dissolved in CH_2Cl_2 (for which these absorptions appear as separated narrow bands), the bands of the polycrystalline thin film show shoulders. As an example, the absorption at 1158 cm^{-1} shows a factor group splitting of about 8 cm^{-1}. For less intense bands, as for example around 1075 cm^{-1}, the band maximum is shifted by only 1 cm^{-1} during rotation of the plane of polarization by 90°. The three components corresponding to B_{1u}, B_{2u} and B_{3u} overlap completely. For this absorption band around 1075 cm^{-1} the intensity nearly does not change by rotating the plane of polarization. This band represents an excellent example of an isotropic-type band (Figure 8c) with a constant dichroic ratio of unity. In other cases the overall intensity of the band is dominated by one of the three components. These bands then show the dichroic behaviour of the dominating component. Besides the imperfect orientation of the polycrystalline layer this explains why for all bands evaluated in our analysis the dichroic ratio never decreases to 0.

Although the maximum factor group splitting is only about 10 cm^{-1} the dynamic coupling nevertheless determines the dichroic behaviour.
As the maximum of the dichroic ratio is observed for the plane of polarization oriented in the xz-plane or in the yz-plane, the axes of the orthorhombic crystal a, b and c coincide with the axes (x,y,z) of the macroscopic system.

3.3 Photopolymerization of Monomeric DSP Films

Photopolymerization at the substrate could be partly achieved for silicon/silica but not for silver substrates. The different relaxations of excited states as they occur for different interactions of DSP with the substrates for different film morphologies determine their different photochemical reactivities. For monochromatic irradiation at the wavelength of the first exciton absorption band ($\tilde{\nu}_{ex}$ = 23000 cm^{-1}) of DSP, the photochemical linking of monomers stops at the trimer with two terminal styrylpyrazyl groups bound to the central pyrazylene ring by cyclobutane groups. The reaction was monitored photokinetically on fused silica plates [Oelkrug et al 1991]. Its final absorption spectrum is shown in Figure 9c. The spectrum is dominated by the absorption bands of the styrylpyrazine chromophore which is non-absorbing for $\tilde{\nu}_{ex}$ < 23000 cm^{-1} so that the trimer cannot be polymerized further at this wavelength. If, however, the energy of irradiation is raised to $\tilde{\nu}_{ex}$ = 27400 cm^{-1}, this reaction on silica proceeds further to form polymers of about 10^2 - 10^3 DSP units.

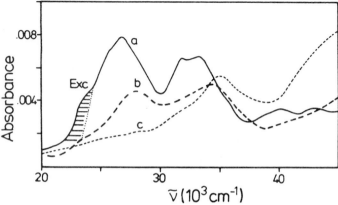

Fig. 9   Optical absorption spectra of DSP and its photoproducts on fused silica; nominal film thickness = 1 nm; a) before irradiation (shaded: first exciton band); b) after irradiation at $\tilde{\nu}$ = 23300 cm^{-1}; c) after irradiation at $\tilde{\nu}$ = 27400 cm^{-1}.

Although a homogeneous polymer film could not be prepared until now by this preparation procedure, the REM pictures of the photoproducts in Figure 10 indicate good order on the macroscopic scale since the islands of Figure 3 are clearly fused together after irradiation with short wavelength.

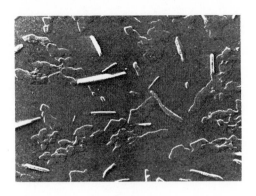

$$\overline{ 2\mu m }$$

Fig. 10   SEM picture of the DSP-polymer on SiO_2/Si.

In addition, the IR-dichroic measurements of the "non-cooled" and "cooled" layers show a higher degree of orientation after photopolymerization but still rotational symmetry around the z-axis. The increasing degree of orientation during photopolymerization can be explained by a link of the monomers across their crystallite edges [Meyer et al. 1978]. This is also in line with results from corresponding SEM-pictures.

ACKNOWLEDGEMENTS

This work was supported by the Bundesministerium für Forschung und Technologie (BMFT project no. TK 0323 0) and the Deutsche Forschungsgemeinschaft (DFG Oe 57/12).

4. LITERATURE

Abraham M, Dütting J, Schreck M, Lege R, Reich S, Oelkrug D and Göpel W 1991 J. Chem. Phys. 94 3235
Braun H G and Wegner G 1983 Makromol. Chem. 184 1103
Bufler J, Abraham M, Bouvet M and Göpel W 1991 J. Chem. Phys. in press
Cartier E, Pfluger P, Pireaux J J and Reivilar M 1987 Appl. Phys. A44 43
Lelay G 1983 Surf. Sci. 32 169
Meyer, W. Lieser G and Wegner G 1978, J.Polym.Sci. Polym.Phys.Ed.16 1365
Nakanashi H, Jones W, Thomas J M, Hasagewa M and Rees W L 1980 Proc. R. Soc. A369 307
Oelkrug D, Lege R, Gauglitz G, Neubauer H-P and Göpel W 1991 in prep.
Otto A, Frank K H and Reihl B 1985 Surf.Sci. 163 140
Otto A and Reihl B 1986 Surf. Sci. 178 635
Philips P J 1990 Rep. Prog. Phys. 53 549
Venables J A, Spiller S D T and Hahnbücken M 1984 Rep. Prog. Phys. 47 339

The chemisorption of polyimide precursors and related molecules on copper

M.R. Ashton, T.S. Jones and N.V. Richardson.

Surface Science Research Centre, University of Liverpool, PO Box 147, Liverpool, L69 3BX, UK.

Abstract

The anhydrides succinic anhydride, phthalic anhydride and pyromellitic dianhydride chemisorbed on Cu{110} have been investigated by high resolution electron energy loss spectroscopy. The anhydride ring is found to open on adsorption with loss of C=O character and bonding to the surface via a carboxylate linkage, characterised by a vibrational band at ca 1410 cm^{-1}, which is strongly dipole active. For the polyimide precursor, pyromellitic dianhydride, only one of the rings is opened to bond to the surface. In all cases, the carboxylate unit is aligned perpendicular to the copper surface.

1. Introduction

Polyimides are formed in the condensation reaction of polyimide precursors, a di-anhydride and a di-amine (Figure 1), via the formation of a polyamic acid and subsequent high temperature imidisation (Figure 2). They are an important class of polymers with high thermal stability which find widespread use in the aerospace and electronics industries.

As part of a larger programme of work, aimed at improving our understanding of the bonding of polyimides to other materials, we have used high resolution electron energy loss spectroscopy (HREELS) and X-ray photoelectron spectroscopy (XPS) to study the adsorption of the polyimide precursors, m-phenylenediamine (m-PDA) and pyromellitic dianhydride (PDMA), I, on well-defined, clean Ni{110} and Cu{110} surfaces. Two simpler anhydrides (phthalic, II, and succinic anhydrides, III) and a simpler amine, aniline, (Figure 1) were also adsorbed on these crystals and their observed behaviour is used to assist our understanding of the larger, polymer precursors.

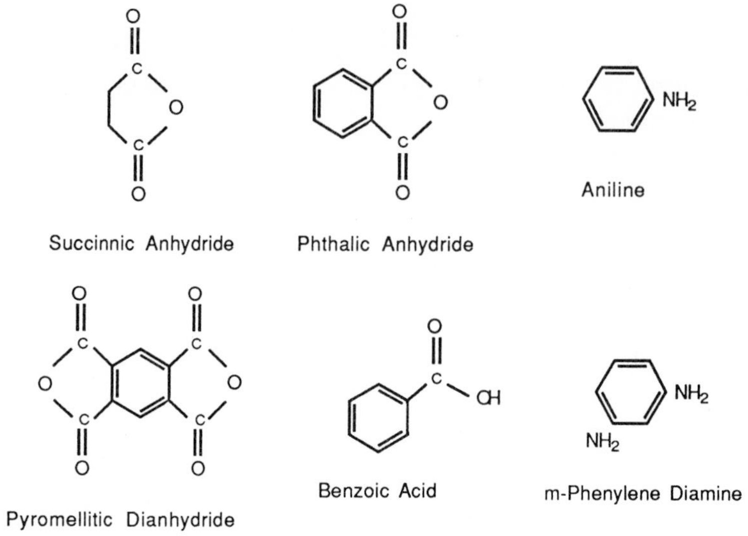

Figure 1. Polyimide precursors and model compounds.

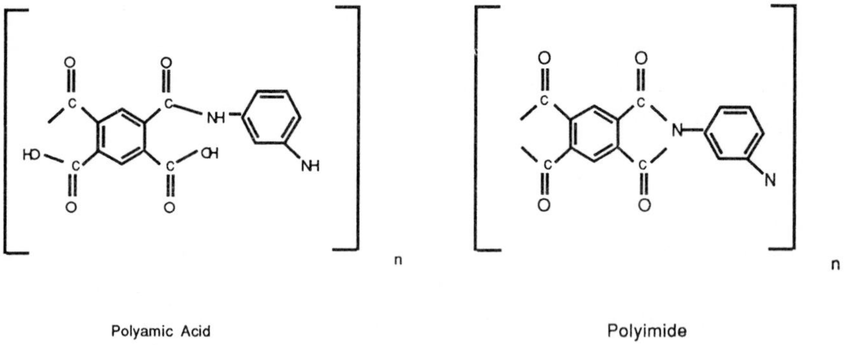

Figure 2. Two step formation of a polyimide

Further insight into the chemisorption process of the anhydrides has been gained by studying the closely related, benzoic acid molecule adsorbed on the copper crystal. In some cases, both deuterated and fully protonated compounds were used. In this particular report, we concentrate on the room temperature interaction of the anhydrides, I-III, with Cu{110} studied by HREELS. For comparison a brief reference will also be made to HREEL spectra of some of these molecules condensed on a cold (170K) substrate.

Some effort has previously been devoted to the adsorption of polyimides [1-3] and even of some of the individual precursors [4-6] onto a variety of substrates, using XPS and several vibrational spectroscopic techniques, including HREELS, IR and Raman spectroscopy. As yet, little attention has been devoted to the adsorption of the simpler anhydrides [7].

2. Experimental

Experiments were carried out in an ultra-high vacuum (UHV) system (base pressure <5 x 10^{-11} mbar) equipped with HREELS, low energy electron diffraction (LEED), Auger electron spectroscopy (AES) and mass spectrometer. The HREEL spectrometer (VSW Scientific Instruments Ltd.) consists of a fixed monochromator and rotatable analyser, both 180° hemispheres with four element lenses. For the rather disordered systems described in this paper, the energy resolution was routinely 8-10 meV (64-80 cm^{-1}) full width half maximum (FWHM). HREEL spectra were collected with incident angles of 60° and off-specular data was collected at an emission angle of 40°, all angles measured relative to the surface normal.

The Cu{110} crystal was mechanically polished, chemically etched, and then cleaned in UHV using standard argon ion sputtering and annealing (ca 800K) procedures. Surface cleanliness was monitored by the appearance of a sharp p(1 x 1) LEED pattern, by the absence of any loss peaks in the tail of the elastic HREELS peak, and by the absence of Auger lines other than those of the metal substrate. Carbon, oxygen and nitrogen impurity levels were certainly less than a 2% of a monolayer.

An independently pumped (base pressure ca 2 x 10^{-7} mbar) side chamber, which could be isolated from the main chamber, was used to house the solid samples before deposition. Succinic anhydride (Aldrich Chemical Co. Ltd., 99+% Aldrich Gold Label), phthalic anhydride (Aldrich, 99+% purity) and pryomellitic dianhydride (Lancaster Synthesis Ltd., 97% purity) were deposited in this way. Each sample in turn was contained in a ceramic dish and heated using external heating tape while being pumped. Pressures in the dosing line typically rose to 2 x 10^{-4} mbar upon heating. Temperatures were monitored externally using chromel-alumel thermocouple wire, and source temperatures were kept at approximately 25°C for succinic anhydride, 30°C for phthalic anhydride and 75°C for PMDA. Each sample was loaded as quickly as possible into the dosing chamber to minimise exposure to moisture in the air. This is particularly important for the anhydrides which hydrolyse easily to the corresponding carboxylic acid in air, although subsequent heating causes reclosure of the anhydride ring. The samples were continuously heated and pumped for several hours before opening the valve to the main chamber for dosing. Typical dosing pressures in the main chamber were 1-5 x 10^{-8} mbar for succinic anhydride, 2 x 10^{-8} mbar for phthalic anhydride and 2 x 10^{-7} mbar for PMDA. Following a typical dose, up to several hours are required to recover the base pressure because of slow desorption of the samples from the walls of the vacuum chamber. This effect makes it very difficult to study the coverage dependence of adsorption of these compounds.

In order to produce condensed layers of each molecule, liquid nitrogen was used to cool the substrate to <u>ca</u> 170K. The crystal was flashed above 275K to desorb any condensed water immediately prior to adsorption of the intended sample.

3. Results and Discussion
Because of space limitations, no attempt will be made to give full assignments of each spectrum but attention will be directed to the main features of the spectra to address the chemisorption interaction and the orientation of the adsorbed species.

3.1 Succinic anhydride
The HREEL spectrum for a condensed layer of succinic anhydride molecules is shown in figure 3. There is a strong feature at 1810 cm^{-1} with a shoulder at 1880 cm^{-1} both in the C=O stretching region, which we assign to the asymmetric and symmetric C=O stretching modes respectively of the intact anhydride unit. The corresponding gas-phase molecule has bands at 1812 cm^{-1} and 1872 cm^{-1}. Other bands in the spectrum also correlate well with features in the spectrum of gaseous succinic anhydride.

Specular and off-specular spectra were recorded after exposing the Cu{110} crystal to 1 - 5 x 10^{-8} mbar succinic anhydride for 5 minutes at 300K (figure 4). The frequencies are given in Table 1. There are major changes compared with the condensed phase spectrum. In particular, there are **no** bands in the C=O stretching region. There is some evidence of intensity changes between specular and off-specular spectra indicative of dipolar scattering into the specular direction although no ordered LEED pattern could be observed. In particular, bands at 340 cm^{-1}, 620 cm^{-1} and 1420 cm^{-1} show significantly increased intensity in the specular direction. The 1420 cm^{-1} band dominates the specular spectrum and has no counterpart in the spectrum of the condensed species. Strong bands in this region of the spectrum are reminiscent of carboxylate symmetric O-C-O stretching vibrations. This suggests that anhydride ring opening occurs, a -CO$_2^-$ unit is formed and a CO unit may be lost to the gas phase. (The EELS shows that chemisorbed CO is not present and indeed it would not be expected at room temperature on a copper sample). The other dipole active bands can then be identified as the Cu-O stretching vibration (340 cm^{-1}) and the O-C-O bending mode (620 cm^{-1}) each of which has its counterpart in the simplest chemisorbed, carboxlate species, namely the formate species adsorbed on Ni{110}, for example [8]. The dipole activity indicates that the -CO$_2^-$ plane is perpendicular to the metal substrate.

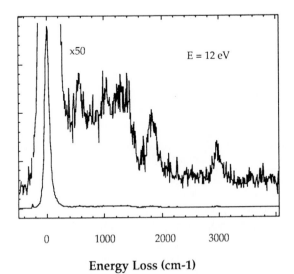

Figure 3. HREEL spectrum of a condensed (170K) multilayer of succinic anhydride.

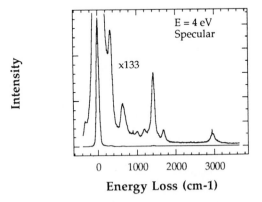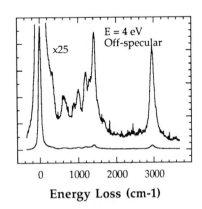

Figure 4. HREEL spectra of succinic anhydride chemisorbed on Cu{110} at room temperature.

Table 1. The vibrational frequencies (cm^{-1}) of succinic anhydride chemisorbed on Cu{110}. Frequencies marked in bold correspond to strongly dipole active bands.

Specular	Off-specular	Assignment
340	325	Cu-O stretch (sym.)
620	615	O-C-O bend
710 sh	695 sh	C-H rock
	870	C-C' stretch
1000	1000	C-H bend
1185	1195	?
1320 sh	1315 sh	?
1420	1410	O-C-O stretch (sym.)
1515 sh	1530 sh	O-C-O stretch (asym.)
1685	1685	?
2950	2950	C-H stretch

In support of the major bonding rearrangement implied by this assignment, it is particularly relevant that the spectra of the chemisorbed species, in both specular and off-specular directions, show no evidence of residual C=O features in the 1800-1850 cm^{-1} region. The weaker feature at 1685 cm^{-1} may be associated with the remaining C=O unit, if it is not fully detached from the molecule. If so, it corresponds to a C=O unit which is not part of a small ring system.

3.2 Phthalic anhydride

Figure 5 shows specular and off-specular HREEL spectra taken after the Cu{110} substrate was exposed to 2 x 10^{-8} mbar of phthalic anhydride at 300K for 20 minutes. The frequencies are quoted in Table 2. Again, as for the succinic anhydride adsorbed at 300K, there is **no** evidence of any C=O stretch at 1800-1900 cm^{-1}. There is, however, again a band at 1405 cm^{-1}. This dominates the specular spectrum while being more weakly present off specular, and is clearly a strongly dipole active mode. We assign this to the symmetric O-C-O stretch of a carboxylate species. The band at 760 cm^{-1} is assigned as the O-C-O bending mode. This is present in both spectra, but is relatively more intense with respect to the other spectral features for the specular spectrum. The Cu-O stretching vibration seems to be at somewhat higher frequency (435 cm^{-1}) than was the case for the succinic anhydride adsorption. Other features in the spectrum are compatible with the existence of the aromatic ring and C-H bending and stretching modes. It has been noted previously that the in-phase, out-of-plane bending mode of

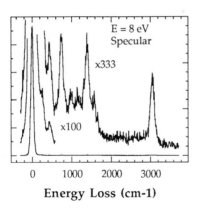
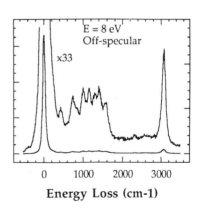

Figure 4. HREEL spectra of phthalic anhydride chemisorbed on Cu{110} at room temperature.

aromatic ring H atoms (670 cm^{-1} in benzene) has a large dynamic dipole moment giving a high intensity to this feature if the aromatic ring is aligned parallel to the surface. Indeed, the dominance of this band in the specularly obtained spectra of many aromatic compounds has been used to determine the molecular orientation at surfaces [9-12]. Unfortunately, the band appears in the 700-850 cm^{-1} region, exactly where we anticipate and have assigned the O-C-O deformation mode. The feature would, however, be weak for an aromatic ring perpendicular to the surface. Experiments with the deuterated molecule could resolve this issue and these will be carried out in the future. Improved resolution would also be an advantage!

Table 2. The vibrational frequencies (cm^{-1}) of phthalic anhydride chemisorbed on Cu{110}. Frequencies marked in bold correspond to strongly dipole active bands.

Specular	Off-specular	Assignment
235		Cu-Cu stretch
435	425	Cu-O stretch (sym.)
760	735	O-C-O bend
	845 sh	C-H out-of-plane bend
1015	1000	C-C ring mode
1160	1155	C-H in-plane bend
	1305	C-C ring mode
1405	1410	O-C-O stretch (sym.)
1550 sh	1565	O-C-O stretch (asym.)
1640 sh?		C-C ring breathing mode
3070	3075	C-H stretch

The HREEL spectrum, (not shown), for a condensed layer of phthalic anhydride deposited at a substrate temperature of 170K has major intensity in the C=O stretching region (1760 cm^{-1}) and relatively little intensity in the 1400 cm^{-1} region. There is a good correlation with all features of the gas phase spectrum confirming the molecular form of the phthalic anhydride after adsorption at 170K.

3.3 PMDA

Having looked at the model compounds, we can turn to the the polyimide precursor itself. The molecular anhydride in this case has two anhydride rings. Figure 6 shows spectra taken following the adsorption of PMDA on Cu{110} at 300K while that for the condensed molecular layer is shown in figure 7. Similar loss features can be seen in both spectra taken at room temperature, although some changes in relative band intensities can be distinguished. The frequencies are given in Table 3.

Table 3. The vibrational frequencies (cm^{-1}) of pyromellitic dianhydride chemisorbed on Cu{110}. Frequencies marked in bold correspond to strongly dipole active bands.

Specular	Off-specular	Assignment
340	340	Cu-O stretch (sym.)
450 sh		?
685	685	O-C-O bend
790 sh		C-H bend
945	930	C-H bend
1250	1235 sh	C-O-C stretch
1435	1420	O-C-O stretch (sym.)
1590 sh	1605 sh	O-C-O stretch (asym.)
1840	1855	C=O stretch
2930 sh	2920	C-H stretch?
3050	3060	C-H stretch

Again, because of the size and complexity of the molecule, a full band assignment is not attempted here. In spectra taken at 300 and 170K, a band can be seen at ca 1820 cm^{-1}, in the C=O stretching region. This was absent for succinic and phthalic anhydride at 300K. However, there is also the band at 1420 cm^{-1} which

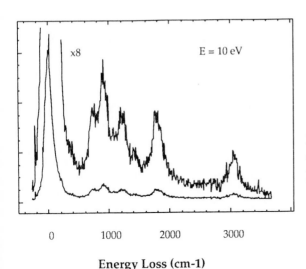

Figure 6. HREEL spectrum of a condensed (170K) multilayer of pyromellitic anhydride.

Energy Loss (cm-1)

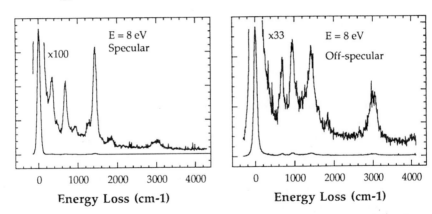

Figure 7. HREEL spectra of pyromellitic dianhydride chemisorbed on Cu{110} at room temperature.

was present for the other, smaller anhydrides. This band is again assigned to the C-O stretch of a carboxylate unit. Since not all of the C=O character is lost, this suggests that only one of the anhydride groups interacts with the surface. XPS results are consistent with this conclusion [5,6]. The feature at 1250 cm^{-1} is assigned to the symmetric C-O-C stretching vibration of the intact anhydride unit. The deformation mode of the carboxylate unit appears at 685 cm^{-1} and the Cu-O stretch at 340 cm^{-1} exactly as in phthalic anhydride. Both these latter modes show strong dipole activity as expected for a perpendicularly aligned -CO$_2^-$ unit. While there is a band at 790 cm^{-1}, which we assign to the in-phase, out-of-plane C-H bend, this is not a dipole active mode of vibration. This, therefore, implies that the plane of the benzene part of the molecule does not lie parallel to the surface but is aligned nearly perpendicular to the surface. This again would imply that bonding to the surface is through only one of the two anhydride groups for steric reasons.

4. Conclusions

Adsorption at 300K leads to significant anhydride ring perturbation such that the C=O stretching intensity is either significantly lower (PMDA) than would have been expected if the anhydrides had adsorbed molecularly or entirely absent (succinic and phthalic anhydrides). A band at ca 1410 cm^{-1} appears for each anhydride, which is reminiscent of the symmetric C-O stretch of a carboxylate group. This band is highly dipole active in all cases and leads us to conclude that the carboxylate group is perpendicular to the surface. The PMDA spectra show that 1420 cm^{-1} feature but also a C=O stretch which implies that some of the carbonyl of the parent molecule is still intact after reaction with the metal surface. The most likely explanation is that bonding to the surface takes place through opening of one of the anhydride units while the other remains intact and directed away from the surface. The aromatic systems have benzene rings which are **not** parallel to the surface.

5. Acknowledgements

Courtaulds Coatings plc are thanked for the award of a studentship to MRA.

6. References

1. Y. Takahasi, M. Ijima, K. Inagawa and A. Itoh, Vacuum, 28 (1985) 440; J. Vac. Sci. Technol., A5 (1987) 2253.
2. R.J. Salem, F.O. Sequeda, J. Duran, W.Y. Lee and R.M. Yang, J. Vac. Sci. Technol., A4 (1986) 369.
3. R.N. Lamb, J. Baxter, C.W. Kong and W.N. Unertl, Langmuir, 4 (1988) 249.
4. J.J. Pireaux, C. Gregoir, P.A. Thiry, R. Caudano and T.C. Clarke, J. Vac. Sci. Technol., A5 (1987) 598.
5. M. Grunze and R.N. Lamb, Surf. Sci., 204 (1988) 183; M. Grunze, W.N. Unertl, S. Ginarajan and J. French, Mat. Res. Soc. Symp. Proc., 108 (1988) 189.
6. T.S. Jones, M.R. Ashton, N.V. Richardson, R.G. Mack and W.N. Unertl, J. Vac. Sci. Technol., A8 (3) (1990) 2370.
7. M.R. Ashton, T.S. Jones, N.V. Richardson, R.G. Mack and W. N. Unertl, J.Elec. Spec. Rel. Phenom, 54/55 (1990) 1133.
8. T.S. Jones, M.R. Ashton and N.V. Richardson, J. Chem. Phys., 90 (1989) 7564.
9. N.V. Richardson and P. Hofmann, Vacuum, 33 (1983) 793.
10. N.V. Richardson, Vacuum 33 (1983) 787.
11. J.E. Demuth, K. Christmann and P. N. Sanda, Chem. Phys. Letts., 76 (1980) 201.
12. H. Ibach, S. Lehwald and J.E. Demuth, Surf. Sci., 78 /(1978) 577.

UPS studies of the electronic structure of merocyanine dye/AgBr interfaces

Kazuhiko SEKI[a] and Tadaaki TANI[b]

[a]Department of Chemistry, Faculty of Science, Nagoya University, Nagoya 464, Japan
[b]Ashigara Research Laboratories, Fuji Photofilm Co., Ltd., Minami-Ashigara, Kanagawa 250-01, Japan

ABSTRACT: The electronic structure of AgBr/Dye interfaces has been studied by UV photoelectron spectroscopy for several merocyanine dye films vacuum deposited on AgBr films. The obtained interfacial energy diagrams, composed of the vacuum levels, Fermi level, the top of the valence band and the bottom of the conduction band of AgBr, and the highest occupied and lowest unoccupied molecular orbitals of the dyes, were consistent with the observed abilities of the dyes for spectral sensitization. The factors affecting the interfacial energy structure are discussed.

1.INTRODUCTION

Spectral sensitization is one of the key phenomena in silver halide photography, which utilizes the photoreaction of silver halide (AgX) particles (James 1977). As described below, this phenomenon is closely related with the electronic structure of the interface between silver halide and adsorbed organic dye. Thus its examination will be useful in understanding the electronic structures and electronic phenomena in more complex polymer/solid interfaces.

At first we briefly explain the photoreaction in silver halide. When unsensitized silver halide is irradiated with photons with higher energy than its bandgap, an electron is excited into the conduction band (arrow I in Fig. 1(a)). This electron is captured in a shallow trap at the surface, and combines with interstitial Ag^+ ion to form a neutral Ag^0 atom. When this process is repeated four times or more at the same site, a stable cluster Ag_4, called latent image center, is formed. It is a deep electron trap, and acts as a catalyst for reducing the whole AgX particle to silver by receiving and transferring electrons from reducing developer to AgX. Thus visible image can be formed with very small number of photons.

However, the energy region of this intrinsic sensitivity covers only a small fraction of the visible region, as can be seen from the bandgaps of AgBr (2.60 eV; 480 nm) and AgCl (3.0 eV; 410 nm). Therefore even black and white photography cannot be performed with silver halide only.

Spectral sensitization is used to overcome this difficulty. Small amount of a dye, of the order of monolayer, is adsorbed on the surface of AgX particle, and latent image center can be formed by light absorbed by the dye. By choosing the dye, the sensitive region can be extended to the whole visible region or even to the infrared. Further, sharp absorption of J aggregates of dyes on AgX enables the reduction of their sensitive range into three principal colors.

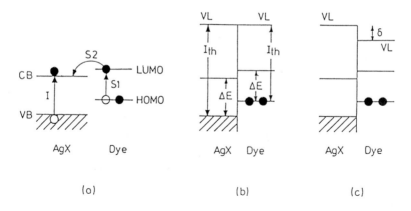

Fig. 1.   (a) Sensitization of a silver halide (AgX). CB and VB: the bottom of the conduction band and the top of the valence band of AgX, respectively. HOMO and LUMO: the highest occupied and lowest unoccupied molecular orbitals of a sensitizing dye, respectively. Arrow I: electron excitation in AgX in the intrinsic sensitivity region. Arrows S1 and S2 electron excitation in the dye and its transfer to AgX in the electron transfer mechanism of spectral sensitization. (b) Traditional way of estimating the interfacial energy diagram assuming a common vacuum level. I_{th}: threshold ionization energy; ΔE: gap between CB and VB in AgX and LUMO and HOMO in the dye. (c) More realistic interfacial energy diagram allowing for a shift δ between the vacuum levels.

The mechanism of this phenomenon has been the subject of long debate (James 1977), but now it seems to be accepted that the electron transfer mechanism, depicted in Fig. 1(a) with the arrows S1 and S2, is the dominant mechanism (Tani 1990). The highest occupied molecular orbital (HOMO) and the lowest unoccupied molecular orbital (LUMO) of the dye are orbitals delocalized over the whole conjugated system (James 1977). By the photoabsorption of the dye, an electron is excited from HOMO to LUMO of a dye molecule, and injected into the conduction band of the silver halide, participating the latent image center formation.

In this mechanism, the relative location of the LUMO and the bottom of the conduction band (CB) of AgX is very important, and a dye with LUMO below CB is expected to be a poor sensitizer. This idea has been confirmed by studying the sensitizing efficiency as a function of reduction potential of dyes in solution, which is a measure of the LUMO energy (Tani et al. 1990). An abrupt change of the sensitizing efficiency at a about -1.4 V (vs. SCE) has been observed.

However, the reduction potential measures isolated dye molecules in solution, and only relative energy values are obtained. Thus direct information has been desired about possibly aggregated dyes on AgBr surface in absolute energy scale. There have been several measurements of interfacial energy diagrams by combining the results for the two components at the interface (Vilesov 1960, Vilesov and Terenin, 1960, Nelson 1961, Akimov et al. 1967). As shown in Fig. 1(b), ionization threshold energy I_{th} and the separation ΔE between the highest occupied and the lowest unoccupied states were determined from photoemission and optical measurements, respectively, and the energies of CB and LUMO were deduced from them. The interfacial energy level diagram was constructed by aligning the vacuum levels. However, the assumption of vacuum level alignment is doubtful, and the deduced relative location of LUMO and CB were inconsistent with the observed sensitizing abilities.

In this work, we directly examined the interfacial energy level diagrams by UV photoelectron spectroscopy (UPS) for merocyanine dyes deposited onto AgBr films, and examined their correlation with the spectral sensitizing abilities of the dyes.

2. EXPERIMENTAL

The five merocyanine dyes in Fig. 2 were studied. They were synthesized according to the method of Brooker et al.(1951). The UPS apparatus is schematically shown in Fig. 3. It is a combination of a rare gas discharge lamp and a measurement chamber with differential pumping part between them. The experiments were performed in the following manner:

(1) UPS measurement of an Ag

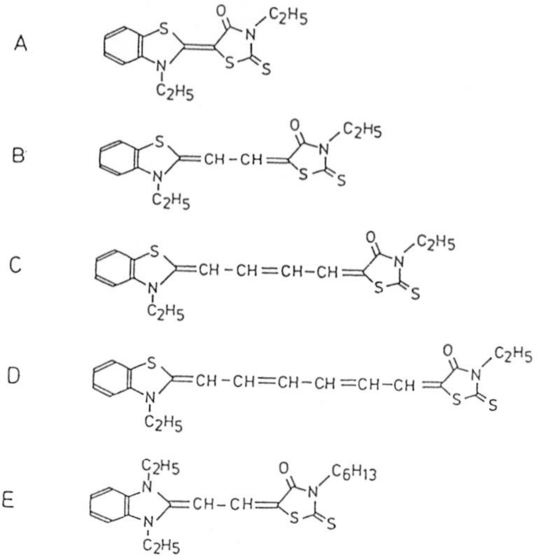

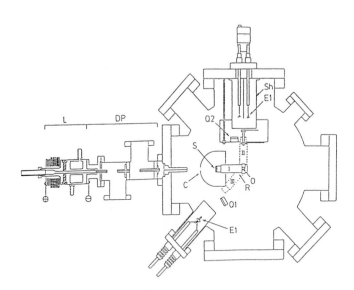

Fig. 2. Dyes studied in this work.

Fig. 3. UPS spectrometer used in this work. L: discharge lamp, DP: differential pumping part, S: sample, C: collector, R: sample rod, O: rotatable axis, E1 and E2: evaporation sources of AgBr and dye, Q1 and Q2: quartz thickness monitors for AgBr and dye, Sh: AgBr shield, and I: positions of R for UPS measurements, AgBr evaporation, and dye evaporation, respectively.

substrate heat cleaned at 500 °C for 10 h, for determining the Fermi edge, with the sample rod R at the position I.

(2) Vacuum deposition of AgBr film of typical thickness of 7 - 10 nm at the position II, and its UPS measurement at the position I.

(3) Vacuum deposition of a dye onto AgBr film at the position III, with UPS measurements at several stages of deposition at the position III. The film thickness of dye films was calibrated by a photometric analysis after dissolving in N,N'-dimethylformamide.

The typical pressure of the measurement chamber was 1×10^{-8} Torr, but the results of AgBr agreed well with those of Bauer and Spicer (1970, 1976) under ultrahigh vacuum condition, indicating insensitivity of AgBr surface to contamination. The UPS measurements were mainly performed with Ar I resonance discharge (11.7 eV) as the light source. Xe I resonance discharge (8.4 eV) was also used.

The energy analysis of photoelectrons was performed using a retarding-field energy analyzer. The sample was surrounded with a Au-coated collector, and the photoelectrons were detected as photocurrent I_{ph} to the collector. The energy of photoelectrons was analyzed by sweeping the retarding voltage V_r applied between the sample and the collector. The energy distribution curve (UPS spectrum) was obtained as dI_{ph}/dV_r by the ac modulation technique (Eden 1970). The resolution was about 0.2 eV as deduced from the measured Fermi edge of Ag.

Care was taken to prevent exposure of each sample to light or electrons from viewports or ion gauges. There was no indication of the incident light for the UPS measurements, such as the change of the UPS spectra after long measurements.

Auxiliary data of diffuse reflection and spectral sensitivity were measured by use of photographic emulsions composed of suspensions of octahedral and cubic AgBr grains (average edge length: 0.83 and 0.73 m, respectively) in aqueous gelatin solutions, which were prepared by a controlled double jet method (Berry and Skillman 1962). A methanolic solution of a dye was added to the above-stated emulsion, which was agitated, coated on a triacetate cellulose film base, dried, and subjected to the measurements.

3.RESULTS AND DISCUSSION

In Fig. 4 we show the change of the UPS spectra from that of AgBr by the
step-by-step deposition of dye C. The abscissa is the retarding voltage
V_r. The location of the Fermi level $V_r(E_F)$ of Ag on the same scale is
also shown. When we assume the alignment of the Fermi levels of Ag and
AgBr, the energy of the highest occupied electronic state (the top of
the valence band VB for AgBr and HOMO for the dye) relative to the Fermi
level ε_{HO} is given by using the value of V_r for the right-hand onset V_0 as

$$\varepsilon_{HO} = V_0 - V_r(E_F). \qquad (1)$$

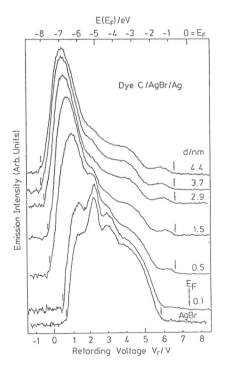

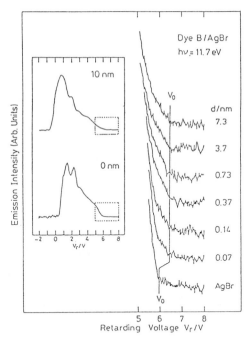

Fig. 4. Change of the UPS
spectra at stepwise deposition
of the dye C on AgBr. The dye
thickness d is shown for each
curve. E_F denotes the location
of the Fermi edge of Ag.

Fig. 5. Change of the right-hand
onset at the stepwise deposition of
the dye B on AgBr. The dye thickness is shown for each curve. The
inserts show the full spectra
before and after deposition.

On the other hand, the work function ϕ, which is the energy of the vacuum level relative to the Fermi level, can be determined from the left-hand onset (V_S) as

$\phi = V_S + h\nu - V_r(E_F).$ (2)

Thus we can trace the energy change of the highest occupied state and the vacuum level relative to the Fermi level of AgBr.

We see that the energy of the HOMO is not dependent on the dye thickness, while the vacuum level is significantly lowered by dye deposition within thickness d of a few nms. Examination of the right-hand onset in thinner thickness region for dye B in Fig. 5 again indicates the thickness independence of the HOMO energy. Similar results were obtained for other dyes.

The lowering of the vacuum level clearly indicates that the assumption of vacuum level alignment in Fig. 1(b) is not valid, and we should use more realistic scheme as shown in Fig. 1(c) with a shift δ between the vacuum

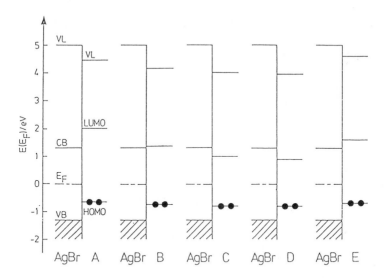

Fig. 6.   Energy level diagrams at AgBr/dye interfaces deduced from the present study. VL: vacuum level, CB and VB: bottom of the conduction band and top of the valence band of AgBr, respectively. E_F: location of the Fermi level deduced from the Fermi edge of the Ag substrate, HOMO and LUMO: the highest occupied and the lowest unoccupied molecular orbitals of dye, respectively.

TABLE 1.   Energy parameters of AgBr and dyes (in eV).   See also Figs. 1 and 6.

Compd.	I_{th} [a]	ε_{HO}^{b}	ϕ^{c}	ΔE^{d}	ε_{LU}^{e}	δ^{f}
AgBr	6.6	1.3	5.3	2.6	4.0	—
Dye A	5.1$_5$	0.6$_5$	4.4$_5$	2.64	2.0	0.5$_5$
Dye B	4.9$_5$	0.7$_5$	4.1$_5$	2.11	1.3$_5$	0.8$_5$
Dye C	4.8	0.8	4.0	1.82	1.0	1.0
Dye D	4.7$_5$	0.8	3.9$_5$	1.70	0.9	1.0$_5$
Dye E	5.3	0.7	4.6	2.30	1.6	0.4

[a] Ionization threshold energy.
[b] Energy of the highest occupied state relative to the Fermi level of the Ag substrate deduced using Eq. (1).
[c] Work function determined assuming the coincidence of the Fermi levels using Eq. (2).
[d] Energy separation between the highest occupied and lowest unoccupied states.
[e] Energy of the lowest unoccupied state relative to the Fermi level of the Ag substrate deduced as $\varepsilon_{LU} = \varepsilon_{HO} + \Delta E$.
[f] Shift of the vacuum level at the AgBr/dye interface.

levels.   Nielsen (1974) pointed out the possibility of this shift, but AgBr/dye systems were not experimentally studied.

In Fig. 6 we show such energy level diagrams deduced from the UPS measurements for thick dye films.   The energies of CB of AgBr and the LUMO of the dyes were deduced assuming that these levels are separated from VB and HOMO by the optical excitation energy ΔE.   The values of ΔE for the dyes were obtained by the diffuse reflection measurements of dyed AgBr emulsions,   which gave very similar results for the cubic and octahedral grains of AgBr.   The numerical data of the location of the energy levels and the energy parameters appearing in Figs. 1, which determine the energy levels, are summarized in Table 1.

The Fermi level of AgBr, assuming its coincidence with that of Ag substrate, is around the midpoint of the bandgap.   If we take account the small upward bending of the bands of AgBr at the surface (Slifkin 1989), the Fermi level is slightly above the middle of the bandgap.

As seen from Fig. 6 and Table 1, the LUMO of dyes A, B, and E are higher than CB, while the LUMO of dyes C and D are below CB, suggesting that the former are efficient sensitizers, while the latter are poor sensitizers.   Note that the conclusion is entirely changed when we assume the alignment of the vacuum levels as in Fig. 1(b). In this case even dyes C and D are expected to be good sensitizers.

We also note the possible distribution of ionization energies of dye molecules (James 1977, Duke 1978). There may be several kinds of sites on AgBr surface, and bulk and surface molecules have different ionization energies in thick films (Salaneck 1978, Harada et al. 1984). The coupling with phonons at photoionization also slightly change the ionization energy (Duke 1978). The peak at the right end of the UPS spectrum of dye C in Fig. 4 gives the feeling of such distribution for thick films.

Thus we should regard the LUMO energy in Fig. 6 as the higher limit, with the peak of distribution a few tenth of eV below the LUMO line. Taking account of this factor, the dye B may be at the border of good and poor sensitizer.

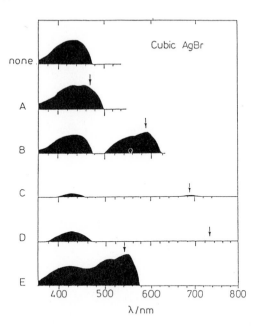

To compare with these results, in Fig. 7 we show the wedge spectrogram of dyed and undyed emulsions with cubic AgBr grains.   Each film coated with an emulsion layer was exposed by a spectrograph with a wedge in such a way that the intensity of the light incident to the film exponentially decreased at any wavelength from its bottom to top, and was developed to give a wedge spectrogram where photo-sensitive region was darkened. The observed results correspond well to those expected from Fig.

Fig. 7. Spectral sensitivity of undyed and dyed photographic emulsion with cubic AgBr grains. Arrows indicate the absorption peak energies of dyes on AgBr.

6. Emulsions with octahedral AgBr grains gave similar results.

Thus the sensitizing behavior of the dyes were well explained in the framework of electron transfer mechanism using the observed interfacial electronic structure. The failure of the previous estimation of the energy level diagrams is due to the large lowering of the vacuum level on dye deposition.

The large lowering of the vacuum level means the existence of an electric double layer at the interface. Since most of the vacuum level shift occurs within dye thickness of 1 nm or so, interaction between the first or a few layers on AgBr surface seem predominantly contribute to this double layer formation. Possible origins of this double layer are:

(1) electron transfer from dye molecules to AgBr,

(2) orientation of permanent and induced dipoles of dye molecules, and

(3) surface rearrangement AgBr surface by dye deposition.

We will briefly examine these factors.

As for (1), we can consider the possibility of Fermi level equalization at least for thick films. If this is true, the energy of the HOMO H_O of a relative to the Fermi level of metal substrate should not change regardless of the existence or nonexistence of AgBr layer in between. Actually the examination of dye B/Ag system gave the same value of H_O as that for dye B/AgBr/Ag system. For further establishing this possibility, studies of other systems and clarification of the origin of charge carriers (e.g. impurity or photoexcitation by to the UPS lamp) should be necessary. For very thin films, possible electron flow should be predominantly controlled by microscopic interaction between AgBr and dye molecules on it.

As for (2), the double layer should be formed with molecules with the positive and negative parts of a molecule at the AgBr and vacuum sides, respectively. Our recent examination of molecular orientation by soft X-ray absorption suggests that merocyanine dyes of the presently studied type adsorb on AgBr at the thioketone (=S) part. The resonance structure of merocyanines indicate that this part should be positively charged, molecular dipole seem to contribute to the double layer formation at least in the initial stage of deposition.

As for (3), presently we have not much information, and further studies are necessary.

4.CONCLUSIONS

We have directly studied the electronic structures of AgBr/merocyanine dye interfaces by UV photoelectron spectroscopy. The obtained electronic energy diagrams were in good consistency with the observed photographic sensitizing ability of the dyes in terms of electron transfer mechanism, resolving one of the difficulties of this mechanism proposed by the traditional way of estimating the interfacial electronic structure. The present study has demonstrated the importance of direct measurements of interfaces. Factors affecting the interfacial electronic structures were discussed.

It is obvious that this kind of studies can also be extended to other photosensitizing systems. Even within the field of silver halide photography, changes from AgBr to AgCl and AgI, from merocyanines to cyanines etc. are possible. In any case, more clear understanding of the interfacial electronic structure and electronic processes should be obtained by the combination with various techniques, including the methods for studying the microscopic structures of interfaces.

ACKNOWLEDGMENTS

We thank Messrs. Y. Kobayashi and H. Yanagi and Prof. T. Ohta for their collaboration. Thanks are also due to Professors H. Inokuchi and H. Kanzaki for their kind interest and encouragements.

This work was partly supported by the Grant-in-Aid for Scientific Research (No. 03640410 and No. 03NP0301 "Intelligent Molecular Systems with Controllec Functionality") from the Ministry of Education, Science, and Culture of Japan.

REFERENCES

Akimov I A, Bentsas V M, Vilesov F I and Terenin A N 1967 Phys. Stat. Solidi 20 771
Bauer R S and Spicer W E 1970 Phys. Rev. Lett. 25 283
Bauer R S and Spicer W E 1976 Phys. Rev. B14 4529
Berry C R and Skillman P C 1962 Photogr. Sci. Eng. 6 159
Brooker L G S, Keyes G H, Sprague R H, Van Dyke R H, Van Lare E, Van Zandt G, White F L, Cressman H W J and Dent S G, 1951 J. Am. Chem. Soc.73 5332
Duke C B 1978 Surf. Sci. 70 674
Eden R C 1970 Rev. Sci. Instrum. 41 252
Harada Y, Ozaki H and Ohno K 1984 Phys. Rev. Lett. 52 2269
James T H (Ed.) 1977 Theory of Photographic Processes, 4th Ed. (MacMillan, New York)
Nelson R C 1961 J. Opt. Soc. Am. 51 1186
Nielsen P 1974 Photogr. Sci. Eng. 18 186
Salaneck W R 1978 Phys. Rev. Lett. 40 60
Slifkin L 1989 Superionic Solids and Solid Electrolytes (Lasker and Chandre S Eds.) (Academic Press, New York) p. 407
Tani T 1990 J. Imag. Sci. 34 143
Tani T, Suzumoto T and Ohzeki K 1990 J. Phys. Chem. 94 1298
Vilesov F I 1960 Dokl. Akad. Nauk SSSR 132 632
Vilesov F I and Terenin A N 1960 Dokl. Akad. Nauk SSSR 133 1060

Morphology of electronic structure of thin, ordered polyimide mono-layers prepared by the Langmuir–Blodgett technique

H Sotobayashi, T Schedel-Niedrig, M Keil, T Schilling, B Tesche and A M Bradshaw

Fritz-Haber-Institut der Max-Planck-Gesellschaft
Faradayweg 4 - 6, W-1000 Berlin 33, Germany

ABSTRACT: Monolayers of two different polyimides, the rod-like PMDA-PDA and the zig-zag PMDA-ODA, have been prepared on graphite and oxidised silicon surfaces using the Langmuir-Blodgett technique of Imai, Kakimoto and co-workers. Scanning tunneling microscopy indicates that both polyimides on graphite are characterized by a high degree of two-dimensional order. X-ray absorption spectroscopy at the C, N and O K-edges has been used to obtain information on the orientation of the different moieties in the polymer chain. With the help of corresponding spectra from various model compounds some insight can be gained into the origin of the resonances at the C and N K-edges.

1. INTRODUCTION

Of the various approaches to the problem of producing long-range order in macro-molecular materials that based on the Langmuir-Blodgett [LB] technique appears to be one of the most promising. With this method oriented monolayers of am-phiphilic molecules at the air/water interface can be transferred to a solid surface (Roberts 1985). Thus Wegner and co-workers (Tieke *et al.* 1979), for example, have prepared polydiacetylene multilayers from LB films of various long chain mono-carbonic acids with subsequent exposure to UV radiation. After polymerisation both the layer structure and the packing in each individual layer were preserved. To obtain good thermal and mechanical stability, however, it is desirable to remove the long chain components. Kakimoto *et al.* (1986) have recently attacked this problem from a new angle. They showed that mono- and multilayers of PMDA-ODA polyimide can be prepared on solid surfaces using a LB-technique which allows subsequent removal of the long alkyl chains. (PMDA-ODA = pyromellitic dianhydride/oxydianiline (Mittal 1984).) In a recent study (Schedel-Niedrig *et al.* 1991) we have used core and valence level photoemission to prove that such layers adsorbed on highly oriented pyrolitic graphite (HOPG) are imidised to a similar extent as thicker samples prepared by spin-coating or by the molecular co-deposition method. Studies with the scanning tunneling microscopy (STM) also indicated that the monolayers are characterised by a high degree of two-dimensional order over areas as large as 100 nm x 100 nm (Sotobayashi *et al.* 1990). Further, a preliminary x-ray absorption study at the N K-edge indicated that the PMDA and ODA sub-units were probably co-planar and adsorbed parallel to the graphite surface (Schedel-Niedrig *et al.* 1991). There was, however, still some doubt as to the exact interpretation of the STM pictures, particularly as to the polymer chain conformation.

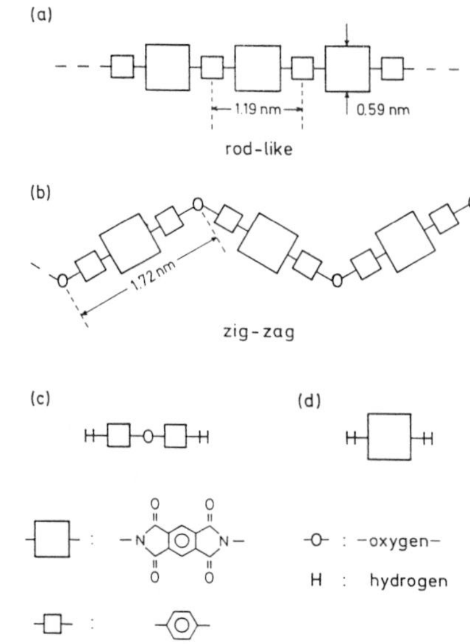

Fig. 1: Molecular structure of two polyimides and related model compounds:
(a) rod-like PMDA-PDA (pyromellitic dianhydride/p-phenylene diamine)
(b) zig-zag PMDA-ODA (pyromellitic dianhydride/oxydianiline)
(c) DPE (diphenyl ether)
(d) PMDI (pyromellitic diimide).
The dimensions have been calculated with the Chem-X set of molecular modelling programs.

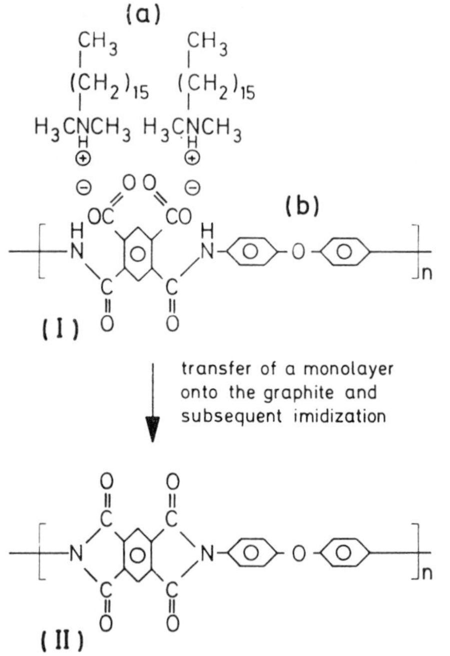

Fig. 2: Schematic representation of the preparation of a monolayer film of PMDA-ODA polyimide using the LB technique:
(I) polyamic acid alkylamine salt,
(a) N,N-dimethylhexadecylamine,
(b) poly (pyromellitamic acid);
(II) polyimide.

The purpose of the present study has been to address the problem of the STM images by extending the investigations to the rod-like PMDA-PDA polyimide and to understand the origin of the resonances in the x-ray absorption spectrum of the PMDA-ODA polyimide by measuring the spectra of certain molecules which might be regarded as the constituent building blocks of the two polymers. (PMDA-PDA = pyromellitic dianhydride/p-phenylene diamine.) The extended chain conformation of the PMDA-PDA polyimide is characterized by a simple rod-like structure (Fig. 1a), whereas the extended chain conformation of the PMDA-ODA polyimide has a bent C-O-C bond at the ether oxygen giving rise to a zig-zag structure (Fig. 1b), in addition to a possible rotation about the C-N bond linking the PMDA and ODA moieties (Karayazan 1972).

2. EXPERIMENTAL

Since the normal starting material for the preparation of polyimide, polyamic acid, possesses only hydrophilic carboxyl groups in the polymer backbone, it does not form an ordered monolayer on the surface of water. As shown in Figure 2, Kakimoto *et al.* (1986) prepared an LB film of a precursor consisting of a polyamic acid alkylamine salt (I) which could then be deposited onto a substrate. Removal of the long alkyl chain and concomitant imidisation was achieved by thermal or chemical means to produce a monolayer film of polyimide (II).

In the present studies both highly oriented pyrolytic graphite (HOPG) and silicon covered with a thin native oxide film 15 Å thick (Henderson 1972), which provide hydrophobic and hydrophilic surfaces, respectively, were used as substrates. The monolayers of two polyimides, (a) the rod-like PMDA-PDA polyimide (Fig. 1a) and (b) the zig-zag PMDA-PDA polyimide (Fig. 1b), were prepared using a circular LB trough and transferred at a constant film pressure of 25 mN/m, a transfer speed of 7 mm/min and a constant temperature of ~25 °C. The details of the subsequent procedure have been given earlier (Sotobayashi 1990).

A commercial STM (NanoScope II; Digital Instruments, Inc. Santa Barbara, CA) with a 0.5 μm scan head was used for imaging the layers on graphite. These experiments were performed in the constant current mode with a Ni tip or a Pt-Ir tip in air. The X-ray photoemission data were taken with a commercial ESCA spectrometer (Leybold-Heraeus LHS12-SCD) using monochromatised AlKα radiation. X-ray absorption spectroscopy was performed in a double-chamber UHV system (base pressur $<2x10^{-10}$ mbar in both chambers) on two different grazing incidence monochromators (SX-700 IV and HE-TGM 1) at the Berlin synchrotron radiation source BESSY. Spectra were recorded in the partial yield mode at a resolution of ~0.5 eV (1.0 eV) at the C K-edge, and of ~0.7 eV (1.4 eV) at the N K-edge. (The values in brackets refer to the HE-TGM). After preparation of the monolayers on graphite and introduction into the UHV system, the samples were heated ~400 °C for 10 minutes in order to drive off fully the remaining water (Hahn 1984). The monolayers on oxidised silicon were only heated to 200 °C because of evidence for decomposition at high temperatures. Thick films of the molecules pyromellitic diimide (PMDI) (Fig. 1d) and diphenyl ether (DPE) (Fig. 1c) were prepared by evaporation onto a cold Cu substrate. Film thickness could be monitored with a quarz crystal balance.

3. X-RAY PHOTOEMISSION

Figure 3 shows oxygen and nitrogen 1s core level spectra (a) for a monolayer of PMDA-ODA polyimide on HOPG and (b) for a spin-coated polyimide film (~20 nm) on an aluminum-coated silicon wafer by Anderson *et al.* (1988). Binding energies were measured relative to the C1s line of the graphite substrate. According to these and earlier authors (Buchwalter 1984, Grunze 1988) the component of lower binding energy OI ($E_B = 532.1$ eV) is due to the carbonyl oxygen and that at higher binding energy OII ($E_B = 533.2$ eV) to the ether oxygen. The component O_S is related to the presence of the graphite substrate and becomes more intense at higher in situ annealing temperatures. It is obvious that the single N1s line at 400.8 eV (NI) is due to the imide nitrogen. As far as peak shape and binding energy are concerned, there is very good agreement between our monolayer spectra prepared by the LB technique and the spectra taken for a thick spin-coated film and indicates that we are dealing with essentially the same material.

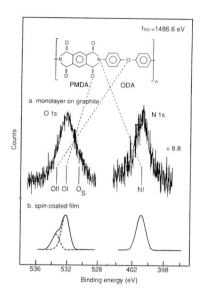

Fig. 3: O1s and N1s core level photoemission lines from PMDA-ODA polyimide. (a) Monolayer on HOPG, (b) Fully cured spin-coated film after Anderson *et al.* (1988).

4. SCANNING TUNNELING MICROSCOPY

Fig. 4a shows a raw topographic image (30 nm x 30 nm) of a monolayer of the rod-like PMDA-PDA polyimide on a HOPG surface. The smaller field of view image (10 nm x 10 nm, Fig. 4b) shows a regular array structure with periodicities of ~1.2 and ~ 0.6 nm in two orthogonal directions. On the basis of the molecular dimensions shown in Fig. 1a, we suggest that the direction of the polymer backbone is coincident with the axis showing ~1.2 nm periodicity and that the distance between the chains is ~0.6 nm. The chain direction corresponds roughly to the transfer direction. Note also a certain degree of "smearing" of the image perpendicular to the chains which is in the scan direction and indicates some interaction between the tip and the film. The observation of this more or less regular pattern which is consistent with both the conformation and molecular dimensions of the polymer is quite significant. The PMDA-PDA data reinforce our view that it is in principle possible with the STM to identify ordered structures in such layers and that the images are consistent with the molecular dimensions.

Figure 5a shows a raw topographic image of a monolayer of the zig-zag PMDA-ODA polyimide on a HOPG surface for a field of view of 100 nm x 100 nm. The micrograph differs slightly from the one published previously due to different film preparation and different tunneling conditions (Sotobayashi 1990). Nonetheless it shows a considerable degree of two dimensional order which is found over large areas of the substrate. The smaller field of view image (4 nm x 4 nm, Fig. 5b) shows more detailed structural features, but these are difficult to interpret satisfactorily. We suggest that the direction of the polymer backbone is coincident with the rows of bright spots and that the distance between the chains is ~1.8 nm.

We thus obtain the remarkable result that the two-dimensional order established in the LB trough is transferred to the polyimide films despite deposition on the substrate, subsequent imidisation and removal of the long alkyl chains.

(a)

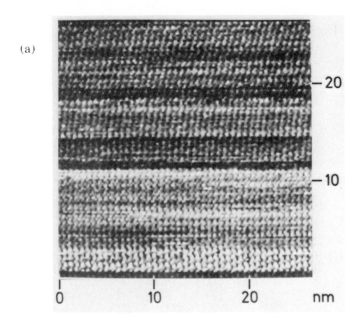

(b)

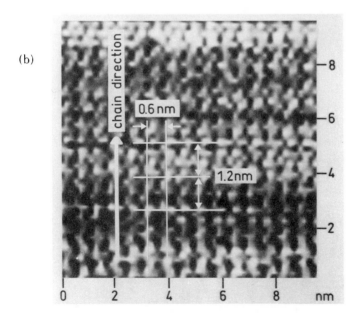

Fig. 4: (a) Raw topographical STM image of a monolayer of the rod-like PMDA-PDA polyimide on a HOPG substrate in air in the constant current mode; area 30 nm x 30 nm; current setting 2 nA; sample bias 10.1 mV. (b) A smaller field of view STM image of (a); area 10 nm x 10 nm.

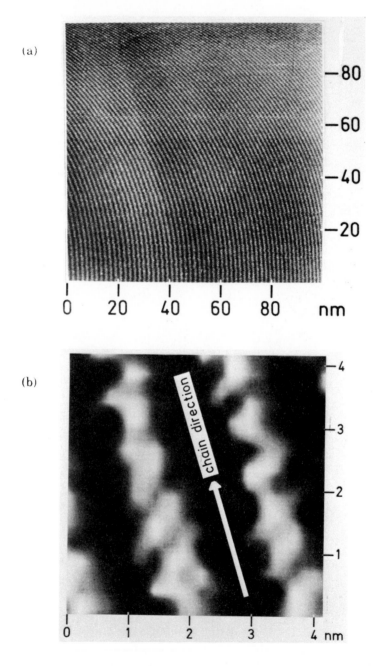

Fig. 5: (a) Raw topographical STM image of a monolayer of the zig-zag PMDA-
ODA polyimide on a HOPG substrate in air in the constant current mode; area 100
nm x 100 nm; current setting 3 nA; sample bias -12.2 mV. (b) A smaller field of
view STM image of (a); area 4 nm x 4 nm.

5. X-RAY ABSORPTION NEAR EDGE SPECTROSCOPY (XANES)

Carbon, nitrogen and oxygen K-edge absorption spectra of adsorbed molecules can be readily measured using synchrotron radiation, not in a conventional absorption geometry, but rather using the electron yield technique (Stöhr *et al.* 1982). The method is not only a probe of unoccupied valence states but also a useful tool for determining molecular orientation (Stöhr *et al.* 1982). This is due to the polarisa-tion dependence of the various resonances, i.e.the variation in their intensity as a function of the orientation of the E vector relative to some symmetry element of the molecule (Bradshaw and Somers 1990). In a simple model, the transitions at the C, N and O K-edges can be grouped into two types: π^* resonances below the ionisation limit, which are polarised perpendicular to the ring planes of the mole-cule or polymer unit, and σ^* resonances occurring above the threshold which are polarised parallel to those planes. In the case of polyimides we might expect that the π orbitals associated with the carbonyl and phenylene groups in both PMDA and ODA or PDA units interact via the nitrogen lone pair and are delocalised over the whole monomer unit if the ring systems are co-planar. An interruption of the conjugation would occur if there is torsion about the C-N bond between the PMDA and the ODA or PDA ring planes.

5.1. The Carbon K-edge Spectra

The carbon K-edge of a single PMDA-ODA polyimide monolayer on oxidised sili-con is shown in Fig. 6 for an angle between the E vector of the light and the sur-face normal of $\theta_E = 25°$. The sharp resonances (features 1 to 7) are assigned to tran-sitions into the π system, in general agreement with calculation (Meyer 1989). They exhibit relatively small changes in intensity as a function of θ_E (spectra not shown), indicating that the polymer is adsorbed such that the PMDA and ODA moieties have fixed, but possibly different, orientations relative to the surface or, alternatively, that the orientation of both is random. In both cases a twisting about the C-N bond could therefore occur. The monolayer spectrum of Fig. 6 is vir-tually identical with the corresponding spectrum from a thick spin-coated film (Fig. 7a) reproduced from Jordan-Sweet *et al.* (1988). The assignment of the ob-served resonances is aided by the spectra of the molecules PMDI and DPE shown in Figs. 7b and c, respectively. These were taken for thick (~15 Å: PMDI; ~30 Å: DPE) films evaporated onto a single crystal copper substrate. Returning to Fig. 6 we note that there is almost a 1:1 correspondence between the all eleven reson-ances identified in the C1s spectrum of the monolayer and the features in the PMDI and DPE spectra. The sharpest, most intense peak at 285.0 is assigned to C $1s{\to}\pi^*$ transitions in both the ODA part (DPE: peak at 285.2 eV designated 2) and the PMDA part (PMDI: peak at 284.8 eV designated 1). The latter appears to con-stitute the low energy shoulder. The second peak (designated 3, 4) also arises from transitions into both π^* systems, as does the next peak which is designated as 5. The shoulder (6) at ~ 291.0 eV is assigned to ODA and the next peak (7) at 291.9 eV to PMDA. The broad features (8, 9) and (10, 11) can be attributed to both the ODA part and the PMDA part and are almost certainly due to σ-type resonances. The fact that the spectrum of Fig. 6 is given essentially by the sum of the spectra b and c in Fig. 7 suggests that there is very little interaction between the two moieties in the polyimide monolayer. It may also mean that there is a torsional angle about the C-N bond, which in turn would be compatible with the polarisation depen-dence. However, it will be seen from the N 1s data for a PMDA-ODA monolayer on graphite reported below that such a conclusion cannot be drawn with certainty. It is not possible to measure carbon K-edge absorption spectra for polyimides on

HOPG due to the very strong, polarisation-dependent resonances of the substrate (Rosenberg 1986).

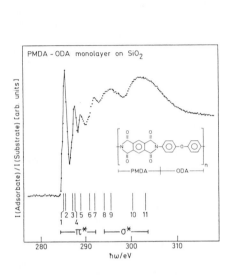

Fig. 6: The carbon K-edge spectrum of a monolayer of PMDA-ODA polyimide on oxidised silicon prepared using the LB-technique. The angle between the E vector and the surface normal is 25°.

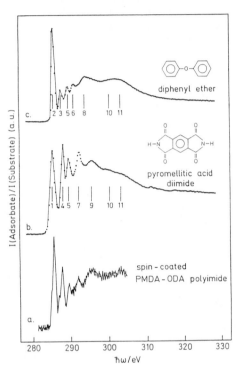

Fig. 7: The carbon K-edge spectra of (a) a thick PMDA-ODA polyimide film prepared using the spin-coating methode [14], (b) a ~15 Å film of pyromellitic acid diimide on Cu{110}, and (c) a ~30 Å film of diphenyl ether on Cu{110}.

5.2. The Nitrogen K-edge Spectra

Fig. 8 shows the nitrogen K-edge spectra of PMDA-ODA polyimide for (a) a monolayer on HOPG and (b) ca. six monolayers successively deposited on HOPG. Spectrum (c) for a spin-coated film is produced from Jordan *et al.* (1988). We note that there is a discrepancy between the PMDA-ODA spectra in our earlier work (Schedel-Niedrig 1991) and the present data in that a third, strong resonance at lower energy was found. This may have been related to incomplete imidisation; some indication for a low energy resonance is still seen in the monolayer spectrum in Fig. 8a. It is reasonable to assign the two lowest energy resonances to transition into the π systems, in general agreement with calculation (Meyer 1989). The similarity of the spectra (b) and (c) in Fig. 8 indicates that we are dealing with essentially the same material and, most importantly, that the degree of imidisation attained in the polyimide chains constituting the monolayer is high.

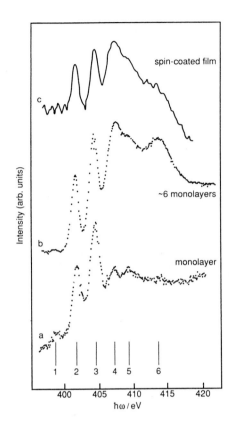

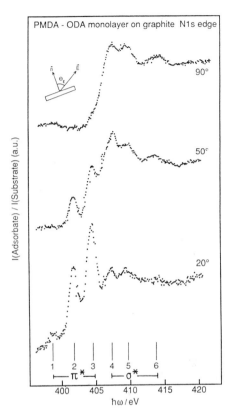

Fig. 8: Nitrogen K-edge spectra of PMDA-ODA polyimide; (a) monolayer on HOPG, $\theta_E = 20°$, (b) ~6 monolayers on HOPG, $\theta_E = 30°$, (c) spin-coated film after Jordan *et al.* (1984).

Fig. 9: The nitrogen K-edge spectra of a single monolayer of PMDA-ODA polyimide on HOPG for angles between the E vector and the surface normal, θ_E, of 20°, 50° and 90°.

Figure 9 and 10 show nitrogen K-edge spectra for monolayers of PMDA-ODA polyimide (Fig. 9) and PMDA-PDA polyimide (Fig. 10) on HOPG as a function of θ_E. Two important observations can be made, the first of which is only apparent by comparison with the corresponding N K-edge spectrum of the model compound PMDI shown in Fig. 11. There is a strong similarity between the spectra and, as in the case of Figs. 6 and 7, an almost 1:1 correspondence is found between the resonances of the monolayers and of the model compound. The spectra of the two polymers thus appear to be dominated by the resonances associated with the imide moiety. However, as in the case of the C 1s spectra, we suspect that the first two peaks contain resonances associated with both PMDA and ODA/PDA moieties. The second observation concerns the variation in intensity as θ_E is varied: There is a strong polarisation dependence with the π* resonances exhibiting a maximum at high angles of incidence. The same result is obtained at the O 1s where the single n* feature (Fig. 12, polarisation dependence not shown) contains contributions from resonances due to the oxygen atoms in both moieties of PMDA-ODA. These results indicate that the PMDA and ODA/PDA moieties in the monolayer are co-planar and oriented parallel to the HOPG surface. Thus for $\theta_E = 90°$, i.e. for the **E** vector parallel to the surface, the intensity of the n* resonances which are polarised perpendicular to the ring planes, goes to zero. Futhermore, the strong similarity with the model compound spectra indicates that - despite the co-planar configuration - there is little or no interaction of the two ring systems, i.e. no n delocalisation across the N atom. The N K-edge XANES and the STM results, taken together, give a consistent picture of the structure of the PMDA-PDA monolayer on graphite: It consists of parallel chains separated by 0.6 nm in which the ring planes are parallel to the surface, as depicted in Fig. 1a. Attempts to reconcile micrographs such as that of Fig. 5 with the XANES information has so far not lead to

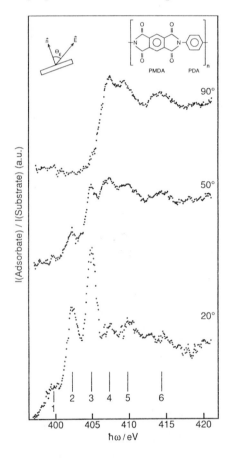

Fig. 10: The nitrogen K-edge spectra of a single monolayer of PMDA-PDA polyimide on HOPG for angles between the E vector and the surface normal, θ_E, of 25°, 50° and 90°.

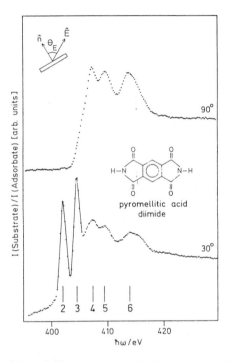

Fig. 11: The nitrogen K-edge spectra of a ~3 Å film of pyromellitic acid diimide on HOPG for angles between the E vector and the surface normal, θ_E, of 30° and 90°. Note the very strong polarisation dependence of the spectra indicating that in this layer the molecule must be oriented almost perfect with its ring plane parallel to the substrate.

any consistent structural model for PMDA-ODA monolayers. One possible reason is that the STM measurements were performed in air and that no heat treatment could be applied as in the case of the samples investigated with XANES in vacuo. We already have some evidence from the appearance of resonance 1 in the nitrogen K edge spectra that such unannealed layers may not be completely imidised. This also applies, however, to the PMDA-PDA samples where the XANES and STM results are comapatible.

5.3 The Oxygen K-edge Spectra

Figure 12 shows the oxygen K-edge spectra of PMDA-ODA polyimide for (a) a monolayer on HOPG, $\theta_E = 20°$, (b) a spin-coated film after Jordan *et al.* (1984). The sharpest, most intense peak at ~532 eV (designated 1), which is assigned to the O 1s→π^* transition at the carbonyl group, exhibits a polarisation dependence (not shown) similar to that at the N edge in Fig. 9, as already noted above. The differences in relative intensity in the two spectra are due to the fact that a $\theta_E = 20°$ geometry for the monolayer has been chosen which gives rise to only weak σ^* resonances.

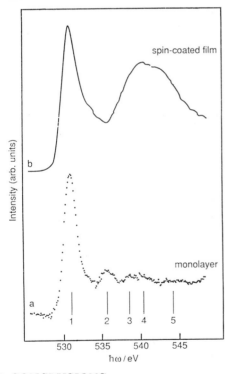

Fig. 12. Oxygen K-edge spectra of PMDA-ODA polyimide, (a) monolayer on HOPG, $\theta_E = 20°$, (b) spin-coated film after Jordan *et al.* (1984).

6. CONCLUSIONS

STM studies reveal that monolayers of both PMDA-PDA and PMDA-ODA on graphite are characterized by a high degree of two-dimensional order. In the case of PMDA-PDA we suggest that the direction of the polymer backbone is coincident with the axis showing ~1.2 nm periodicity and that the intermolecular distance is ~0.6 nm. These data provide confirmation of the simple interpretation of such STM images in terms of molecular models. The corresponding images for PMDA-ODA are, however, still difficult to interpret. Using the polarisation dependence of x-ray absorption spectra both the PMDA-ODA and PMDA-PDA monolayers are found to adsorb parallel to the graphite substrate.

ACKNOWLEDGEMENTS

We are greatly indebted to Prof. Kakimoto and Prof. Imai for preparing the starting materials and to Dr. A. W. Moore (Union Carbide) for supplying the HOPG. This work has been supported by the Bundesministerium für Forschung und Technologie under grant number 05 490 FX B 8 and by the Fonds der Chemischen Industrie.

REFERENCES

Anderson S G, Meyer H M, Gnanorajan S and French J 1988 Mater. Res. Soc. Symp.Proc. 108 189

Bradshaw A M and Somers J 1990 Phys. Scripta T **31** 189

Buchwalter P L and Baise A I 1984 Ref. (Mittal 1984)

Dietz E, Braun W, Bradshaw A M and Johnson R L 1985 Nucl. Instr. Meth. A239 359

Grunze M and Lamb M 1987 J. Vac. Sci. Technol. A5 1695

Hahn P O, Rubloff G W. and Ho P S 1984 J. Vac. Sci. A2 756

Henderson R C 1972 J. Electrochem. Soc.: Solid-State Sci a. Technol. 119 772

Kakimoto M, Suzuki M, Konishi T, Imai Y, Iwamoto M and Hino T 1986 Chem. Lett. 823

Karayazan L G, Tsvankin D Ya, Ginzburg M B, Tuichiev Sh, Korzhavin L N, Frenkel S Ya 1972 Vyskomol. soyed A14 1199 (Polymer Sci. USSR (Engl. Transl.) 14 (1972) 1344)

Jordan-Sweet J L, Covac C A, Goldberg M J and Morar J-F 1988 J. Chem. Phys. 89 2482

Jordan J L, Morar J F, Hughes G, Pollack R A and Himpsel F J 1984 NSLS Annual Rep. Brookhaven National Laboratory p. 125

Meyer H M, Wagener T. J, Weaver J H, Feyereisen M W and Almlöf J 1989. Chem. Phys. Lett. 164 527

Mittal K L (ed.) 1984 Polyimides: Synthesis, Characterization and Applications, Plenum Press, NY

Petersen H 1982 Opt. Commun. 40 402

Roberts G G 1985 Adv. Physics 34 475

Rosenberg R, Love P J and Rehn V 1986 Phys. Rev. B33 4034

Schedel-Niedrig Th, Sotobayashi H, Ortega-Villamil A and Bradshaw A M 1991 Surface Sci. 247 83

Sotobayashi H, Schilling T and Tesche B 1990 Langmuir 6 1246

Stöhr J and Jaeger R 1982 Phys. Rev. B26 4111

Tieke B, Lieser G and Wegner G 1979 J. Polymer Sci. 17 1631

Grafting and growing of poly(N-vinyl-2-pyrrolidone) films on a Pt anode: experimental and theoretical study

E. Léonard-Stibbe, P. Viel, E. Younang[*], M. Defranceschi, G. Lécayon, and J. Delhalle[+]

CEA-Saclay, DSM-DRECAM-SRSIM, F-91191 Gif-sur-Yvette Cedex (France)

([+])Laboratoire de Chimie Théorique Appliquée, Facultés Universitaires N.D. de la Paix
61, rue de Bruxelles B-5000 Namur (Belgium)

(*) Permanent address : Laboratoire de Chimie Physique, Facultés des Sciences, Université de
Yaoundé, BP 812 Yaoundé, Cameroun

ABSTRACT: The electrochemical behaviour of N-vinyl-2-pyrrolidone and its capability to lead to polymer films are studied by cyclic voltammetry in the anodic regime. Thin, insulating, adherent and covering films of poly(N-vinyl-2-pyrrolidone) are obtained on a Pt surface by electropolymerization under anodic polarization. Their electronic and molecular characteristics are studied by UPS and IRRAS spectroscopies.

1. INTRODUCTION

Poly(N-vinyl-2-pyrrolidone) or PVP is a very important specialty polymer with a wealth of uses in various fields : pharmaceuticals, cosmetics, adhesives, detergents, photography, crop protection, oilfield chemicals, etc. (Haaf et al 1985; Linke et al 1987). Physiological harmlessness and physico-chemical properties of PVP, could advantageously be exploited with metallic implants coated with PVP films provided sufficiently dense and homogeneous distributions of chemical bonds can be formed between the substrate and the polymer film to resist mechanical stress, chemical attacks from the environment and to oppose to microbial colonization.

A possible approach towards this goal is to create surface reactive centers on which either to initiate the polymerization of the monomer or to react the functional groups of an already formed polymer. This has recently been used to graft PVP onto carbon fibers by cationic polymerization initiated by carboxylic groups previously created on the surface (Tsubokawa et al 1988). A drawback of such procedure, that could be remedied by electrochemical techniques, is the small number of true chemical bonds formed between the polymer and the surface (Allara et al 1986).

As confirmed in a recent study (Younang et al 1991), the electrochemical polymerization at the cathode, capable of providing films homogeneously covering surfaces of oxidizable metals and resisting redissolution in solvents (Bruno et al 1975; Debesne-Monvernay et al 1978; Lécayon et al 1982, 1984, 1987; Deniau et al 1990, 1991), is impossible in the case of N-vinyl-2-pyrrolidone (VP). The purpose of this work is to investigate the possibilities to graft and grow PVP films at a Pt anode using the same stringent conditions as in cathodically polarized electropolymerizations. Our goal is to obtain thin films of functionalized polymers presenting the same properties as the corresponding polymers in bulk, it is thus different from the electroinitiated polymerizations in the anodic regime of vinylic systems in solution, e.g. styrene (Akbulut et al 1975) or conducting polymers such as polypyrrole, polyaniline, etc..

2. THEORETICAL RESULTS

Possible ways of polymerization of N-vinyl-2-pyrrolidone (VP), a vinylic polymer, initiated by the electrochemical polarization at the anode surface are schematically shown in Figure 1.

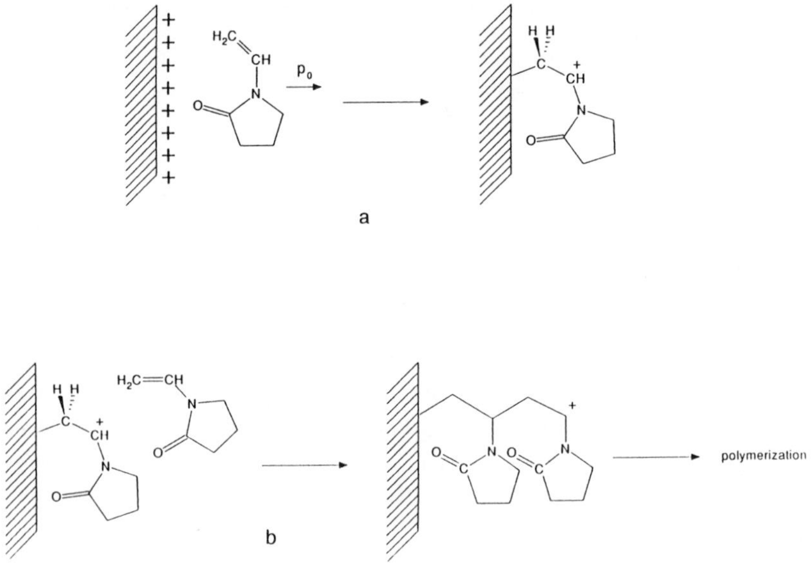

Fig.1. Mechanism leading to electroinitiated chemisorption and polymerization under anodic polarization (the molecule is represented in the synperiplanar form).

Generally, these mechanisms are possible if the substituent is an electron donating group capable of accomodating the positive charge of the cation formed by the reaction of the C=C double bond with an electron deficient site. Ideally, the substituent should also contribute to orient the reactive part ($RCH=CH_2$) of the vinylic moiety towards the anode and preferably should not include sites that could compete with this reactive part.

Theoretical calculations carried out at the ab initio RHF level using the 3-21G basis set (Younang et al 1991) have shown that the HOMO energy of VP, -8.73 eV, is close to the corresponding values of styrene, -8.21 eV, which is known to polymerize electrochemically in solution (Akbulut et al 1975). Thus, from the point of view of its electronic characteristics in the isolated state, VP appears to be a suitable candidate for grafting PVP on a Pt anode.

As shown in Figure 2, VP can adopt two conformations, antiperiplanar (ap) and synperiplanar (sp); the direction of the molecular dipole moment is also indicated in the figure. Theoretical calculations (Younang et al 1991) indicate that the (ap) form is more stable than (sp) by 12.7 kJ.mol^{-1}. The (ap) conformation does not correspond to a suitable orientation of the molecule approaching towards the anode when subject to the double layer electric field, the synperiplanar conformation being more appropriate. However, to go from the (sp) to the (ap) conformation it is necessary that the pyrrolidone ring undergoes a 180° rotation around the $H_2C=HC-N$ bond. A single point calculation at 90° rotation leads to an energy barrier of 15 kJ.mol^{-1}. Electric field intensities and dipole moments of the order of 10^9 V.m^{-1} and 4 D, respectively, correspond to an energy lowering of approximately 8 kJ.mol^{-1}. This should be sufficient to start inducing rotations of the pyrrolidone group around the C-N bond and bring the VP molecule in a suitable conformation for reaction as it approaches the anode surface.

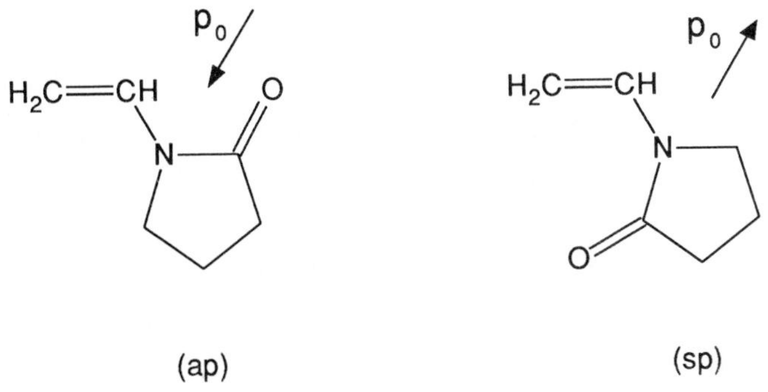

Fig.2. Representation of the antiperiplanar (ap) and synperiplanar (sp) forms of N-vinyl-2-pyrrolidone.

Nitrogen, oxygen and the terminal carbon of the vinylic moiety, $CH_2=CH-$, are nucleophilic centers of VP which can interact with the electron deficient surface sites. Simulation of these surface sites by a proton provides valuable insight on the possible competition between the various nucleophilic centers of VP to interact with the electron deficient surface sites. Total energy results indicate that VPH$_C^+$, the system protonated on C, is more stable than VPH$_O^+$ (system protonated on O) by 11.6 kJ.mol^{-1}, itself being more stable than VPH$_N^+$ (system protonated on N) by 281.5 kJ.mol^{-1}, thus the electrophilic attack should preferentially take place on the terminal carbon CH_2=CH- of the vinylic moiety of the molecule. This follows the Markovnikov rule which states that with unsymmetrical alkenes the proton adds to the less substituted carbon of the double bond so as to afford the more stable carbocation. However, one cannot disregard the fact that the proton affinity of the oxygen of the carbonyl group is only 11.6 kJ.mol^{-1} smaller than that of CH_2=CH-.

3. EXPERIMENTAL

All experiments (reactant purification and electrochemical processes) are carried out in glove boxes under recycled argon atmosphere; to keep the amount of residual water vapour and oxygen below 1 and 5 ppm respectively, argon is treated by passage through various purifiers (Lécayon 1981). The monomer (Aldrich 99%) is purified by storage on molecular sieves before fractional distillation under reduced pressure.The solvent (acetonitrile, Fluka >99.5 %) is treated by stirring 24h with potassium permanganate in order to oxidize the traces of unsaturated impurities such as acrylonitrile or allylic alcohol present in the commercial product and then, like the monomer, purified after distillation.The amounts of water in both monomer and solvent distillates are titrated by the Karl Fischer method and found to be less than 5×10^{-4} mol.dm^{-3}. Tetraethylammonium perchlorate (Fluka 99%) is used as supporting electrolyte after drying by permanent storage under reduced pressure at 110°C. The electrolytic solution is prepared by dissolution of the monomer in acetonitrile with 5×10^{-2} mol.dm^{-3} of supporting electrolyte.

The electrochemical behaviour of N-vinyl-2-pyrrolidone at the anode is studied on a working electrode prepared by cathodic sputtering of a 100 nm thick layer of Marz platinum (99.99%) on a 5 cm^2 glass plate.The auxiliary electrode is a platinum sheet and the reference electrode is made using the electrochemical Ag/Ag$^+$ couple to which all potentials are referred. Electrochemical experiments are performed with a PAR 173 potentiostat-galvanostat coupled to a COMPAQ 386S computer.

After electrolysis, the samples are rinsed with tetrahydrofuran in which PVP is not soluble and then analyzed by ultra-violet photoelectron spectroscopy (UPS) and by reflexion-absorption infra-red spectroscopy (IRRAS). The UPS spectra are collected on a Vacuum Generator Escalab MKII spectrometer with the He II line (40.8 eV). The molecular structure of the film is studied by IRRAS with a BRUCKER IFS 66 instrument.

4. ELECTROCHEMICAL BEHAVIOUR OF N-VINYL-2-PYRROLIDONE

In this section the electrochemical behaviour of the electrolytic medium is studied at low and high concentrations of VP during anodic polarization by cyclic voltammetry. First, experiments are performed in a one-compartment cell with a solution of low concentration, C, in the monomer (5×10^{-3} mol.dm^{-3}) to work under pure diffusion control and to prevent the occurence of secondary chemical reactions at the electrode surface and in the electrochemical cell that could interfere with the intrinsic electrochemical behaviour of VP. Then, studies at higher concentrations which are necessary to obtain films are reported.

4.1 Low Monomer Concentration

In Figure 3 are presented a reference current-potential curve recorded on the electrolytic solution without monomer and a forward and return curve for a solution of N-vinyl-2-

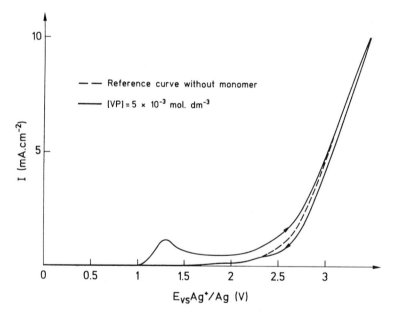

Fig.3. Intensity-potential results on N-vinyl-2-pyrrolidone on a Pt anode
(v= 50mV.s^{-1}).

pyrrolidone ($C = 5 \times 10^{-3}$ mol.dm^{-3}) recorded with a sweep rate equal to 50 mV.s^{-1}. Under these conditions, oxidation of the monomer can be observed at $E_{pa}=1.30$ V, i.e. before the barrier corresponding to the oxidation of the supporting electrolyte; no reduction peak can be observed on the return curve. These observations suggest that N-vinyl-2-pyrrolidone is oxidized by direct and irreversible electron transfer at the anode. This is in good agreement with experimental data previously obtained by Akbulut et al (1975) in their study of electroinitiated polymerization of vinyl monomers and corroborate our theoretical results (Younang et al 1991).

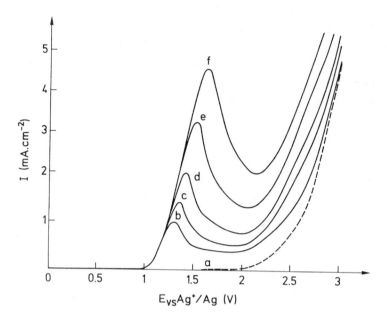

Fig.4. Influence of sweep rate on the oxidation peak of the monomer (C = 5×10^{-3} mol.dm^{-3}): (a) reference curve without monomer v= 50 mV.s^{-1} ; (b) v=50 mV.s^{-1} ; (c) v= 100 mV.s^{-1} ; (d) v= 200 mV.s^{-1} ; (e) v= 500 mV.s^{-1} ; (f) v= 1000 mV.s^{-1}.

In order to further specify the nature of the oxidation peak for this concentration, additional voltammetric studies are performed at different sweep rates. Experimental curves are presented in Figure 4.

TABLE I. Randles-Sevick equation parameters.

entry	C (mol.dm^{-3})	σ (mS.cm-2)	Ep (V)	Ip (mA.cm^{-2})	Ip/C
1	5×10^{-3}	6.5	1.30	0.27	540
2	5×10^{-2}	6.4	1.60	2.75	55
3	10^{-1}	6.3	1.65	4.50	40.5
4	2.5	4.4	1.55	4.05	1.62
5	5	2.6	1.5	3.50	0.70

For an irreversible system, the peak current is characterized by the Randles-Sevick equation : $I_p = k'v^{1/2}C$. In our case, a correlation between the square root of the sweep rate and the intensity shows that peak current increases linearly with $v^{1/2}$; the results are listed in Table I. This behaviour indicates that, at low concentration ($C = 5 \times 10^{-3}$ mol.dm^{-3}), the oxidation process is controlled by the diffusion of electroactive species in the electrochemical cell.

4.2 High Monomer Concentrations

Electrochemical experiments aimed to study the influence of the concentration in the monomer on the formation of films are now reported. Indeed, previous work in the cathodic regime (Durig et al 1971) showed that the formation of an electroinitiated polymer film requires concentrations in the monomer 10 to 10^2 times that of the concentration in the supporting

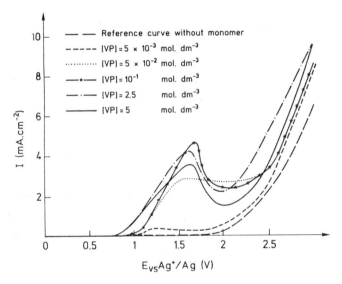

Fig.5. Dependence of the electrochemical behaviour on the concentration (v=5 mV.s^{-1}).

electrolyte. A series of voltammograms are recorded for concentrations varying between 5×10^{-3} and 5 mol.dm^{-3} with a sweep rate of 5 mV.s^{-1}; the corresponding curves are presented in Figure 5.

In all cases the oxidation peak appears before the barrier corresponding to the oxidation of the supporting electrolyte which confirms the electroactivity of the molecule. Table II shows the evolution of E_{pa} and I_{pa}/C with respect to concentration. A significant deviation from the ideal behaviour (E_{pa} independent of C and I_{pa}/C constant) can be noted : the peak intensity decreases as the concentration increases which is incompatible with a diffusion-controlled process.

TABLE II. Influence of the concentration on E_{pa} and I_{pa}.

entry	v (mV.s^{-1})	v$^{1/2}$	Ep (V)	Ip (mA.cm^{-2})	Ip/v$^{1/2}$
1	50	7.07	1.30	1.00	0.14
2	100	10.00	1.35	1.54	0.14
3	200	14.14	1.37	2.05	0.14
4	500	22.36	1.50	3.10	0.14
5	1000	31.62	1.60	4.15	0.13

For concentrations ranging from 2.5 to 5 mol.dm^{-3} a change of colour can be observed in the cell after electrochemical polarization as well as the presence of a thin brown-coloured film on working electrode. They reveal the existence of secondary reactions occurring both in the electrolytic medium and at the electrode surface. All of these observations point to the existence of important perturbations in the reactive medium when the concentration in monomer increases which make very difficult any interpretation of the evolution of the electrochemical characteristics. It must also be noted that experiments are perturbed by the products of the secondary reactions in the reactive medium; this probably modifies the diffusion of electroactive species because of the evolution of viscosity in the reactive medium. Furthermore, the conductibility of the solution largely decreases when the monomer concentration increases.

5. SPECTROSCOPIC STUDY OF THE ANODE SURFACE

In order to study the various steps of the reaction leading to the formation of a polymer film, a series of electrolyses is realized by the voltammetry technique by stopping the sweep at different potentials determined on the I/E curve presented in Figure 5. This is done in the highly-concentrated solution of monomer (5 mol.dm^{-3}) to increase the probability of secondary electrode reactions responsible for the polymer film formation.

The passivation of the surface is studied by electrolyses stopped at 0.5 , 1.5 , 2 and 2.5 V, in order to evidence the formation of a PVP film after oxidation of the monomer, by direct electron-transfer at the anode and not by other species such as ClO$_4^-$. After polarization the samples are rinced with tetrahydrofuran and analyzed spectroscopically. The films synthetized by our electrochemical technique are compared with a film obtained by dipping a Pt electrode in a solution of commercial polymer (Aldrich, average MW=10000 , synthetized by the Reppe's process) in acetonitrile.

The evolution of the density of occupied states is followed by UPS measurements. The spectra are shown in Figure 6. The spectra of surfaces after anodic polarization are compared with the initial Pt surface before and after ionic etching (Ar^+ , 3 kV , 8 μA , 10mn). All spectra are referenced to the Fermi level of the bare metal surface obtained by ionic etching (Figure 6a).

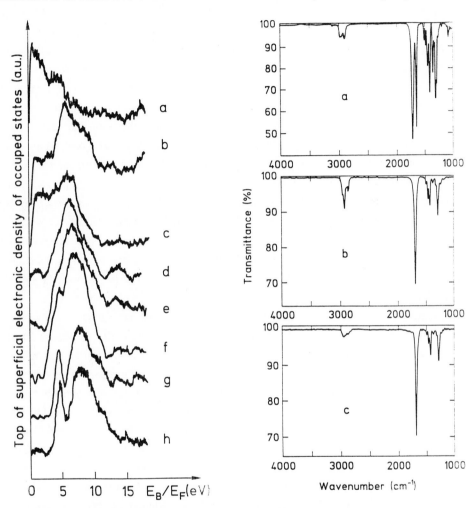

Fig.6. UPS spectra of the Pt surface : (a) bare Pt surface ; (b) initial working surface ; (c)-(f) surface obtained after linear sweep potantial stopped respectively at 0.5, 1.5 , 2 and 2.5 V ; (g) surface obtained after 10 min of polarization at 1.5 V (h) surface of a Pt coated by dipping in an a solution of Aldrich polymer.

Fig.7. IRRAS spectrum of the : (a) Aldrich monomer ; (b) film obtained after 10 min of anodic polarization at 1.7 V (c) film obtained by dipping the surface in a solution of Aldrich polymer.

The spectrum is characterized by a peak centred at 0.6 eV below the Fermi level and attributed to the $5d^8$ levels of the metal : this is in good agreement with the results obtained by Wendelt (1982). The spectrum of the initial working surface (Figure 6b) shows an intense structure centred at 6.4 eV below the Fermi level and a weaker intensity of the $5d^8$ peak can be noticed. This evolution can be attributed to a surface contamination by the laboratory atmosphere. After polarization at 0.5 V (Figure 6c) an exaltation of the $5d^8$ levels can be noted. At 1.5 V (Figure 6d) a more significant evolution is noted : an increase by 2.8 eV of the electronic work function which supports the hypothesis of a surface modification. The spectrum is still dominated by the structure corresponding to hydrocarbon contaminants. Nevertheless, a shoulder at 4.8 eV becomes visible on the side of the main structure at 6.4 eV. The spectra obtained for 2.0 and 2.5 V (Figure 6e to 6f) corroborate this evolution : the increase of the work function stays equal to 2.8 eV and confirms the hypothesis of the formation of a covering and insulating coating on the surface; the shoulder becomes a peak which can be attributed to the oxygen lone pair by comparison with the spectrum of the monomer analyzed by Woydt et al (1989). As the initial main structure at 6.5 eV decreases, a poorly resolved structure emerges at 8.5 eV, which is tentatively assigned to the π_{CO}, π_{CH2} and σ_{CH2} levels of the pyrrolidone ring. The spectrum recorded after potential-controlled electrolysis (+1.5 V, 10 min) and shown inFigure 6g is in good agreement with those obtained by voltamperometric studies. Finally, the Pt surface covered with a PVP film obtained after dipping in a solution of commercial polymer (Figure 6h) presents an electronic structure comparable with the preceding observations .

In order to confirm the nature of the film formed at the electrode after anodic polarization and to study some eventual defects in the molecular structure of the films obtained by our electrochemical technique, the surface obtained after potential-controlled electrolysis is compared in a first step with the IR spectrum of the monomer and in a second step with a film of commercial polymer obtained by dipping. The IRRAS spectra are presented in Figure 7. After polarization at 1.7 V during 600 s , the anode is coated with a thin (30 nm as determined by spectroscopic ellipsometry), homogeneous brown-coloured film. It is important to notice that it has been impossible to obtain thick films. The film obtained by electrolysis presents the characteristic bands of the molecular structure of PVP. The assignment of its spectral bands established by comparison to Socrates's work (1980) can be summarized as follows : 1288 cm^{-1}, 1420 , 1462 and 1493 cm^{-1} (CH def. of cyclic CH$_2$ groups), 1694 cm^{-1} (CO str.). The characteristic bands of the monomer corresponding to the vinylic structure have totally disappeared : 1333 cm^{-1} (CH def.) , 1629 cm^{-1} (CC str.) and weak bands over 3000 cm^{-1} (CH str. of CH and CH$_2$). No defect in the molecular structure can be noticed.

6. DISCUSSION

As expected from the calculated HOMO enrrgy (ε_{HOMO}= -8.73 eV), N-vinyl-2-pyrrolidone has been shown to be electroactive under anodic polarization. From a mechanistic point of view, we propose that the formation of a stable radical-cation by direct interaction between the vinylic extremity of the monomer and the anode is promoted by a possible delocalization of the positive charge on the molecular skeleton.The theoretical simulation of the interaction with the electrode by protonation of the monomer shows that the most reactive site is the less substituted vinylic carbon ($CH_2=$), according to Markovnikov's rule.

The fact that PVP is soluble in the reactive medium suggests that the film is built up at the surface from oxidized and chemisorbed monomer molecules. The grafting mechanism of the oxidized monomer can be schematized as follows : after activation in the double layer by the electric field, the molecule reacts with the superficial reactive sites of the electrode to form a chemisorbed carbocation (Figure 1a) .Then a secondary electrode reaction leads to the formation of a polymer film by a cationic mechanism (Figure 1b). This hypothesis is reinforced by the UPS results : the modification of the electrode surface by formation of an insulating film is noted after anodic polarization (increase of the work function , presence of a structure corresponding to the pyrrolidone motif). This result is further corroborated by the IRRAS measurements : the molecular structure is characterized by a saturated skeleton functionnalized with hanging pyrrolidone groups. It must be noted that, contrary to the classical cationic polymerization (Mercier 1987), the growth rate of the electrochemically initiated film is very slow. Possible explanations to this is the partial solubility of the polymer in the reactive medium but no polymer is detected by infra-red measurements in the anodic compartment after one hour of polarization at 1.7 V . Another hypothesis concerning the structure of the monomer can be proposed : according to theoretical calculations , the antiperiplanar conformation is the most stable but its dipolar moment is not favourable to oxidation under an anodic electric field.It is possible that the activation energy necessary for setting the molecule in the adequate geometry restrains the process. Finally, one cannot exclude the occurence of a radical-cation coupling mechanism which would lead to a lowering of the polymerization process (Waltman et al, 1986).

7. CONCLUSION

Theoretical previsions concerning the possibility for N-vinyl-2-pyrrolidone to form a polymer film under anodic polarization are verified by experimental results. Electrochemical studies show that the film results from a direct interaction between the monomer and the electrode surface. The molecular structure of the grafted polymer is similar to that of PVP obtained by classical polymerization methods. These results show that the process involves carbocation-type reaction intermediates : this is in good agreement with theoretical simulation of the behaviour of the different molecular sites potentially reactive under oxidation.
Complementary theoretical and experimental studies will be realized in order to further specify

the relative contribution of these different sites, the possible dependence of the polymerization on the ability for the pyrrolidone substituent to rotate around the C-N bond, and to definitly clarify the origin of the low growth of the polymer.Finally, the fact that the film thickness can be of the order of 50 nm, in spite of the known solubility of PVP in the reactive medium used, raises the problem of knowing the chain organization at the molecular level for the electropolymerized PVP. Studies are now in progress to get some insight into these questions (chain length and cross-links).

Acknowledgments. The authors acknowledge with appreciation the support of this work within the scientific exchange agreements between France and the Communauté française de Belgique (1991), project n° 91.23. E. L.-S. is grateful to the IRAP (Institut de Recherche Appliquée sur le Polymères, Le Mans, France) and the CCI (Chambre de Commerce et d'Industrie du Mans et de la Sarthe, France) for a Ph.D. stipend and their interest in this work. E.Y. thanks the CEA for financial support that made possible his research visit at the DSM/SRSIM Sacaly (France).

References

Akbulut U, Fernandez J E, and Birke R L 1975 J. Polym. Sci. Polym. Chem. Ed. 13 133
Allara D L, Fowkes F M, Noolandi J, Rubloff G W, and Tirell M V 1986 Materials Science and Engineering, **83** 213
Boiziau C and Lécayon G 1986 Scanning Electron Microsc. **6** 207
Bruno F, Phan M C and Dubois J E 1975 J. Chim. Phys.**72** 490
Deniau G, Lécayon G, Viel P, Zalczer G, Boiziau C, Hennico G and Delhalle J 1990 J. Chem. Soc. Perkin Trans. **2** 1433
Desbene-Monvernay A and Dubois J E 1978 J. Electroanal. Chem. **89** 149
Durig J R, Tong C K, Hawley C W, and Bragin J 1971 J. Phys. Chem. **75** 44
Haaf F, Sanner A, and Straub F 1985 Polymer J. **1** 143
Lécayon G 1981 rapport CEA N2181
Lécayon G, Bouizem Y, Le Gressus C, Reynaud C, Boiziau C and Juret C 1982 Chem. Phys. **91** 506
Lécayon G 1984 J. Phys. C2 **45** 713
Lécayon G, Viel P, Le Gressus C and Boiziau C 1987 Scanning Microsc. **1** 85
Linke W and Vogel G M 1987 Polymer News **12** 232
Mercier J P 1983 Polymérisation des Monomères Vinyliques (Lausanne: Presses Polytechniques Romandes)
Reppe W 1954 Polyvinylpyrrolidon,monographie zu "Angewandte Chemie" n°66, Weinheim/Bergstr.
Socrates G 1980 Infrared Characteristic Group Frequencies, (Chichester: Wiley)
Tsubokawa N, Maruyama H, and Stone Y 1988 J. Macromol. Sci.-Chem. **A25** 171
Waltman R J, Bargon 1986 J Can. J. Chem. **64** 76
Wandelt K 1982 Surface Science Report Vol.2
Younang E, Léonard-Stibbe E, Viel P, Defranceschi M, Lécayon G, and Delhalle J, "Prospective theoretical and experimental study towards electrochemically grafted poly(N-vinyl-2-pyrrolidone) films on metallic surfaces" Mol. Eng. to be published

Paper presented at First International Conference, Namur, Belgium
2–6 September 1991: Section I

Heat-induced chemical evolution of polyamic acid–metal interfaces

Fabio Iacona°, Giovanni Marletta*, Marco Garilli§, Orazio Puglisi* and Salvatore Pignataro*

°C.N.R.-I.Me.Te.M., viale A.Doria 6, 95125 Catania (Italy);
§Consorzio Catania Ricerche, viale A.Doria 8, 95125 Catania (Italy);
*Dipartimento di Scienze Chimiche dell'Università, viale A.Doria 6, 95125 Catania (Italy).

ABSTRACT: In this paper we report the XPS study of the heat-induced chemical evolution of the interface between thin polyamic acid (PAA) films (thickness 25±5 Å) and oxidized metal surfaces. Depending on the nature of the metal surface, the process of thermal curing may lead to a partial degradation of the polymer, in competition with the reaction of imidization, the extent of the phenomenon being correlated with the isoelectric point of the metal surface (IEPS). Furthermore, also catastrophic degradation processes have been observed, due to the instability of some oxidized metal surfaces in the employed curing conditions.

1. INTRODUCTION

Polyimides (PI) are widely used in a number of advanced technologies including microelectronics, corrosion and radiation protection, aerospace and wire insulation. Most of these applications involve the formation of reliable and durable interfaces with metal surfaces, by depositing polyamic acid (PAA) solutions onto surfaces which are often covered by a more or less thick native oxide layer.

The adhesion performances of these joints are dependent on the chemistry and physics of the involved surfaces, as well as on the stresses in the polymer film and substrate (Buchwalter 1990). Among all these factors, a particular importance is attributed to the formation of chemical bonds at interfaces. Accordingly, in the last years a great effort has been directed towards the knowledge of the interfacial chemistry of several polyimide-on-metal (Buchwalter 1990) and metal-on-polyimide systems (Ho 1989).

However, since the polyimide film is applied as a PAA solution, i.e., in the uncured state, chemical reactions may occur at the interface not only as the characteristic acid-base reactions expected for these systems (Bolger 1983), but also during the subsequent solvent removal and curing steps. In particular, the reactions responsible of the evolution of the polymer-solid interfaces during these steps has been the object of several papers (Burrell et al. 1989, Flament et al. 1990, Kelley et al. 1987, Russat 1988, Stewart et al. 1989, Kim et al. 1988).

In the last years we focussed our attention to the study of this type of problems (Garilli et al. 1989, Iacona et al. 1991). We found that in a PAA/metal system the interaction starts at room temperature, mainly consisting in acid-base reactions, and particularly in salification of the carboxylic groups, prompting the occurrence of very complex reaction patterns during the curing process, including different degrees of modification of the polymer and the metal at the interface (Iacona et al. 1991).

In this paper we show that can be recognized two different kinds of interfacial reactions induced by curing, depending either on the thermal stability of the oxidized metal surfaces, or on the strength of the preliminary acid-base interaction between PAA and native oxide.

2. EXPERIMENTAL

The investigated polyamic acid was Pyralin 2555 from DuPont de Nemours, consisting mainly of a 3,4,3',4'-benzophenonetetracarboxylic dianhydride-4,4' oxydianiline (BTDA-ODA); thin films of PAA were obtained by spin coating, at 5000 rpm for 45 seconds, a solution prepared by 1:50 dilution in N-methyl-2-pyrrolidone of the commercial product. After the coating step the PAA films were heated at 90°C for 5 minutes to remove the residual solvent. The thickness of these PAA films was estimated at 25 ± 5 Å by means of angle resolved XPS measurements. The substrates for spin coating were bare or metallized 5 inch silicon wafers; the metallized wafers were prepared by e-beam deposition of Cr, Ni and Ti (thickness 3500 ± 500 Å). In the Ni and Ti cases we also studied the surfaces obtained by heating in air at 300°C for 30 minutes the as deposited ones.

XPS analysis was performed in a Kratos ES 300 electron spectrometer (operating pressure 1×10^{-9} torr), equipped with a dual anode (Mg/Al), a hemispherical analyser and a preparation chamber, where we performed the "in situ" curing of the PAA films, by heating at 200°C for 2 hours and at 300°C for 30 minutes. XPS peak analysis was performed by means of computer program of gaussian peak fitting with inelastic background subtraction.

3. RESULTS AND DISCUSSION

All the studied metal surfaces are covered by thin layers of oxides and hydroxides, whose acid-base properties can be expressed by using the isoelectric point of the surface (IEPS). The IEPS has been extensively used by Bolger (1983), Buchwalter (1987) and Oh et al. (1990) to describe the interaction between oxide surfaces and polymers. The IEPS values reported by Parks (1965) for the surfaces studied in the present work are: 2.2 for SiO_2, 5.0 for TiO_2, 7.0 for Cr_2O_3 and 10.4 for NiO. On the other hand PI, in the form of its soluble precursor PAA, is an acidic polymer, due to the presence of -COOH groups (estimated pK_a = 3.0, Buchwalter 1987).

We have previously shown (Iacona et al. 1991) that when PAA comes into contact with basic surfaces like oxidized Cr and Ni, the C 1s XPS peak shows some new features, which can be explained in terms of the formation of carboxylate groups, while no significant modification can be observed with a strong acidic surface like oxidized Si. Spectroscopic evidence of the formation of carboxylate at the PAA/metal interface has been previously reported by Kelley et al.(1987) for the case of Cu and by Flament et al. (1990) for the case of Al.

This reaction has been proposed by Kim et al. (1988) to be a necessary step of the curing-induced formation of oxide precipitates in PAA films on oxidized Cu and Ni surfaces, while in the case of oxidized Cr it is not able to prompt the subsequent formation of precipitates, due to the higher corrosion resistance of Cr_2O_3. This work mainly concerned the curing-induced degradation of the metallic phase. On the other hand, the occurrence of this interfacial reaction may prompt characteristic degradation processes of the chemical structure of the cured polymer depending on the nature of the metal surface. This is shown in detail in fig.1, which reports typical C 1s and N 1s XPS peaks for a fully cured BTDA-ODA thick film (about 1 μm thickness, fig.1a), compared with those obtained for cured thin films (about 25 Å thickness) of the same polymer deposited on a silicon wafer (fig.1b), on a titanium surface obtained by heating in air at 300°C for 30 minutes the as deposited one (fig.1c), and on an as deposited chromium surface (fig.1d). The main features of the XPS spectrum of a fully cured BTDA-ODA has been already extensively discussed by Leary et al. (1979) and Buchwalter and al. (1984). We focus our attention on the main carbonylic component in the C 1s peak (about 288.7 eV B.E.) and on the low B.E. component (399.2-.3 eV) in the N 1s peak, assigned by Buchwalter et al. (1984) to =C=N- groups present in the isoimidic structure, or to =C=N- groups present in linear structures by Toth et al.(1986) and Anderson et al. (1988). In the cases of Ti and Cr substrates we observe four main effects: 1) a decrease in the intensity of the the carbonylic component;

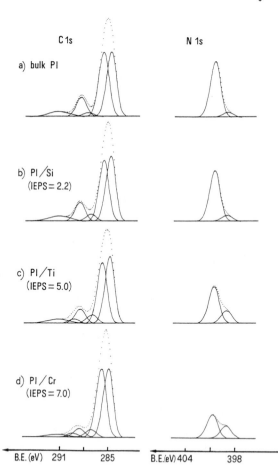

Fig. 1. C 1s and N 1s XPS peaks for: a) bulk PI; b) PI on Si; c) PI on Ti; d) PI on Cr.

2) the increase of the intensity of the N 1s component at about 399.2 eV;
3) the formation of a component at about 289.4 eV in the C 1s peaks, due
to -COOH groups (that should be absent in a fully cured PI) and 4) the occurrence of nitrogen loss (all peaks in fig.1 are area normalized in order
to evidence this effect).

All these features indicate the occurrence of a partial degradation of the
polymer in the interfacial region. This effect can be explained by taking
into account the interaction between the -COOH groups of the polymer and
the oxidized metal species, that inhibits the closure of the imidic rings,
leaving the polymer in the polyamidic form. The presence of typical polyamidic features in thin PAA films deposited and cured on Cu has been previously revealed by means of FTIR by Stewart et al. (1989) and Burrell et
al. (1989). The polyamidic form is by far less thermally stable than the
imidic form and it may undergo remarkable degradation processes already at
the typical curing temperatures. The main decomposition pathways, involving hydrolysis of amidic bonds, decarbonylation and decarboxylation, formation of =C=N- groups and further complex reactions leading to the loss
of small nitrogen-containing molecules, have been previously discussed by
Iacona et al. (1991).

Fig.1 suggests that the extent of degradation of the polymer is proportional to the IEPS of the metal surface with which it is in contact; indeed,
in the Si case, that is characterized by a very low IEPS value (2.2), no
relevant differences can be observed with respect to a bulk-like PI, while
the spectroscopic evidences of degradation are more marked in the Cr case
(IEPS = 7.0), than in the Ti case (IEPS = 5.0).

The spectra reported in fig.1 for Ti refer to a surface obtained by heating in air at 300°C for 30' the as deposited one. This treatment has been
performed in order to "stabilize" this surface, because we found that it
is remarkably reduced when heated in ultra high vacuum (U.H.V.). A similar
effect has been already reported by Garilli et al. (1989) and Iacona et
al. (1991) for Ni and it is illustrated in fig.2. Fig.2a reports Ni $2p_{3/2}$
and O 1s XPS peaks for an as deposited Ni surface. This surface is strongly oxidized, while containing metallic component, as shown by the complex
shape of the Ni $2p_{3/2}$ peak, with a prevalent contribution of hydroxyl-con-
taining species (component at about 531.4 eV in the O 1s peak) with re-

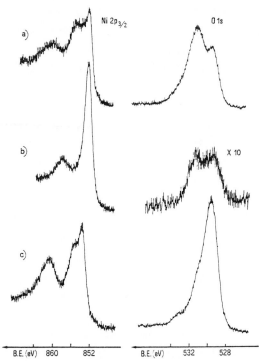

Fig. 2. Ni $2p_{3/2}$ and O 1s XPS peaks for: a) as deposited Ni; b) Ni after heating in U.H.V.; c) Ni after heating in air.

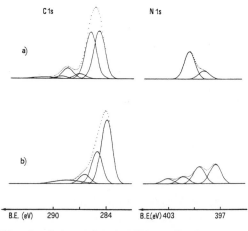

Fig. 3. C 1s and N 1s XPS peaks for: a) PI on Ni heated in air; b) PI on as deposited Ni.

spect to NiO (component at about 529.5 eV). If this surface is heated in U.H.V., the almost complete reduction of the oxidized species occurs; indeed we obtain a Ni $2p_{3/2}$ peak almost totally due to metallic Ni, while only a small amount of oxygen is still detected (see fig.2b). The occurrence of this phenomenon can be avoided by heating in air for 30 minutes at 300°C the as deposited Ni surface. This new Ni surface (see fig.2c) has a thicker oxide layer (no metal peak can be detected), with only a minor contribution of hydroxylic species (the shoulder on the high B.E. side of the Ni $2p_{3/2}$ peak is mainly due to the multiplet splitting), and is stable if heated in U.H.V..

The different reactivity of these two oxidized Ni surfaces towards a PAA overlayer is shown in fig.3. The "stabilized" Ni surface behaves similarly to Cr, i.e. induces only the partial decomposition of the polymer (see fig.

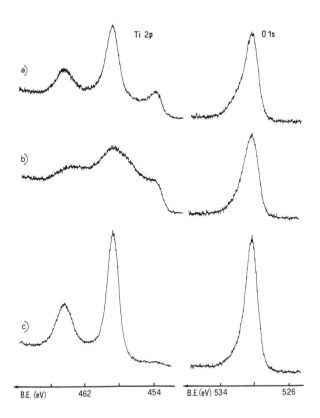

a)

b)

c)

Ti 2p   O 1s

B.E. (eV)   462   454   B.E.(eV) 534   526

Fig. 4. Ti 2p and O 1s XPS peaks for: a)
as deposited Ti; b) Ti after heating in
U.H.V.; c) Ti after heating in air.

3a). On the other hand, the effect of the instability of the as deposited Ni surface on the PAA film is catastrophic: the release of oxygen from the metal surface induces the total decomposition of the polymer, leaving only a very thin layer, essentially composed by amorphous carbon, with a relatively high content of different nitrogen-containing species, probably including metal nitrides (see fig. 3b).

In the case of Ti the chemical modification induced by the heating is less marked.

Fig.4a shows the Ti 2p and the O 1s XPS peaks for an as deposited Ti surface. This surface is characterized by the presence of a layer of TiO_2 (about 458.7 eV B.E. for the $2p_{3/2}$ component), thin enough to allow the detection of metallic Ti (about 453.7 eV B.E. for the $2p_{3/2}$ component); it is to note that the $2p_{3/2}$ component of TiO_2 overlaps with the $2p_{1/2}$ component of Ti. Furthermore, the shoulder on the high B.E. side in the O 1s peak indicates the presence of hydroxylic species. After the heating in U.H.V. new Ti species besides Ti and TiO_2 are present on the surface (see fig.4b), presumably oxides in a lower oxidation state. Accordingly, the B.E.s reported by Chan et al. (1990) are 454.2-.4 eV for TiO and 455.6-456.8 eV for Ti_2O_3. The heating in air at 300°C for 30 minutes of the as deposited surface (see fig.4c) produces a thicker oxide layer (no metallic

Ti can be detected), with a remarkably reduced content of hydroxylic species (see O 1s peak). As in the Ni case, this surface is stable if heated in U.H.V.. Also in this case the modification undergone by the Ti surface greatly influences the curing of a PAA overlayer. Indeed, while a PAA film in contact with the "stabilized" Ti surface undergoes only minor effects of decomposition (see fig.1c), the as deposited surface induces an extensive decomposition of the polymer, even if the phenomenon is less "violent" than in the Ni case, leaving a small but detectable amount of unaltered imidic rings.

The instability of naturally passivated Ni surfaces under heating in U.H.V. conditions remains to be explained, also if Cochran et al. (1985) have already reported that NiO (as powder pellets) was reduced by heating in U.H.V., and attributed this effect to a reaction with carbonaceous (possibly hydrocarbons) contaminants. In contrast with this observation, it is to note that our "stabilized" NiO surface does not show appreciably changes in the same heating conditions. Nevertheless, we may argue that when PAA is deposited on this Ni surface it behaves like the "contamination layer" proposed by Cochran et al., promoting the oxide reduction and in turn, the decomposition of the polymer. In the case of Ti the degradation pathway involves only the partial reduction of TiO_2 (or more probably of the hydroxylated species) to intermediate oxidation states like Ti^{3+} or Ti^{2+}.

In both cases, a major role seems to be played by the presence of an high concentration of hydroxylic species on the naturally passivated surfaces. This hypothesis is confirmed by the fact that both the "stabilized" Ni and Ti surfaces contain only small amount of these species.

4. CONCLUSIONS

We have shown that the interface of PAA with oxidized metal surfaces at typical curing temperatures can be the site of relevant chemical reactions, leading to a more or less extended degradation of the polymer as well as of the metallic-like phase. It is to stress that these degradation pathways are very different from typical imidization reaction occurring in the polymer bulk at the same temperatures. The relevant point is that in the interfacial region the imidization process is inhibited to some extent, so that alternative degradation pathways are preferred, characteristic of thermally unstable polyamidic forms. In this sense, we can say that the heat resistance of polyimides is completely lost at the interface region. We can classify the degradation reactions in two distinct categories:

a) Reactions that occur when both the polymer and the inorganic phases are thermally stable. In this case the reactivity is mainly driven by acid-base interactions and in particular by the difference between the IEPS of the oxidized metal surface and the constant of acidity or basicity of the polymer.

b) Reactions that occur when the inorganic phase is not thermally stable. In this case the acid-base interactions are still present, but the overwhelming contribution due to the "catastrophic" reactions induced by the decomposition of the oxidized metal surface makes them negligible.

ACKNOWLEDGMENTS

MURST (Rome) and P.F. "Chimica Fine e Secondaria" (CNR) are gratefully acknowledged for financial support. Dr. A. Porto (SGS-Thomson, Catania) is gratefully acknowledged for his help with electron-beam deposition facility.

REFERENCES

Anderson S.G., Meyer III H.M., Atanasoska Lj. and Weaver J.H., 1988 J. Vac. Sci. Technol. **A6** 38.

Bolger J.C., 1983 "Adhesion Aspects of Polymeric Coatings", ed. K.L.Mittal (New York: Plenum) pp.3-18.

Buchwalter L.P., 1987 J. Adhesion Sci. Technol. **1** 341.

Buchwalter L.P., 1990 J. Adhesion Sci. Technol. **4** 697.

Buchwalter L.P. and Baise A.I., 1984 "Polyimides, Synthesis, Characterization and Applications", ed. K.L.Mittal (New York: Plenum) vol.1, pp.537-545.

Burrell M.C., Codella P.J., Fontana J.A. and Chera J.J, 1989 J. Vac. Sci. Technol. **A7** 1778.

Chan C.-M., Trigwell S. and Duerig T., 1990 Surf. Interface Anal. **15** 349.

Cochran S.J. and Larkins F.P., 1985 J. Chem. Soc. Faraday. Trans. I **81** 2179.

Flament O., Russat J. and Druet E., 1990 J. Adhesion Sci. Technol. **4** 109.

Garilli M., Marletta G., Puglisi O., Oliveri C., Magro C. and Ferla G., 1989 "Interfaces Between Polymers, Metals and Ceramics", MRS Symp. Proc., eds. B.M.DeKoven, A.J.Gellman and R.Rosenberg, (Pittsburgh: MRS) vol.153, pp.273-278.

Ho P.S., 1989 Appl. Surf. Sci. **41/42** 559.

Iacona F., Garilli M., Marletta G., Puglisi O. and Pignataro S., 1991 J. Mater. Res. **6** 861.

Kelley K., Ishino Y. and Ishida H., 1987 Thin Solids Films **154** 271.

Kim Y.-H., Kim J., Walker G.F., Feger C. and Kowalczyk S.P., 1988 J. Adhesion Sci. Technol. **2** 95.

Leary H.J. and Campbell D.S., 1979 Surf. Interface Anal. **1** 75.

Oh T.S., Buchwalter L.P. and Kim J., 1990 J. Adhesion Sci. Technol. **4** 303.

Parks G.A., 1965 Chem. Rev. **52** 177.

Russat J., 1988 Surf. Interface Anal. **11** 414.

Stewart W.C., Leu J. and Jensen K.F., 1989 "Interfaces Between Polymers, Metals and Ceramics", MRS Symp. Proc., eds. B.M.DeKoven, A.J.Gellman and R.Rosenberg, (Pittsburgh: MRS) vol.153, pp.285-290.

Toth A., Bertoti I., Székely T., Sazanov J.N., Antonova T.A., Shchukarev A.V. and Gribanov A.V., 1986 Surf. Interface Anal. **8** 261.

Electrochemical and electrical study of the n-TiO$_2$/pbT heterojunction

L.Torsi, C.Malitesta, F.Palmisano, L.Sabbatini, and P.G.Zambonin

Dipartimento di Chimica - Universita', 4 Trav. 200 Re David -70126- BARI (I)

ABSTRACT: Polibithiophene (pbT) thin films have been electrochemically deposited for the first time on n-type titanium dioxide (TiO$_2$) prepared by anodic oxidation of a Ti sheet. The linear sweep voltammetry measurements using the TiO$_2$/pbT heterojunction as working electrode as well as the measured I-V characteristic show its rectifying behaviour. In addition estimates of the barrier height of the n-TiO$_2$/p-pbT heterojunction, as well as the pbT work function are given.

1. INTRODUCTION

The investigation of polymer films grafted onto electrode surfaces is one of the most active research areas in the field of contemporary electrochemistry (Murray 1984) and this investigation is even more interesting when semiconducting electrodes are used (Noufi *et al* 1980). At the same time organic polymers are very promising 'for replacing the inorganic semiconductor in device fabrication.

Among conducting polymers polybithiophene seems to be suitable to be used for device due to its stability both in its neutral and p-doped states (Tourillon 1986). Nevertheless up to now, to our knowledge, no investigation of polybithiophene / inorganic semiconductor heterojunction exists.

In this work the synthesis of a heterojunction by photoelectrochemical deposition of pbT directly onto n-type titanium dioxide is reported, and the characterization of this heterojunction both from an electrochemical and an electrical point of view is described.

2. EXPERIMENTAL

The n-type part of the heterojunction is a sodium doped titanium dioxide (TiO$_2$) film, 300-400 Å thick, prepared by anodic oxidation of a titanium sheet (Peraldo Bicelli *et al* 1986). X-ray analysis gives a polycrystalline rutile structure and the carrier density is in the range of 10^{18}-10^{19} cm^{-3} giving a conductivity of about 10^{-4} ohm^{-1} cm^{-1}. The measured TiO$_2$ band gap is 2.99 eV (Peraldo Bicelli *et al* 1986).

The p-type part is a polybithiophene (pbT) film electrochemically deposited in its oxidized p-doped form direcly onto the n-type TiO_2. Polymerization of bithiophene onto the n-type TiO_2 sheet was carried out illuminating the sheet (Bard *et al* 1980) with an Osram XBO 150W/1 xenon lamp. Photoelectropolymerization was performed in a single-compartment three-electrode cell. The working electrode was a Ti/TiO_2 sheet. The bottom of the cell was closed by a teflon screw cap which sealed the Ti/TiO_2 sheet leaving 0.5 cm^2 of the TiO_2 face in contact with the solution. The TiO_2 face was illuminated through an optical window placed on the top of the cell. The counter electrode was a Pt sheet; the reference electrode was Ag/Ag^+ (0.1 mol dm^{-3}) in MeCN (+0.47 V vs NHE). Polymerization was performed at a constant current density of 800 μA cm^{-2} from a solution containing 5×10^{-3} mol dm^{-3} bithiophene and 0.1 mol dm^{-3} tetrabutylammonium hexafluorophosphate (TBAPF) in MeCN. Deposition on illuminated TiO_2 occurs at less anodic potential than on Pt, as it might be expected for a photoassisted process. The film thickness estimated from the charge transferred during the polymerization (Tourillon *et al* 1987) had a typical value d=500Å. All the electrochemical measurements were performed using an EG&G Princeton Applied Research Model 273 potentiostat/galvanostat, driven by a microcomputer. The polymer film was contacted using a platinum probe and I-V measurements were performed in air at room temperature using the EG&G potentiostat.

3. RESULTS

Figure 1 shows the linear sweep cyclic voltammograms at a sweep rate of 30mV/sec for the pbT coated n-TiO_2 electrode in MeCN solution containing only supporting electrolyte, in the range from -0.8 V to 1.1 V vs. Ag/Ag^+ .In this figure the main feature is a cathodic peak at about -0.5 V vs. Ag/Ag^+, but it is also visible an anodic current, with the onset at 0.5 V vs. Ag/Ag^+, spread over a wide range of potential. Passing from the first cycle to the third one there is a decrease of both the anodic and the cathodic contribution.

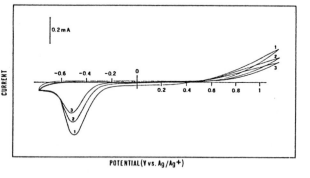

Fig. 1. Linear sweep cyclic voltammograms (s.r.= 30mV/sec) of pbT/TiO_2 working electrode in 0.1 mol dm^{-3} TBAPF in MeCN.

In figure 2, the same linear sweep cyclic voltammetry experiment in the range from -1.6 V and -2.4 vs. Ag/Ag⁺ is shown; it is recognizable a redox process peaked at about -2.1 V vs. Ag/Ag⁺. The curve presented is recovered at all the subsequent cycle.

Figure 3 reports a diagram of the TiO_2(n-type)/pbT heterojunction energy band levels allignement respect to vacuum, before and after contact at open circuit. The pbT band edges energy levels have been deduced by taking into accout that, as demonstrated both in the works of Yoshino *et al* (1988) and Hillman *et al* (1990), these coincide with the peak potentials of cyclic voltammograms on condition that the same energy reference is taken. The voltammetric peak potentials of the two redox couple "oxidized pbT pristine / reduced pbT" (i.e. p-type pbT / neutral pbT) and "n-type pbT / neutral pbT" are respectively at +0.45 V vs Ag/Ag⁺ (Malitesta 1988) and at -2.2 V vs Ag/Ag⁺ (Mastragostino *et al* 1989), and the position of the Ag/Ag⁺ redox couple energy level is 5.0 eV vs vacuum, as the energy level of the redox couple H^+/H_2 is -4.5 vs vacuum (Peraldo Bicelli *et al* 1981). According to these facts the pbT valence and conduction bands levels respect to vacuum are respectively -5.45 and -2.8 eV.

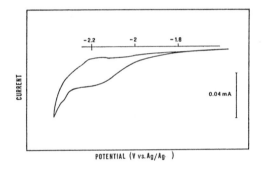

Fig. 2. Cyclic voltammogram (s.r.=30mV/sec) of pbT/TiO_2 working electrode in 0.1 mol dm⁻³ TBAPF in MeCN.

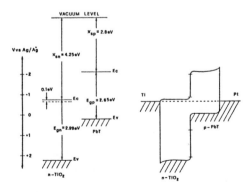

Fig. 3. Energy band levels alignement diagram, before and after contact at open circuit, of a heterojunction between p-type pbT and n-type TiO_2

The pbT/TiO$_2$ heterojunction I-V characteristic in the range from -2V to +2V at room temperature is shown in figure 4. The pbT film was left in its oxidized p-doped form and the I-V measurements were carried out in air. The knee voltage is at about 0.5 V and the rectification ratio at 3V is about 10^3. The Ti/n-TiO$_2$ and Pt/p-pbT metal-semiconductor contacts were verified to have an ohmic behaviour. This is in agreement with the theorethical prediction for ohmic contacts between a metal and n-type or p-type semiconductor since Ti has E$_f$=-4.1 eV (Riviere 1967) higher than TiO$_2$, and Pt has a very low E$_f$ (-5.36 eV) (Riviere 1967).

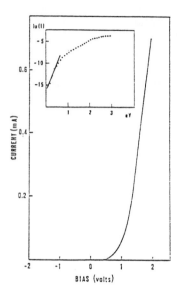

Fig. 4. Current - Voltage characteristic of p-pbT / n-TiO$_2$ heterojunction. Inset : ln(I) vs. V plot.

4. DISCUSSION

The rectifying behaviour of the n-TiO$_2$/p-pbT heterojunction, expected because of the energy band levels allignment (figure 3), is readily apparent from the electrochemical and electrical results presented in figure 1 and figure 4 respectively. In fact, regarding the I-V measurements of the heterojunction in air (figure 4), when a positive bias is applied to p-pbT, the junction is forward biased and a current of majority carriers flows through the interface. At low bias, i.e. less than 0.8 V the I-V curve follows the semiclassical diode characteristic, as lnI vs. V has a linear shape (inset of figure 4). The well known I-V semiclassical law is:

$$I = I_0 \exp{(qV/nKT - 1)}$$

where I$_0$, the saturation current , is given by:

$$I_0 = A \; A^* \; T^2 \exp{(-qX_b/KT)}$$

where A is the effective junction area, A* is the Richardson constant (120 A/K^2 cm^2 for free electrodes), X$_b$ is the barrier height and n is the diode quality factor; the other symbols have the usual meaning. At high forward bias, the lnI-V plot is no longer linear because the current is limited from both the resistence of TiO$_2$ and pbT (Inganàs et al 1983). From the linear interpolation of the lnI-V plot in the range 0-0.8 V we estimate a X$_b$ value of about 0.8 V and n=5. Since TiO$_2$ E$_f$ is at -4.35 eV and the estimate of X$_b$ is about 0.8 V, the pbT work function results to be close to -5.15 eV.

The barrier height value is in agreement with that deducible in figure 3, if one takes into account that the Fermi Energy Level (E_f) is 0.3 eV above E_V (see works by Chung *et al. 1984* and Stockert *et al* 1991).

An estimate of X_b can be obtained also from the linear sweep cyclic voltammetry experiments of figure 1. In fact the thermodynamic interpretation of the E_f level in a semiconductor allows to correlate the electronic properties of the semiconductor with those of the redox couple in solution (Bard 1980). In the present case the redox couple is the pbT electroactive film deposited on the working electrode (Malitesta 1988). The peak potential of the redox couple "p-type pbT/ neutral pbT" is about +0.45 V vs. Ag/Ag$^+$ (Malitesta 1988), and it is significantly more positive than the Flat Band Potential (V_{fb}) of TiO$_2$ (V_{fb}=-0.6V vs. Ag/Ag$^+$ Peraldo Bicelli *et al* 1985). The reduction potential of pbT when grafted onto TiO$_2$ electrode is -0.5V vs. Ag/Ag$^+$ (see figure 1). The absolute value of the cathodic shift with respect to bare platinum is therefore 0.95V, which is comparable with the value of 0.8V found for the barrier height. The little discrepancy could be ascribed to the fact that approximation is made that the voltammetric peak potential is equal to the redox standard potential. This is true on condition that (Hillman 1990): i) the voltammetric current is totally Faradaic (in our case a capacitive contribute cannot be excluded), ii) the peak itself is symmetric in shape.

The absolute cathodic shift can be explained on considering that even if the reduction process can take place (is thermodinamically possible) when the electrode potential is equal or more negative than the standard potential, the electrode kinetics may be slow. The process becomes faster and faster as the electrode potential reaches the value of the V_{fb} (Frank *et al* 1975). As far as the reoxidation process is considered, it is carried out by a weak dark current of minority carriers (holes h$^+$), and the rate of the process is potential independent (Frank *et al* 1975). When the cycle is repeated, the reduction peak current becomes lower and lower since the previous oxidation was not complete.

As to the redox couple "n-type pbT - neutral pbT", its peak potential is at about -2.2 V vs Ag/Ag$^+$ (Mastragostino *et al* 1989). This process is quite reversible even in the case of pbT on TiO$_2$, as shown in figure 2, and occurs also at the same potential of pbT on Pt. This results can be easily explained if one considers that the level -2.2 V vs Ag/Ag$^+$ falls in the TiO$_2$ conduction band and, therefore the electrode behavior is metal-like (Frank et al 1975).

5. CONCLUSIONS

To summarize, we have successfully prepared a rectifying polymer-semiconductor heterojunction by electrodepositing pbT in its oxidized (p-type) form directly onto n-type polycrystalline TiO$_2$. The rectifying characteristic is investigated both electrochemically and electrically and the results are compared with the outcome of standard semiclassical models for semiconductor heterojunctions. A barrier height of about 0.95V (electrochemical estimate) and 0.8V (electrical estimate), was calculated. A first estimate of the pbT work function, has been also obtained.

ACKNOWLEDGMENT

Prof. L. Peraldo Bicelli (Politecnico-Milano) and dr. A. Guerrieri (Universita'-Bari) are greatly acknowledged for stimulating discussions.

REFERENCES

Bard A J and Faulkner L R 1980 *Electrochemical Methods, foundamentals and applications* (New York: John Wiley & Sons) pp 636-6
Chung T C, Kaufman J H, Heeger A J and Wudl F 1984 *Phys. Rev. B* **30** 702
Frank S N and Bard A J 1975 *J. Am. Chem. Soc.* **97** 7427
Hillman A R and Mallen E F, 1990 *Electroanal Chem* **281** 109
Inganàs O, Skotheim T and Lunstròm I 1983 J. Appl. Phys. **54** 3636
Malitesta C 1988 PhD Thesis pp 219
Mastragostino M and Soddu L 1989 Electrochim. Acta **35** 463
Murray R W 1984 *Chemically Modified Electrodes in Electoanalytical Chemistry* ed A.J. Bard (New York: Marcel Dekker) **13** pp 191-368
Noufi R, Tench D and Warren L F 1980 J. Electrochem Soc. **127** 2310
Peraldo Bicelli L and Scrosati B, 1981 *La Chimica e l'Industria* **63** 172
Peraldo Bicelli L and Razzini G 1985 Int. J. Hydrogen Energy **10** 645
Peraldo Bicelli L, Pedeferri P and Razzini G 1986 *Int . J. Hydrogen Energy* **11** 647
Riviere J C 1967 A.E.R.E. Harwell (U.K.) - Internal Report
Stockert D, Kessel R and Schultze J W 1991 *Synth. Met. Proceeding Int. Conf ICSM '90.*
Tourillon G, 1986 *Handbook of Conducting Polymers* ed. A. Skotheim (New York: Marcell Dekker) **1** pp 293
Tourillon G, Dartyge E, Fontaine A, Garret R, Sagurtun M, Xu P and Williams G P,1987 Europhys. Lett. **4** 1391
Yoshino K, Onoda M, Manda Y and Yokoyama M, 1988 *Jap J. Appl. Phys.* **27** L1606.

Paper presented at First International Conference, Namur, Belgium
2-6 September 1991: Section I

Polymers on mica

Walter R. Caseri, Ronald A. Shelden and Ulrich W. Suter

Institut für Polymere, ETH Zentrum, CH-8049 Zürich

ABSTRACT: With a view to obtaining thin polymer film- and supramolecular structures on mica, the ion exchange adsorption of ionic initiators, ionic monomers, and ionic polymers on mica sheets and powder surfaces was investigated. To facilitate studies, an ultrahigh surface area mica powder was prepared by a new process. Surface polymerizations with styrene were attempted in order to obtain bound polymers. Toluene and THF insoluble polymer fractions of high molecular weight were obtained and evidence is presented suggesting that this was polystyrene bound to the mica by ionic groups.

1. INTRODUCTION

Mica is a common filler for polymers that can increase stiffness, improve dimensional stability or lower cost (Bajaj *et al* 1989, Rochette *et al* 1988, Lusis *et al* 1973, Favis *et al* 1983, Shepherd *et al* 1974, Theng 1979, Trotignon *et al* 1986). Composites of polymers and mica are also technically important high voltage insulating materials. In all these applications, the interface between polymer and filler can play an important role in determining mechanical properties and the chemical and environmental stability of the composite systems. Thus the nature of the bond between polymers and mica is an important theme and possibilities for modification or improvement of this bond are of considerable interest.

Moreover, since mica has a well defined sheet-like crystal structure, and is in fact available in the form of crystalline sheets, it is also a potentially valuable substrate for building thin polymer film and supramolecular structures with applications in, for example, microelectronics and sensor technology.

For these reasons, we have been interested in techniques for binding polymers to mica and assembling supramolecular structures on mica, and in this paper we report on our investigations with a particular mica mineral, muscovite.

Mica is a layered material, consisting of aluminum and magnesium stabilized silicate layers with a thickness of about 1.0 nm held together by potassium ions (Bailey 1984, Theng 1974). It is available in sheets from about 10 to 100 cm^2 in area, or in powder form. The mineral is commercially cleaved by grinding or wet pulping assisted sometimes by chemical treatment such as the Bardet process (1951). After cleavage, exchangeable cations are present on the surface (Gaines 1957).

Two types of muscovite mica were used in the present work. One was a commercial filler obtained from Plüss Stauffer AG, Oftringen, Switzerland, designated M2/1 by the supplier, and referred to hereinafter as mica A. The other was a very high surface area material (hereinafter mica B) obtained by a new chemical cleavage process we recently developed, described in Section 3.

Mica powders obtained by previously known cleavage processes have particle thicknesses of at least about 100 silicate layers. They typically have specific surface areas of 2 to 5 m^2/g. For many investigations it would be useful to have micas with higher specific surface areas, e.g. for taking IR spectra and DSC or TGA measurements of adsorbed molecules. Moreover, ultrahigh surface area micas might prove useful for the preparation of high performance composite materials. We have therefore sought to prepare, and have succeeded in preparing, micas with, so far as we know, previously unattained high surface areas.

The surface areas of the mica A and mica B that we used were respectively 5.2 m^2/g and 98.3 m^2/g as determined by the methylene blue adsorption method (Blake and Ralston 1985, Giles and Trivedi 1969, Hang and Brindley 1970).

2. TECHNIQUES FOR BONDING POLYMERS TO MICA

Just as with glass-filled polymers, when polymers are filled with mica, silane coupling agents such as chlorosilanes or alkoxysilanes are often used to improve interfacial bonding (Theng 1979, Debnath *et al* 1989, Boaira and Chaffey 1977, Rochette *et al* 1988, Favis *et al* 1983). We have rather chosen another route to binding polymers to mica, making use of the ion exchangability of the cationic surface ions. It is well known that mica's surface potassium ions can exchange with other inorganic and organic cations (Theng 1974) as well as cationic polymers (Theng 1979, Fröhlich 1983). This capability allows a number of techniques to be applied to generate polymer chains bound ionically to the mica surface.

We have considered three approaches: 1) attach ionic initiators to the surface and use the complex to initiate polymerization. This should lead to terminally bound polymer chains. 2) attach ionic monomers to the surface and use the complex as monomer or comonomer. This should lead to polymer chains bound to the surface by one or more ionic units. 3) attach preformed polymers such as polyelectrolytes or ionically terminated polymers. We have so far made use of each of these approaches, and will summarize some of our results in the following sections.

We attribute method 1), using ionic initiators, basically to Dekking (1965, 1967). Dekking reported the preparation of a number of vinyl polymers bound to the silicate clay minerals bentonite and kaolin. He prepared these bound polymers by attaching, presumedly via ion exchange, the free radical initiator 2,2'-azobisisobutyramidine hydrochloride (hereinafter AIBA) and then using the resulting material as a polymerization initiator. We have employed Dekking's technique with mica. In contrast to mica, bentonite is an expanding clay and consists of small crystallites with a very irregular form (Cairns-Smith 1982). The ion exchangability of kaolin is due to structural defects, the exchangeable ions lying mainly at the edges and corners of particles (Dekking 1965). Dekking carried out the polymerizations in bulk and in water emulsion. While polymer fractions were obtained which were insoluble in the usual solvents for those polymers, no definitive evidence was presented that the polymers were ionically bound to the clays, nor were the insoluble polymers characterized.

We attribute method 2) to Hauser (1946,1953). Hauser attached polymerizable organic salts to bentonite via ion exchange and used the resulting material as a comonomer.

Method 3), binding polymers having ionic groups, can be attributed to the numerous early studies on the adsorption of polyelectrolytes on micas and clays (see for example Theng 1979).

To use methods 1) and 2), it is necessary to understand the adsorption of certain organic salts on mica. In Section 4, we discuss the adsorption of AIBA (ionic initiator) and a vinyl pyridine salt (ionic monomer). In Section 5, we consider polymerization results using these salts. In Section 6 we discuss work with polyelectrolytes.

3. MICA WITH ULTRAHIGH SPECIFIC SURFACE AREA

The mica starting material that was used for preparing mica B was a waste product (mica fines) obtained in the production of muscovite pulp for insulating "papers". The waste product was obtained from Isola, Breitenbach (Switzerland). The muscovite pulp was designated by them as Mica 21. The waste product had a specific surface area of 3.4 m^2/g. According to Isola, Mica 21 is the powdered form of mica that most closely retains the structure of the natural sheet material.

It was hypothesized that it might be possible to cleave mica by introducing an excess of small ions, previously dissolved in water, into the potassium interlayers leading to a type of inert salt effect which might weaken the attractive force between the silicate layers. This effect, however, if observed, would depend not only on ion size, but also on solubility, charge and coordinated water sphere.

Following this hypothesis, mica was treated with a $LiNO_3$-water system and increases were obtained in the specific surface area of the mica of up to two orders of magnitude. The efficacy of the process depended strongly on the $LiNO_3$/water ratio and the process temperature. At 130 °C, highest surface areas were obtained with a $LiNO_3$/water weight ratio of 1.7 to 1.8 (reaction time 46 h), corresponding to about 2 moles water per mole $LiNO_3$. Therefore, the process took place rather in a $LiNO_3$ hydrate melt than in a $LiNO_3$ solution. The specific surface area increased with increasing temperature (Figure 1). At 190 °C, a specific surface area of 295 m^2/g was obtained, corresponding to an average of 2 to 3 silicate layers per particle.

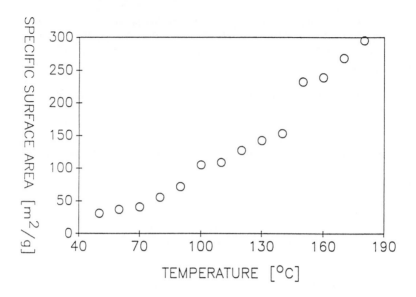

Fig.1. Specific surface area versus process temperature for muscovite. Reaction time 46 h, LiNO$_3$/water weight ratio of 1.72.

Particles were deposited from an aqueous suspension onto a silicon wafer and their profiles were recorded with a surface profilometer (vertical resolution 0.5 nm, horizontal resolution 40 nm). Since particles below ca. 1 μm diameter or ca. 100 nm thickness cannot be seen on the video microscope of the profilometer they are only detected accidentally during scanning. Therefore, particle size distributions obtained by this method are only qualitative. However, it was clear that the particle thicknesses were drastically reduced by this process. The particle diameters were also reduced (see Figures 2a and 2b), so the aspect ratio (diameter/thickness) increased only slightly. Most of the particles of the micas with a specific surface area above 100 m^2/g were smaller than about 10 μm in diameter and had an average aspect ratio of about 250. This value, about 250, can be compared with that reported for a commercial mica filler, about 30 (Boaira and Chaffey 1977). Since high aspect ratio is an important determinant in the strength and modulus of mica-filled polymers (Shepherd *et al* 1973), the retained high aspect ratio of mica B is an important characteristic.

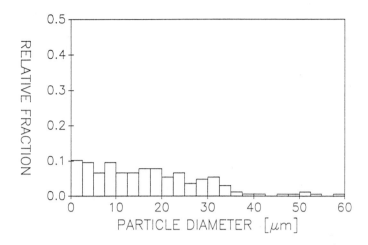

Fig. 2a. Particle diameter distribution before cleavage (mica B).

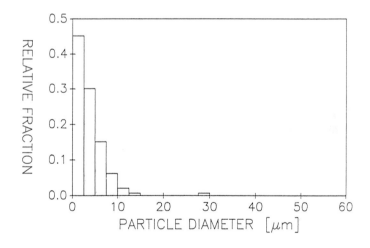

Fig. 2b. Particle diameter distribution after cleavage (mica B).

After cleavage with $LiNO_3$, the surface ions are believed to be predominantly lithium (see Section 4). The silicate structure itself appeared to be little damaged, if indeed it was damaged at all, by the cleavage process. The IR spectra of the micas before and after processing were quite similar. In particular, the O-H stretching frequency bands (ca. 3620 - 3650 cm[-1]) and the absorption bands between 400 and 1200 cm[-1] did not change significantly following cleavage.

Further details can be found in an upcoming paper (Caseri *et al*, in press).

4. ADSORPTION OF ORGANIC IONS

Experiments were performed with the mono-cation salts N-methyl-4-vinylpyridinium methylsulfate (VPQS), N-methyl-4-ethylpyridinium methylsulfate (EPQS), N-methyl-4-methyl pyridinium methylsulfate (MPQS), N-dodecylpyridinium chloride (NDPCl) and the bi-cation 2,2'-azobisisobutyramidine hydrochloride (AIBA). The pyridinium salts were particularly useful for our experiments because of their strong UV absorption. The vinyl salt was selected for its polymerizability, the ethyl, methyl and dodecyl salts as model salts for comparison.

Aqueous solutions of the organic salts were added to the micas, the mixtures stirred for several hours and left overnight to equilibrate. The amount of salt adsorbed was determined from the change in salt concentration of the supernatant solution as measured by UV absorption. For the mono-cation salts, it was assumed that a simple ion-exchange equilibrium was applicable. It can then be shown that:

$$\mu/(S-\mu)=K \; c/(c_0-c) \qquad (1)$$

where μ is the concentration of the inorganic salt on the mica surface, S is the total ion exchange capacity of the surface, c_0 is the initial organic salt concentration in solution, c is the organic salt concentration in the solution at equilibrium, and K is the equilibrium constant for the adsorption reaction.

Equation 1 can be rearranged to give:

$$1/\mu=(1/S)+(1/KS)(c_0-c)/c \qquad (2)$$

Thus a plot of $1/\mu$ versus $(c_0-c)/c$ should give a straight line with slope $1/(KS)$ and intercept $1/S$. Plots were obtained from our experimental adsorption data and one of these, that for VPQS and mica A, is shown in Figure 3.

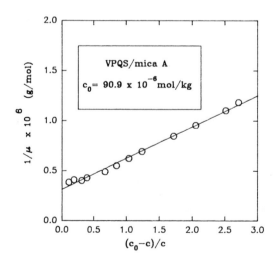

Fig. 3. $1/\mu$ vs $(c_0-c)/c$ for VPQS/mica A.

The values of K and S were obtained by linear regression and are shown in Table 1. The expected inorganic ion exchange capacities based on the measured surface area and the reported literature value (Gaines 1957) for the monovalent inorganic ion capacity per unit of surface area for muscovite are also shown. It can be seen that VPQS, EPQS and MPQS gave, for mica A, S values that are several times smaller than the expected inorganic ion capacities. With NDPCl, adsorption values closer to the expected inorganic ion exchange capacities were obtained with both micas. The differences between the observed values of S and the expected ion capacities can perhaps be attributed to the space requirements of the adsorbed ions which can vary depending on whether or not the ions lie flat on the surface.

	mica A		mica B	
Salt	S µmol/g	K	S µmol/g	K
VPQS VPQS (1)	3.20 3.97	1.00 1.71	289	6.71
EPQS	2.55	1.20	404	.68
MPQS	2.22	1.08		
NDPCL	12.55	3.17		
AIBA actual adsorption	9.1-15.7		231-248	
Expected inorganic ion capacity	18.1 (2)		342 (3)	

(1) mica A treated with lithium salts to exchange surface ions
(2) based on 3.48 µmol/m^2 (Gaines 1957) x 5.2 m^2/g = 18.1 µmol/g
(3) based on 3.48 µmol/m^2 (Gaines 1957) x 98.3 m^2/g = 342 µmol/g

Table 1. Adsorption of organic salts on micas A and B.

For AIBA, being bi-cationic, equations 1 and 2 are not applicable. For this salt, the actual adsorption values on the two micas are shown in Table 1, rather than the K values. The amounts of AIBA adsorbed corresponded very roughly to the expected inorganic ion capacities. It should be born in mind that, since AIBA is bi-cationic, only half as many molecules should be adsorbed by ion exchange at capacity as would be the case with a mono-cationic salt.

An important question, for AIBA, is the degree to which the salt deprotonates in water to give the corresponding bis-amidine and HCl. We measured the pK values and found for both amidine groups a value of 10.53. This corresponds to more than 99.99 % of the amidine groups in the protonated form, in the concentration ranges used to treat the micas (about 4 to 40 x 10^{-3} mol/kg).

As expected for ion-exchange with lithium surface ions, it was found for mica B using ion-chromatography that roughly two (1.85 ± 0.17) lithium ions were released per AIBA molecule adsorbed.

Upon adsorption of NDPCl on freshly cleaved mica sheets, the surface characteristics as determined by wetting angle experiments changed from hydrophilic (about 3° with water) to significantly more hydrophobic (about 40 ° with water). VPQS-treated mica gave a wetting angle of about 15°. When a droplet of dilute aqueous NDPCl (255 μmol/l) was placed on the mica sheet, the droplet first spread out, reflecting the good wetting of the mica by the very dilute solution and then, within seconds, the droplet began to draw together reflecting the increasingly hydrophobic character of the surface as the organic ions were adsorbed. The change in contact angle from about 3° at the start to about 48° at the end, took 2 to 3 minutes.

5. POLYMERIZATION

The micas were equilibrated with solutions containing the salt to be adsorbed, then filtered (mica A) or centrifuged (mica B) and washed 2 to 3 times with water and methanol. The micas were then transferred to polymerization tubes and dried overnight under vacuum. Freshly distilled styrene (inhibitor removed) was added, and polymerization carried out under nitrogen for 1 to 3 days at 60 °C. At the end of the polymerization period, the contents of the tubes were poured into methanol, resulting in a mixture of mica and polymer.

An amount of polymer could be removed by repeated extraction with toluene that corresponded roughly to that expected for thermal polymerization at 60°, roughly 2.5 % per day of the styrene charged (Rubens and Boyer 1952). As shown by TGA, polymer equal to 0.5% to 8% by weight of the mica remained with the mica after repeated extraction. Carbon analysis gave somewhat lower values e.g. 5% compared with 8% from TGA analysis. The TGA plot for the 8 wt % sample is shown in Figure 4. By comparing the TGA plots with those of polystyrene and of AIBA residues, we conclude that the observed weight loss at 400-500°C is attributable to the former and the loss around 350° attributable to the latter.

The amount of insoluble polymer varied considerably from run to run, even under ostensibly similar conditions. However, it appears that the amount of insoluble polymer per gram of mica was several times greater for mica B than for mica A, and increased proportionally or more than proportionally with the time of polymerization over a three day period.

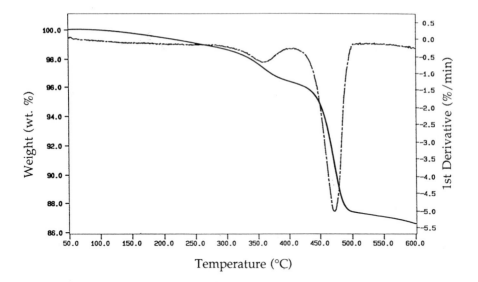

Fig. 4. Weight loss (%) versus temperature for toluene-insoluble polymer/mica B in nitrogen at 20°C/min.

When untreated mica was used in place of the AIBA-treated mica, the amount of soluble polymer was similar, but no toluene-insoluble polymer was produced.

The toluene-insoluble polymer appears, from TGA, IR and NMR studies (see below), to be polystyrene. An IR spectrum of the solid mica B-polymer complex is shown in Figure 5. The following assignments have been made: bands at 3600-3650 cm^{-1}, mica OH-stretch; broad bands at 3000-3500 cm^{-1}, water OH stretch and amidino NH stretch; 3061 and 3028 cm^{-1}, polystyrene aromatic CH stretch; 2925 and 2852 cm^{-1}, polystyrene aliphatic CH stretch; 1696 cm^{-1}, water OH bending and amidino NH bending; 1601, 1493 and 1452 cm^{-1}, polystyrene aromatic ring skeletal vibrations. The small absorption bands at 1515, 1506, 1393 and 1372 cm^{-1} are attributed to the residue of the AIBA, as they are also found in the spectra of micas treated with AIBA solutions and then heated at 60° C for a day.

The toluene-insoluble polymer could be dissolved in a THF solution containing lithium chloride but not in THF alone (THF normally dissolves polystyrene). This was shown by TGA on the remaining solids and NMR on the solution. This observation is consistent with the polymer chains being bound to the mica by one (one end) or two (both ends) cationic groups; ion-exchange of the lithium ions for bound cationic polymer end groups would take place followed by solution of the "freed" polystyrene in the THF.

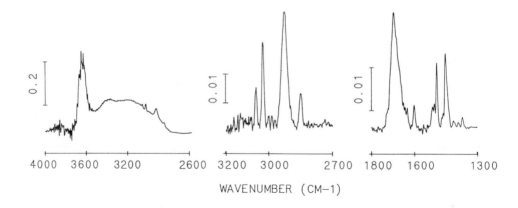

WAVENUMBER (CM–1)

Fig. 5. Infrared spectrum of toluene-insoluble polymer/mica B powder between KBr plates.

The molecular weights of the toluene-insoluble polymer were also obtained by GPC and light scattering in THF-LiCl solution. The molecular weights appear to be well above a million. Given the amount of toluene-insoluble polymer per gram of mica, at most a few per cent, it appears that only a very small fraction of AIBA molecules initiate chains of significant length. Most of the free radicals produced from the rest of the initiator presumably recouple on the surface producing low molecular weight species.

We measured the half life of the initiator adsorbed on mica B in an argon atmosphere, using mass spectroscopy, and found it to be about one hour at 60 °C. The lifetime of the chains is presumable on the order of a few seconds or less. The fact that the mass of "bound" polymer appears to increase over a polymerization period of 1 to 3 days therefore requires further study.

In some polymerization runs, high molecular weight polystyrene was added to the styrene monomer (producing 5 wt % polystyrene solution) to test the possibility that thermally produced polystyrene was being trapped by "bound" polymer or being grafted to the AIBA-mica complex. No significant difference in the amount of insoluble polymer formed was observed. Hence entrapment or grafting apparently did not occur. In other runs, the added polystyrene was dissolved in toluene and no styrene monomer was present. In this case, no insoluble polymer resulted, again showing that grafting did not occur.

Experiments were performed using AIBA-treated mica sheets instead of powders in styrene polymerization. The resulting sheets after extraction with toluene gave wetting angles with water of about 25°. This is intermediate between the 3° obtained with freshly cleaved mica and the 84° obtained with mica sheets coated with about 20 μm of polystyrene by evaporation of polystyrene solution.

Preliminary experiments on the copolymerization of styrene with the vinyl pyridinium salt (VPQS) adsorbed on the mica also gave toluene-insoluble polymer on the order of 0.5 to 1 wt. % of mica A. This polymer has not yet been characterized.

6. ADSORPTION OF IONIC POLYMERS

Since cationic polymers and anionic polymers react in solution to form insoluble polyelectrolyte complexes, we thought it would be possible to create alternating layers of these polymers on mica surfaces by sequenced adsorption. This is similar to what has been reported for SiO_2 surfaces modified with organic cations (Decher and Hong 1991). They could obtain at least 100 layers in a controlled manner. The sequenced adsorption of polyelectrolytes could be an alternative to the Langmuir-Blodgett technique in the preparation of polymer films with controlled layer types and thicknesses. In contrast to the Langmuir-Blodgett technique, however, sequenced adsorption is not restricted to sheets but can in principle also be performed with powders.

We tried to obtain alternating layers of polymers using poly(diallylammoniumchloride) (hereinafter PDAC) as cationic polymer and poly(styrenesulfonicacid Na-salt) (hereinafter PSSNa) as anionic polymer. The cationic PDAC readily adsorbed on mica B as confirmed by IR spectroscopy. Also, a freshly cleaved mica sheet treated with a solution of PDAC gave a contact angle of 14° with water, compared with 3° obtained with an untreated freshly cleaved sheet. In contrast, the anionic polymer PSSNa did not adsorb from aqueous solutions on mica B, even over 30 days, as shown by IR spectroscopy of the mica and UV spectroscopy of the supernatant solution.

Surprisingly, however, the PSSNa, also did not adsorb significantly on mica previously treated with PDAC. This was confirmed by IR spectroscopic analysis of the powders and UV spectroscopic analysis of the supernatant solutions. This, in spite of the fact that the surface properties of the mica were clearly altered by the adsorption of PDAC. This was shown not only by the contact angle experiments previously mentioned, but also by the behavior of PDAC-treated powders contacted with methylene blue solutions. PDAC-treated mica adsorbed only 0.35% as much as did untreated mica. Presumably, the PDAC is adsorbed on the mica in such a way that virtually all the cationic groups are on the surface and virtually none are exposed in loops or tails to complex the PSSNa. Similarly, the methylene blue obviously "sees" a surface for which it has only a small affinity. Nevertheless, we are exploring other approaches to the formation of these potentially alternating layers on mica because we feel the method has considerable potential for application in a wide variety of fields.

We have also adsorbed on mica B about 6 wt % of a water soluble oligomer bearing a cationic end group (ethoquad C/25 obtained from AKZO, Düren, FRG). It is a quaternary ammonium salt with two OH-terminated polyethylene oxide chains (total of 15 ethylene oxide units), a primarily C_{12} fatty acid residue, and a methyl group as the organic radicals (MW ca. 700 g/mol). Adsorbed on mica sheets, a contact angle of about 53° resulted with water. We are now in the process of synthesizing ionically terminated polymers of higher molecular weight to explore this simple method of obtaining bound polymers.

7. REFERENCES

Bailey S W 1984 *Reviews in Mineralogy Vol. 13*. (Blacksburg: Virginia Polytechnic Institute & State University) pp 1-12

Bajaj P, Jha N K and Jha R K 1989 *Polym. Eng. Sci.* **29** 557

Bardet J J 1951 *US Pat 2549880*

Blake P, Ralston J 1985 *Colloids Surf.* **15** 101

Boaira M S and Chaffey C E 1977 *Polym. Eng. Sci.* **17** 715

Cairns-Smith, A G 1982 *Genetic Takeover and the Mineral Origins of Life* (Cambridge: Cambridge University Press) pp 234-6

Caseri W R, Shelden R A and Suter U W in press *J. Colloid Polym. Sci.*

Debnath S, Sadhan K D and Dipak K 1989 *J. Appl. Polym. Sci.* **37** 1449

Decher G and Hong J D 1991 *presented at Makromolekulares Kolloquium Freiburg, Germany, held 28 February - 2 March*

Dekking H G G 1965 *J. Appl. Polym. Sci.* **9** 1641

Dekking H G G 1967 *J. Appl. Polym. Sci.* **11** 23

Favis B D, Leclerc M and Prud'homme R E 1983 *J. Appl. Polym. Sci.* **28** 3565

Fröhlich H P 1983 *PhD Thesis* (University Köln) pp 3-99

Gaines G L 1957 *J. Phys. Chem.* **61** 1408

Giles C H and Trivedi A T 1969 *Chem. Ind.* **40** 1426

Hang P T and Brindley G W 1970 *Clays Clay Miner.* **18** 203

Hauser E 1946 *US Pat 2,401,348 (June 4, 1946)* cited in Dekking 1967

Hauser E 1953 *US Pat 2,651,619 (Sept. 8, 1953)* cited in Dekking 1967

Lusis J, Woodhams T and Xanthos M 1973 *Polym. Eng. Sci.* **13** 139

Rochette A, Choplin L and Tanguy P A 1988 *Polym. Composites* **9** 419

Rubens L C and Boyer R F 1952 in *Styrene, Its Polymers, Copolymers, and Derivatives* eds R H Boundy, R F Boyer and S M Stoesser (Reinhold: New York) p 217

Shepherd P D, Golemba F J and Maine F W 1974 *Adv. Chem. Ser.* **134** 41

Theng B K G 1974 *The Chemistry of Clay-Organic Reactions* (London: Hilger) pp 1-16, pp 211-38

Theng B K G 1979 *Formation and Properties of Clay-Polymer Complexes* (Amsterdam: Elsevier) pp 109-54

Trotignon J P, Verdu J, de Boissard R and de Vallois A 1986 *Polymer Composites* ed B Sedlàcek (Berlin: Walter de Gruyter) pp 191-83.

ACKNOWLEDMENTS

Our thanks go to Martin Colussi who performed the numerous TGA analyses. We also gratefully acknowledge financial support for this work through the (Swiss) Kommission zur Förderung der wissenschaftlichen Forschung (KWF) as well as stimulating discussions with several members of the Research Divisions of Isola, Breitenbach (Switzerland) and Asea Brown Boveri, Baden (Switzerland).

Fundamental concepts in the interfacial adhesion of laminated safety glass

D. J. David

Monsanto Chemical Company, 730 Worcester St., Springfield, MA 01151, U.S.A.

ABSTRACT: In attempting to understand the adhesive forces between glass and plasticized poly (vinyl butyral) interlayer (PVBI), molecular modelling was used to examine the nature of the glass and PVBI surfaces. Plasma treated PVBI and untreated PVBI were evaluated for their adhesion to glass. Differences in adhesion levels of the untreated PVBI and the plasma treated PVBI are discussed in terms of XPS results. Mechanisms for the adhesion of PVBI to glass are proposed and are considered in light of experimental and characterization results. Results support the concept that interfacial adhesion in the interphase layer is a result of interaction of the glass silanol groups and the residual hydroxyl groups of PVBI.

1. INTRODUCTION

Laminated safety glass finds applications in architectural and automotive uses. In automotive applications it is used in windshields throughout the world to protect the driver and passengers from serious injury. In architectural applications, it protects pedestrians on the streets below from falling glass in the event of glass breakage, provides protection in skylights, and reduces the sound transmission from outside noise sources.

Safety glass consists of two pieces of glass laminated with plasticized poly (vinyl butyral). In both architectural and automotive applications adhesion is a key issue. Much effort has been expended over the years to understand and control adhesion to desirable levels. This paper will attempt to address some of the factors that generally influence glass - plasticized poly (vinyl butyral) adhesion.

We have seen many improvements in automotive comfort, styling, and safety over the years. This can be better appreciated by focussing on any one of hundreds of components. One essential component that provides both comfort and safety is the windshield What is generally not appreciated is the fact that the control of adhesion for this application is critical to provide the necessary passenger safety.

The automotive windshield has evolved from one that offered little protection from rocks and foreign objects and lack of safety features to one that provides excellent passenger safety. Laminated glass evolved from use of cellulose nitrate as an adhesive to the use of poly (vinyl butyral) interlayer (PVBI) which appeared over fifty years ago and provided major improvement in performance and durability.

Improvements have continued over the years with the next major improvement occurring in the sixties with the introduction of the "high penetration resistance" windshield. This design introduced controlled adhesion of the interlayer to the glass and substantially improved the impact resistance via improvement of the energy absorption characteristics of the laminate. Thus, in present day construction, the adhesion of the interlayer to glass is held within a narrow range. In the event of an accident and passenger impact with the windshield, the adhesion is sufficient to prevent major face lacerations from shards of glass coming loose while simultaneously providing energy absorption to prevent major head injuries.

These safety features are monitored by measuring adhesion and impact properties of test laminates and closely monitoring the adhesion of randomly selected production windshields.

2. INTERLAYER COMPOSITION

The plastic interlayer used in laminated glass consists of plasticized poly (vinyl butyral) (PVBI). This polymer is prepared by acetalization of polyvinyl alcohol with butyraldehyde, as illustrated in Figure 1.

The concentrations shown are given in weight percentages for the commercial ranges of importance. It will be noted that there is generally present a small amount of residual acetate which remains from incomplete hydrolysis of polyvinyl acetate. Thus the interlayer consists of a properly plasticized terpolymer. But for practical purposes, it is considered to be a copolymer.

The amount and type of plasticizer is highly dependent upon the application. Automotive formulations are different from formulations used for architectural applications and these are in turn different from those used for aircraft applications.

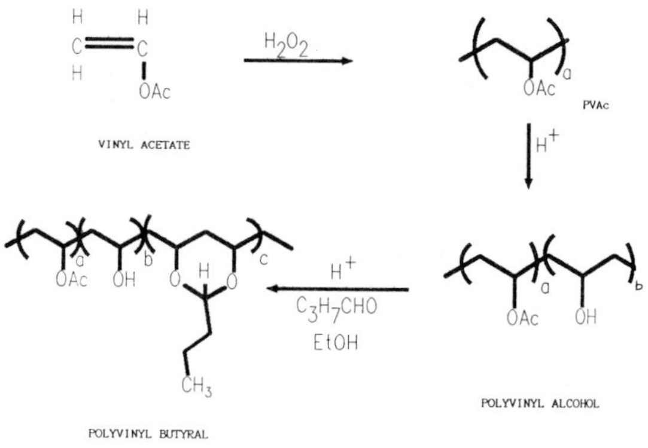

Fig. 1. Synthesis of poly (vinyl butyral)
a = 2-8%, b = 16-28%, c = 64-82%

3.   ELEMENTS OF ADHESION

In order to understand the mechanisms which account for the adhesion of poly (vinyl butyral) (PVB) to glass, the bulk and surface composition of each material must be considered. Extensive reviews of the infrared analysis and surface chemistry of high area silicas have been published[11,18,21]. Using attenuated total reflectance (ATR) infrared spectroscopy, hydroxyl groups at the surface of glass specimens can be monitored[3,4,5].

From these studies, glass can be represented schematically as shown in Figure 2. In Figure 2a, it can be seen that glass consists of a silica tetrahedron network with alkali "modifiers". The surface of the glass undergoes hydration with water to produce the configuration shown in Figure 2b. In fact, the literature indicates that a layer of "silica gel" develops that can be on the order of 500-1000Å thick and for greater thicknesses, the glass will be hazy. Bershstein et al.[3,4,5] concluded that the hydrolysis of siloxane linkages occurs preferentially at weakened $\equiv Si - O - Si \equiv$ sites and that the concentration of these sites sets a limit upon the number which can be hydroxylated. These and other aspects have been summarized by Pantano[30].

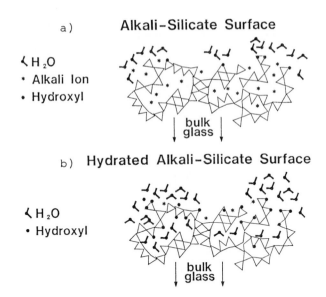

a) **Alkali-Silicate Surface**

⟨ H₂O
* Alkali Ion
• Hydroxyl

bulk glass

b) **Hydrated Alkali-Silicate Surface**

⟨ H₂O
• Hydroxyl

bulk glass

Fig. 2. Schematic Representation of Silicate Glass Surface Structures Containing Silanol
Groups and Adsorbed Molecular Water
a = Alkali-Silicate Surface under Ambient Conditions
b = Alkali-Silicate Surface Hydrated in Water

It is recognized that glass surfaces are covered with physically adsorbed water which is often referred to as water of hydration. It is the underlying chemisorbed water, often referred to as the hydroxyl or silanol group, which is the more important form. This chemisorption occurs at the siloxane linkages $\equiv Si - O - Si \equiv$, as mentioned above, to produce the silanol group, which can occur in two configurations which can be distinguished, as shown in Figure 3.

From Figure 3, it can be appreciated that the concentration and configuration of silanol groups determine the capacity for physical adsorption of water and other materials which are capable of forming hydrogen bonds. Although physical adsorption of molecular water occurs primarily at the vicinal silanol groups, multilayer adsorption and clustering of molecular water occur via intermolecular hydrogen bonding of the water molecules. In general, one would expect that other materials possessing groups capable of interacting or hydrogen bonding with the active glass sites would adhere to the glass surface. Recalling the structure of poly (vinyl butyral) from Figure 1, it can be seen that this polymer has an abundance of hydroxyl groups available for bonding to the active sites on the glass. These active glass sites are mainly silanol groups but this is not to exclude $\equiv Si - O - Si \equiv$ structures.

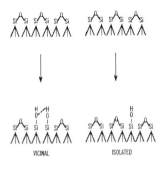

Fig. 3. Illustration of the Vicinal and Isolated Silanol Groups

The models of the polymer and glass structures become key in understanding the adhesion between the two materials in spite of the fact that much of our knowledge of glass surfaces was obtained using high surface area silicas. The interaction of these two materials can be visualized as shown in Figure 4. There can be a whole range of possible configurations but the main point is that these two materials possess the number and types of groups necessary to form strong permanent bonds while maintaining excellent optical quality.

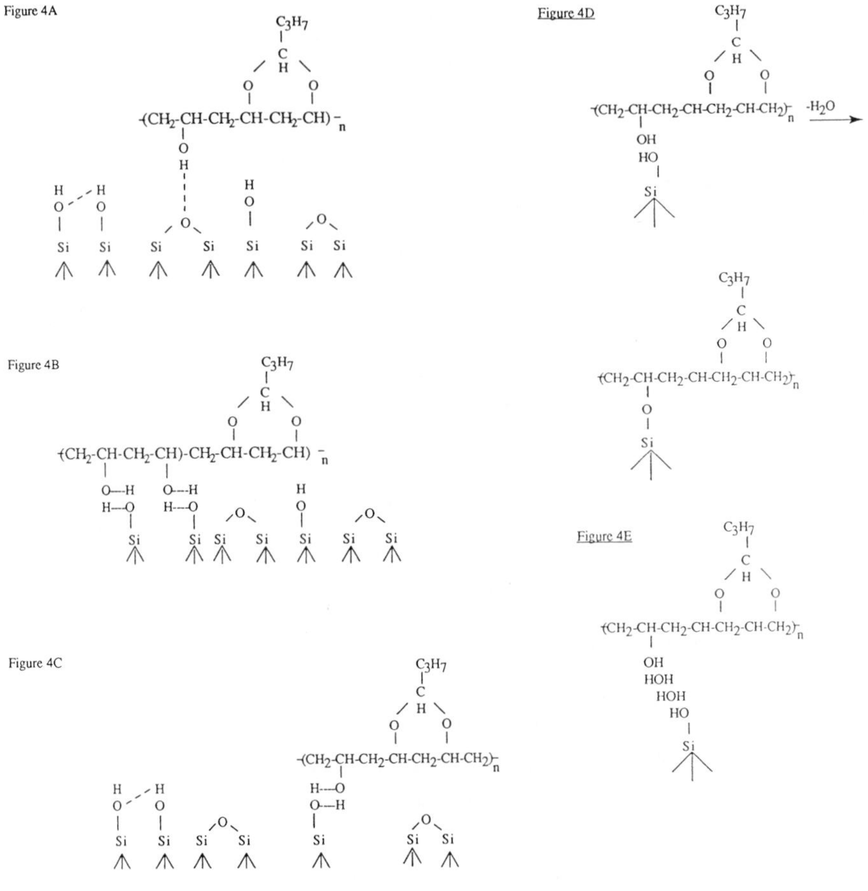

Figure 4A

C_3H_7

$-(CH_2\text{-}CH\text{-}CH_2\text{-}CH\text{-}CH_2\text{-}CH)-_n$

Figure 4B

C_3H_7

$-(CH_2\text{-}CH\text{-}CH_2\text{-}CH)\text{-}CH_2\text{-}CH\text{-}CH_2\text{-}CH)-_n$

Figure 4C

C_3H_7

$-(CH_2\text{-}CH\text{-}CH_2\text{-}CH\text{-}CH_2\text{-}CH\text{-}CH_2)-_n$

Figure 4D

C_3H_7

$-(CH_2\text{-}CH\text{-}CH_2\text{-}CH\text{-}CH_2\text{-}CH\text{-}CH_2)-_n$ $-H_2O \longrightarrow$

C_3H_7

$-(CH_2\text{-}CH\text{-}CH_2\text{-}CH\text{-}CH_2\text{-}CH\text{-}CH_2)-_n$

Figure 4E

C_3H_7

$-(CH_2\text{-}CH\text{-}CH_2\text{-}CH\text{-}CH_2\text{-}CH\text{-}CH_2)-_n$

Fig. 4. Model for Poly (vinyl butyral) - Glass Surface Interaction
   4A = Adhesion via SiO/PVB Hydroxyl Interaction
   4B = Adhesion via Vicinal Silanol/PVB Hydroxyl Interaction
   4C = Adhesion via Isolated Silanol/PVB Hydroxyl Interaction
   4D = Adhesion via Silanol/PVB Hydroxyl Interaction with Water Elimination
   4E = Adhesion via Silanol/PVB Hydroxyl Interaction through Multilayer
        Moisture Interaction

Recent adhesion studies carried out at ambient temperatures in which we have participated[34], have demonstrated that there is a 50% increase in adhesion when the hydroxyl content is increased from 18% to 28% using the same glass substrate. Our own studies on the effects of glass hydration[14] on adhesion have shown that "pristine float" glass that is fully hydrated results in an increase of adhesion of 200% with PVBI of constant hydroxyl level. The general affect of the plasticizer is to lower the level of adhesion depending upon the extent of plasticization so we can conclude that the plasticizer mainly acts as a diluent.

These studies show: the hydroxyl group is the primary bonding group of the PVBI; the silanol group is the major bonding site on the glass; and the effects of glass treatment on adhesion can be far greater than the possible variation of the PVBI hydroxyl content. In order to test whether or not the models shown in Figure 4 are realistic, molecular modelling was carried out. The questions asked were: "What does a realistic glass surface look like and what is the probability of a hydroxyl group(s) on the polymer encountering a silanol group (binding site) on the glass surface?" The silica surface was constructed using 3,4,5,6, and 7 membered rings as described previously[26,2,13,19,32] and summarized by Maniar and Navrotsky[27]. The model surface obtained is shown in Figure 5 a and b and consists of 233 atoms containing 249 bonds and uses a preponderance of the more stable 5-8 fold rings[24]. Figure 5a is a top view of the surface and is similar to conventional views presented in the literature[23,35,12]. It is interesting to note that the top view does not provide a feel for the molecular roughness of the surface nor is this point brought out in classical texts[23]. This can be appreciated from the side surface view shown in Figure 5b.

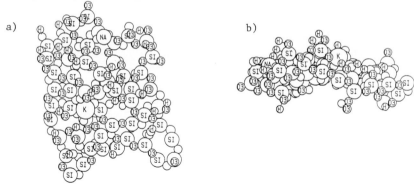

Fig. 5. Model Silica Surface
    a = Top view of silica surface with Na and K incorporated to illustrate differences in atomic size
    b = Side view of surface shown in (a)

Using the x, y and z coordinates of each atom and standard vector mathematics, the distances from each silanol hydrogen to every other silanol hydrogen were determined, eliminating duplications and self measurement distances. These distances were determined only after energy minimization of the molecular configurations. The final results are displayed in the bar graph in Figure 6.

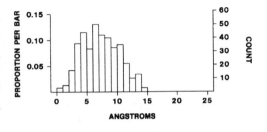

Fig. 6. Silanol hydroxyl distribution on a glass surface

These results show that on a molecular scale, the silanol groups are distributed in regular discrete steps, the envelope of which constitutes a normal distribution curve. Thus the step-wise but continuous distances between silanols indicate that there is a high probability of adjacent hydroxyl groups on the polymer backbone, which are 3.0Å apart, finding silanol binding sites on the glass. It would be even easier for hydroxyl groups along a polymer backbone chain to find not just one but multiple binding sites.

4.   TECHNIQUES OF ADHESION CONTROL

For both automotive and architectural laminates, the adhesion, as measured by some standardized test, must be maintained in the control range for that specific application. In the case of automotive applications, this range can be quite narrow in order to provide the necessary energy absorption to prevent serious injury by preventing head penetration and accompanying skull injury or serious lacerations.

While gross changes can be obtained by drastically altering the glass composition or the polymer composition, in practice neither of these two approaches represent viable commercial alternatives. This issue is generally addressed by adding adhesion control agents that allow "fine tuning" of the adhesion to the desired level without major changes in either polymer or glass composition.

The literature indicates that PVBI adhesion can be controlled using moisture and metal carboxylates such as potassium, cesium, magnesium, calcium and other such salts[20,6,22,29,7,28,25,33,8]. Moisture plays an important role in PVBI adhesion but unfortunately also acts as a plasticizer.

Although moisture can be used to adjust adhesion as shown in Figure 7, since PVBI is known to be hygroscopic, any change in moisture content can produce unwanted effects on adhesion.

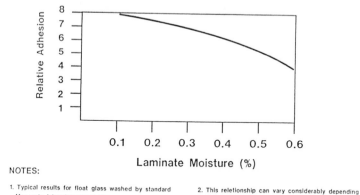

NOTES:

1. Typical results for float glass washed by standard   2. This reletionship can vary considerably depending
   Monsanto laboratory procedure.   on specific laminating conditions, glass type and
                                                              glass washing conditions.

Fig. 7. Relative adhesion vs moisture of plasticized poly (vinyl butyral)

5.   CHARACTERIZATION OF ADHESION INTERFACES

The plasticized PVB is extruded into sheets which are cut to size and laminated at elevated temperatures and pressures for a short period of time to provide an optically clear laminate. During this operation the plasticized PVB flows and bonds to the glass substrates. The lamination process changes the physical appearance and the interfaces and therefore the surface chemistry of the two materials. The resultant glass laminates now contain the encapsulated polymer/plasticizer which presents experimental problems in characterization of the glass/polymer system interface.

Our approach to this dilemma was to utilize specimens that had been prepared for peel adhesion measurements. These specimens consist of strips of plasticized PVB that are laminated to aluminum foil on one side and glass on the other. The interlayer was hand peeled from the glass and the interphase adhesive layer remaining on the glass could then be examined.

One of the techniques that we utilized is XPS. Using XPS the thickness of this interphase layer remaining on the glass can be estimated. This is done by employing the equation:

$$\frac{I_{Si}}{I_{Si*}} = 1 - \frac{I_c}{I_{c*}}$$

where:

I_{Si} = Measured intensity of the Si2p peak

$I_{Si}*$ = $F\alpha Dk\lambda Si$

I_c = Measured intensity of the Cls peak

I_{c*} = Intensity from an infinitely thick polymer layer

From a plot of I_{Si} vs I_c the intercept gave I_{Si*} which was then used in the following equation to calculate the interphase thickness.

$$I_{Si} = F\alpha Dk\ \lambda_{Si}\left(e^{-(\sqrt{2})t/\lambda_{Si}}\right)$$

where: $\lambda_{Si} \sim \lambda_c \left(\lambda_c = 27\text{Å}, \lambda_{Si} = 30\text{Å}\right)$

The derivations and applications have been developed in detail in previous publications[8,9].

The interphase adhesive layer thickness was found to be proportional to peel strength as is illustrated in Figure 8. These results show that the actual adhesive interphase is very thin indeed and that this thin layer is left on the glass surface as a result of peeling the PVBI from the glass. Thus it appears that laminated glass specimens fail cohesively at the glass/polymer interface[9,10].

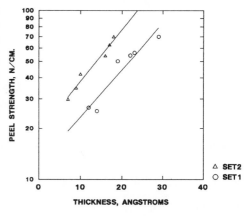

Fig. 8. Peel strength vs interphase adhesive layer thickness

6. MECHANISMS OF ADHESION CONTROL

In a previous study[20] Huntsberger pointed out that potassium acetate and formate are known to be effective adhesion control additives. Experimental work supported the premise that the solubility of the salts and their ability to form stable equilibrium solutions are necessary for these materials to be effective. This results in adhesive performance that is a function of the interfacial aqueous layer. Data were presented to show the importance of the presence of a water layer and the salt concentrations in that interphase layer.

A subsequent XPS study[9] showed that the interphase layer thickness varies from about 7Å to 40Å and that peel strength is directly proportional to the interphase layer thickness and varies depending upon the type of formulation. In a further study[10] it was demonstrated that the measurement of contact angle by using a technique that allows the polar force contributions to be measured, related the polar force contributions to contact angle and contact angle work of adhesion. This was accomplished by measuring the polar force contribution of the interphase layer left on the glass after peeling.

In an attempt to understand the way in which these metal salts control adhesion, we used XPS to examine the glass surface from which the PVBI had been removed by peeling. What we found was that the metal salts diffused to the interphase layer between the glass and the PVBI and in some fashion interfered with adhesion. This is demonstrated in the plot in Figure 9 which shows that peel strength decreases as the metal salt concentration of total alkali content increases per interphase layer thickness. We found that this relationship held independently of how the metal surface concentration was obtained, i.e., by increasing the metal content or increasing the water content. This metal salt concentration enhancement is a result of the thermodynamic driving force due to the initial concentration gradient at the interphase. The diffusion of metal salts continues until the interactions at the interphase have been satisfied and a new concentration equilibrium has been established.

This illustrates that both water and salt content are involved but does not provide information on the precise role of these components.

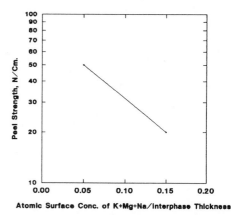

Fig. 9. Saflex® peel strength vs surface metal concentration per interphase thickness

Of a number of possible ways in which the metal ions can specifically interfere with the binding of polymer hydroxyl groups to the glass silanol groups, the following appear to be the more likely candidates.

1. Reaction of metal ions with silanol groups
   A metal carboxylate can undergo an acid base interchange to produce the resulting conjugate acid/base.

$$-SiOH + M^+ + An^- \underset{\leftarrow}{\overset{\rightarrow}{\sim}} SiOM^+ + H^+ + An^-$$

2. Interaction of metal ion and water
   The metal ion can act as the nucleus for water clustering. The type and size of the metal would determine the extent. This would give rise to the effect that Huntsberger[20] has proposed. The relative ion size is illustrated in Figure 5 in order that ion size effect can be appreciated.

3. Interaction of metal ions with polymer hydroxyl groups
   Metal ions may repulse the polymer due to electrostatic interaction or induce polymer hydroxyl clustering and thus serve to prevent polymer glass binding.

7. ALTERNATE TECHNIQUES OF INFLUENCING ADHESION AND MOISTURE SENSITIVITY

Previous work has identified techniques of flame treatment and electrical discharge for the control of adhesion and to eliminate the tendency for rolls of PVBI to stick to themselves[15,16,17]. This tendency of the roll laps to stick to themselves is referred to as blocking. More recently, a low temperature plasma has been used to improve the blocking resistance[31].

In general, one can speculate that the reason the electrical discharge techniques have not been commercially practiced is that acceptable and uniform adhesion control cannot be achieved and there may be a problem with the permanence of the treatment since the electrical discharge does not penetrate as deeply as with plasma. In the present context, deep is a relative term since the advantage of the plasma treatment is that the effects are confined to surface depths only leaving the bulk of the treated polymer in its pristine condition. These surface effects are generally thought to be 1-$10\mu m$ in depth depending upon plasma conditions.

The advantages of using a plasma have been described by Bell[1]. The advantages of using plasmas or glow discharges can be appreciated by recalling that the discharge is characterized by average electron energies of 1-$10eV$ and electron densities of 10^9-10^{12} cm^{-3}. These electron densities are on the order of 10^2 higher than those existing in flames. An additional characteristic of such plasmas is the lack of equilibrium between the electron temperature T_e, and the gas temperature T_g. Typical ratios of T_e/T_g lie in the range of 10-10^2. The absence of thermal equilibrium makes it possible to obtain a plasma in which the gas temperature may be near ambient values at the same time that the electrons are sufficiently energetic to cause the rupture of molecular bonds.

In order to investigate the effects of plasma treatment on PVBI, samples of PVBI were exposed to Argon plasmas for short periods of time. Previous work[15,1] had not characterized the effects of electrical discharge or plasma treatment of PVBI. In this instance we utilized XPS to characterize the effects of argon plasma treatment.

The effects of the plasma on PVBI can be seen by comparing the XPS spectrum of the untreated (Figure 10) PVBI with that of the treated (Figure 11).

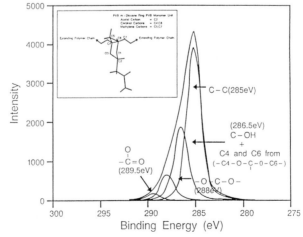

Fig. 10. Cls scan untreated PVBI

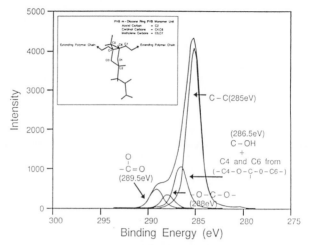

Fig. 11. C1s scan argon plasma treated PVBI

The peak that occurs at approximately 286.5eV is a combination of the response of the hydroxyl carbon and the C4 and C6 carbons as illustrated in the accompanying figure. The peak at 288eV is due to the response from the C2 carbon and has decreased about 1.5 times on the treated vs untreated surfaces. A comparison of the ESCA spectra before and after exposure in the plasma shows that the hydroxyl groups and the ether linkages of the m-dioxane ring have decreased considerably in concentration.

The interpretation is that hydroxyl groups and rings would be cleaved in a manner that would allow formation of carboxyls, the end effect of which would be to lower adhesion and roll blocking tendencies. This appears to be corroborated by the fact that the carboxyl group at 289.5eV has increased approximately three times which is the value it should attain, neglecting major contributions from the plasticizer, and assuming one half the response is from hydroxyl and the other half from the ring carbons. It is reasonable to expect that both reactions would be accompanied by crosslinking. Although plasma energetics are complex and there are a variety of species present, possible plasma reactions which would account for the loss of hydroxyl groups, the increase of carboxyls and cross linking reactions are as follows:

$$P_1OH + Ar^* \rightarrow P_1O\cdot + H\cdot + Ar$$

$$P_2OH + P_3OH + Ar^* \rightarrow P_2 + OH\cdot + P_3 + Ar$$
$$\underset{OH}{|}$$

$$P_1{=}0 + HOH + P_2\underset{\underset{OH}{|}}{-}P_3$$

where P = PVBI molecule

The quantitative effects of plasma exposure are demonstrated in Figure 12 in which adhesion is plotted as a function of moisture content. The plasma treatment has also resulted in substantially less moisture sensitivity as compared to the response of a standard control formulation. There is also a slight difference between the anode and cathode electrodes which vary in polarity with respect to their electrical biasing to ground.

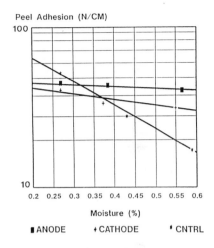

Fig. 12. Peel adhesion vs moisture for plasma-treated PVBI

8.   SUMMARY

The careful control of adhesion of PVBI to glass is essential in providing safety to front seat passengers in an automobile. This is true for preventing intrusion of foreign objects from without and energy absorption of potential head impacts from within. Adhesion occurs as a result of the silanol groups on the glass surface bonding to the hydroxyl groups in the PVBI. The specific formulation of the PVBI determines the adhesion in conjunction with the nature of the glass surface via moisture and other components in the formulation.

The contemporary technique of XPS has proven to be extremely useful for characterizing the thickness and nature of the interfacial adhesive layer which is on the order of 10Å-40Å.

Plasma technology was found to be an alternate technique for modifying adhesion and also provided a relatively moisture insensitive PVBI material. XPS characterization of this material showed that the plasma effectively reduces the concentration of hydroxyl groups and ring ether groups present on the surface. The end effect is to lower adhesion.

LITERATURE CITED

1.   Bell A T 1974 in *Techniques and Applications of Plasma Chemistry* ed Hollahan J R and Bell A T (New York Wiley)
2.   Bell R J and Dean 1971 *P. Philos. Mag.* **25** pp 1381
3.   Bershstein V A and Nikitin V V 1970 *Sov. Phys-Doklady* **15** (2) pp 163
4.   Bershstein V A and Nikitin V V 1974 Proc. of Xth Intl. Cong. Class **9** pp 105
5.   Bershstein V A Novoikov S N and Nikitin V V 1973 *Sov. Phys. Solid State* **15** (2) pp 348
6.   Buckley F T and Nelson J S 1966 (to Monsanto) *Safety Laminates* US Patent 3249487
7.   Buckley F T and Riek R F 1971 (to Monsanto) *Process for Production of Interlayer Safety Glass* US Patent 3556890
8.   Carlson T A and McGuire G E 1972 *J. Electron Spec.* **1** pp 2
9.   David D J and Wittberg T N 1984 *J. Adhesion* **17** pp 231
10. David D J and Misra A 1985 *J. Colloid Interface Sci.* **108** pp 371

11.   Elmer T H and Eaton D L 1980 in *Silylated Surfaces* (New York Gordon & Breach)
12.   Garofalini S H 1989 in *Physics and Chemistry of Glass Surfaces* Proceedings of the Tenth University Conference on Glass Science ed G C Pantano (Elsevier Holland)
13.   Gerber T and Himmel J 1987 *Non-Cryst. Solids*, **92** pp 407
14.   Grover J Monsanto Internal Report April 1991
15.   Hailstone R B 1966 (to E. I. DuPont de Nemours & Co.) *Method of Flame Treating Poly (vinyl butyral) and the Product Thereof* US Patent 3282722
16.   Hailstone R B 1968 (to E. I. DuPont de Nemours & Co.) *Process of Treating Poly (vinyl butyral) Sheeting by an Electric Discharge in Nitrogen to Reduce Blocking* US Patent 3407130
17.   Hailstone R B 1968 (to E. I. DuPont de Nemours & Co.) *Process of Treating Poly (vinyl butyral) Sheeting by an Electrical Discharge to Reduce Blocking* US Patent 3407131
18.   Hair M L 1975 *J. Non-Cryst. Sol.* **19** pp 299
19.   He H 1987 *J. Non-Cryst. Solids* **89** pp 402
20.   Huntsberger J R 1981 *J. Adhesion* **13** pp 107
21.   Iller R K 1979 *The Chemistry of Silica* (New York Wiley)
22.   Inskip H K 1980 (to E. I. DuPont de Nemours & Co.) *Poly (vinyl butyral) Compositions* US Patent 4210705
23.   Kingery W D Bowen H K and Uhlmann D R 1976 *Introduction to Ceramics* (New York Wiley)
24.   Kubichi J D and Lasaga A C 1988 *Am. Mineral* **73** pp 941
25.   Lavin E and Mont G E 1966 (to Monsanto) *Laminated Safety Glass* US Patent 3271235
26.   Liebau F 1987 in *Silicon Chemistry* (New York Ellis Harwood-Halstead)
27.   Maniar P D and Navrotsky A 1989 in *Physics and Chemistry of Glass Surfaces* Proceedings of the Tenth University Conference on Glass Science ed C G Pantano (Elsevier Holland)
28.   Mont G E and Lavin E 1966 (to Monsanto) *Laminated Safety Glass* US Patent 3249488
29.   Moynahan R E 1983   (to E. I. DuPont de Nemours & Co.) *Process for the Preparation of Polyvinylbutyral Sheeting and Adhesion Control* US Patent 4379116
30.   Pantano C G 1985 in *Strength of Inorganic Glass* ed C R Kurkjiam (Plenum New York)
31.   Tanaka M Okamoto T Koyama F and Okazaki H 1984 (to Sekisui Chemical) *Method for Manufacturing A Poly (vinyl butyral) Sheet with Improved Blocking Resistance* Japanese Kokoku Patent Sho 59 -30730
32.   Tadros A Klenin M A and Lucovsky G 1984 *J. Non-Cryst. Solids* **64** pp 215
33.   Technical Guide *Moisture Conditioning of Saflex®* Monsanto Company, St. Louis, MO
34.   Thakker B 1991 PhD Dissertation U. of Minnesota (to be submitted).
35.   Van Vlack L H 1964 *Physical Ceramics for Engineers* (Keading, MA Addison-Wesly)

Multitechnique spectroscopic analysis of plasma treatment carbon fibres

C. LEFEBVRE, J. VERBIST [*]
D. LEONARD, P. BERTRAND [**]
[*] Facultés Universitaires Notre Dame de la Paix, LISE, 61 rue de Bruxelles, 5000 Namur   BELGIUM
[**] Université Catholique de Louvain, PCPM, 1 Place Croix du Sud, 1348 Louvain-la-Neuve
  BELGIUM

ABSTRACT : Carbon fibre surfaces were submitted to three processes : activation by O_2 plasma, AcryloNitrile (AN) plasma polymerization and finally neutralization under AN or Argon atmosphere. The effect of these treatments were evaluated by XPS, Static SIMS and ISS. XPS and ISS techniques give a good characterization of the surface but can not show the fine differences between the processes. The SIMS technique has been used in order to identify more precisely the nature of the physico-chemical changes after the process and bring information about the efficiency of each step of these treatments.

1. INTRODUCTION

Epoxy resins may be reinforced by carbon fibres to yield high strength-to-weigth composite resin systems suitable for remplacement of metal component in the aerospace industry. Plasma treatment of carbon fibres is used for modifying their wetting characteristics. The plasma grafted layer is a flowless highly crosslinked film which has excellent adhesion to the carbon fibres. Various steps have been studied during plasma treatment : plasma activation, plasma polymerization and neutralization. Our aim in this work was to use a combination of techniques sensitive to surface chemistry in order to understand the effects of the plasma.

X-ray Photoelectron Spectroscopy (XPS) technique was used to determine the elemental surface composition of the carbon fibres before and after plasma treatment. Xie and Sherwood (1990), Kozlowski and Sherwood (1986) and Ishitani (1982) showed clearly the potential use of XPS in carbon fibres studies. Specifically, the information from the core-level photoionization peaks yields both qualitative and quantitative data regarding the near surface regions. Qualitatively, beyond elemental identification, the chemical shift gives information about the types of bonding around each type of atom. Quantitatively, the XPS can identify the major constituents and minor impurities.

Secondary Ion Mass Spectrometry (SIMS), thanks to its molecular specificity, could help to know more precisely the physico-chemical changes of the carbon fibre surface during the different parts of the plasma treatment. It is important to underline the difference between

XPS and SIMS sampling depths : about 10 Å for SIMS and 50 Å for XPS. In order to obtain static conditions, assuming that the analyzed material is not damaged, the total ion dose was chosen as low as possible, but sufficient to obtain a good signal to noise ratio in the studied mass range. Hearn and Briggs (1991) have recently published a paper about Time Of Flight SIMS studies of carbon-fibre surfaces. They underline the influence of the different parts of the processing of a commercially electrochemically treated carbon-fibre. A lot of peaks are to be related to different contaminants which are, as it were, a real "memory" of the processing stages.

Ion Scattering Spectrometry (ISS) is even more surface sensitive than SIMS. Thanks to a good choice of the analysis conditions, only the topmost layer is analysed (when the ion goes further in the surface, it has a great probability to be neutralized). In itself, the information derived from only about ISS spectrum could not be really quantitative. It has been used here with the aim of comparing different parts of the treatment.

2. EXPERIMENTAL

Plasma treatments were carried out in an inductively coupled plasma reactor operating at the rf frequency of 13.56 MHz (Figure 1). Carbon fibres are deposited on a glass support introduced in the quartz glow discharge tube (2). The whole chamber is pumped to a pressure of 10^{-3} Torr using the primary pump (4).

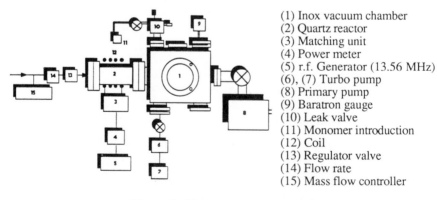

(1) Inox vacuum chamber
(2) Quartz reactor
(3) Matching unit
(4) Power meter
(5) r.f. Generator (13.56 MHz)
(6), (7) Turbo pump
(8) Primary pump
(9) Baratron gauge
(10) Leak valve
(11) Monomer introduction
(12) Coil
(13) Regulator valve
(14) Flow rate
(15) Mass flow controller

Figure 1 : Plasma treatment system

Various steps were studied : *plasma activation* (**A**) is conducted under O_2 with a flow rate fixed at 200 ml/min at 0.06 Torr of pressure, at 10 watt during 2 min. *For plasma* *polymerization* (**P**), the acrylonitrile monomer is placed in a sealed vessel located before the plasma quartz tube. Under a working pressure of 10^{-1} Torr, the liquid is volatilized, providing thus a gazeous source of monomer. The carbon fibres are coated for 10 min at a power of 20 W. The *neutralization step* (**N**) was carried out under AN or Ar atmosphere during 45 min. One sample was drenched in an acetone *bath* for 2 min (**B**).

HTA-3000 carbon fibres (ENKA S.A.) (ex-PAN, unsized and oxided) were plasma treated. Five samples, labelled P_1 to P_5, which represented five levels of surface treatment, constituted the batch of plasma treated fibres :

$$P_1 = A + P + N_{AN} + B$$
$$P_2 = A + P + N_{AN}$$
$$P_3 = A + P + N_{Ar}$$
$$P_4 = \quad P + N_{AN}$$
$$P_5 = A + P$$

X-ray Photoelectron measurements were carried out on an HP 5950 A spectrometer using monochromatized AlK_α radiation. Static SIMS and ISS analysis were performed at Louvain-la-Neuve in the same analysis chamber under ultra high vacuum conditions (residual pressure less than 10^{-9} Torr). The other experimental conditions are given in Table 1.

Table 1 : ISS and SSIMS experimental conditions

	ISS	**Static SIMS**
Spectrometer	Kratos WG-541	Riber Q-156
Type of spectrometer	CMA (Cylindrical MirrorAnalysis)	Quadrupole
Primary Ion beam : - gas	He^+	Xe^+
- energy	2 keV	4 keV
- angle*	0°	75°
Sample current**	10 nA	10 nA
Ion beam rastered area	1.9 x 2.2 mm^2	33 mm^2
Filament current for charge compensation	-	2 A
Scan region	0.3 to 0.8 x E/E_0	50 - 0 amu
Scan time	200 sec	102 sec
Total ion dose (ions/cm^2)	3 x 10^{14}	2 x 10^{13}

* with respect to the surface normal of the samples
** measured into a Faraday cup

The ion gun is mounted coaxially in the Cylindrical Mirror Analyser. Backscattered ions were analysed and detected in a full annular solid angle at 138° with respect to the beam direction. Samples were maintained by the use of scotch tape. PolyAcryloNitrile powder (PAN) (Aldrich Chemicals), used as a reference, was compacted (as much as possible) before being introduced into the chamber. Charge compensation was necessary in the case of SSIMS because we used the negative mode and, in this case, we need to flood the surface with electrons to optimise the observation of negative peaks.

3. RESULTS AND DISCUSSION

3.1. XPS

Surface compositions determined by XPS are presented in Table 2, and Figure 2 illustrates the C1s spectra of untreated carbon fibres, of two groups of plasma-treated fibres (P_1 and P_2) and of reference PAN.

Table 2 : XPS composition percentages of untreated, treated
carbon fibres and PolyAcryloNitrile (PAN).

	%C	%N	%O
untreated fibres	82.8	4.0	13.2
P_1 (APN$_{ANB}$)	78.6	8.7	12.7
P_2 (APN$_{AN}$)	72.6	24.8	2.6
P_3 (APN$_{Ar}$)	70.7	22.2	7.1
P_4 (PN$_{AN}$)	72.8	23.3	3.9
P_5 (AP)	71.6	23.8	4.6
PAN	74.6	25.4	-

The presence of nitrogen on the untreated carbon fibre (4%) is probably due to a small quantity of nitrogen trapped within the fibre after incomplete decomposition of precursors such as PAN. Nevertheless, for the other samples, the amount of nitrogen illustrates the presence of Plasma Polymerized AN (PPAN) of a composition close to the reference PAN one, except for P_1, which is the only one to have been drenched in an acetone bath; this latter process clearly removes part of the plasma coating (%N=8.7). In order to understand the composition and chemical bonds of plasma-treated carbon fibres, the high deconvolution of the untreated fibres and reference PAN C1s spectra was studied (Figure 2). Three types of oxides can be identified: -C-OH, -C=O and -COOH, shifted by 1.5, 3.0, and 4.5 eV respectively from the primary carbon peak of the untreated carbon fibres, and a shake-up shifted at 6.0 eV. The chosen assignments were based on a comparison with chemical shifts proposed by Proctor and Sherwood (1982) and Takahagi and Shimada (1986). The reference PAN studied by Raynaud and Riga (1991) was fitted to two components separated by 1.2 eV. The principal peak was assigned to the >CH-C≡N and >CH-C≡N groups, and the second one to those of CH₂ groups.

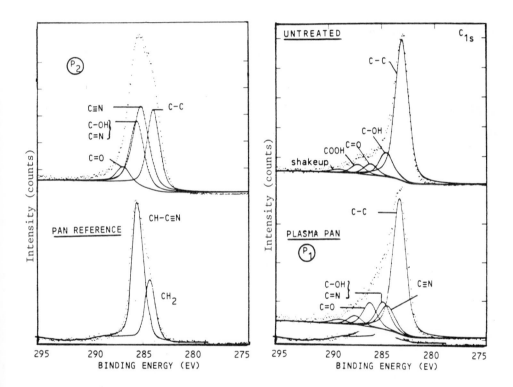

Figure 2 : C1s spectra of carbon fibres a) untreated, b) P_1, c) P_2 and d) PAN.

The C1s spectrum of P_1 shows an increase of C=O groups compared with the untreated fibres. This could results from the activation effect or acetone oxidation. In the case of P_2 to P_5, activation and neutralization steps seem to bring no great changes in the elemental composition (Table 2) and in the C1s lineshape (not shown here). P_2 was thus chosen to represent also P_3, P_4 and P_5 plasma treated carbon fibres. The activation step, used to create reactive sites which could improve the coating adhesion on carbon fibres, is masked by the following processes and can not be detected by XPS : indeed, P_2 and P_4 compositions and deconvolutions are similar. The purpose of the neutralization step was to find out what occurs when the active surface was in presence of different atmospheres as AN (P_2), argon (P_3) or air (P_5). The active surface, resulting from plasma polymerization, reacts with the AN monomer as indicated by a weak increase of the nitrogen concentration (24.8%), and undergoes a rearrangement and an oxidation (7.1% O) under argon atmosphere. The results from P_5, which was let in the air immediately after the polymerization, indicate a weak oxidation of the surface.

As for the plasma polymerization step, we notice that the fit of the C1s spectrum of P_2 sample contains different proportions of functional groups, most probably different types of C/N groups. The analysis of the N1s spectrum of PPAN (P_2) shows an N1s peak broader (FWHM = 1.7-1.8 eV) than for the reference PAN (FWHM = 1.3 eV), and slightly asymmetrical with a new shoulder at higher binding energy (Figure 3).

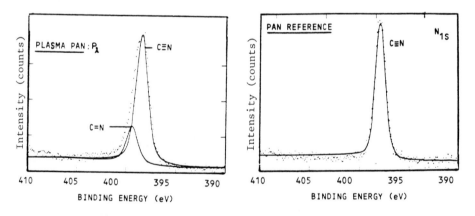

Figure 3 : N1s spectra of P_2 treated fibres and PAN.

The shoulder peak could be belonged to the nitrogen of the -C=N group. This appearance of imine groups is in agreement with the study of Lefebvre (1991) on IR absorption spectra of PPAN on carbon fibres which reveals the =N-H absorption band at 3360 cm^{-1}. It was very difficult to integrate their contribution in the C1s fits. Indeed, the chemically shifted region of C/O and C/N bondings is similar, which renders the interpretation difficult. Nevertheless, we decided to fit to three components separated by 1.2, 1.5 and 3.1 eV from the principal component, attributed to [C≡N], [C-OH and C=N] and [C=O] groups respectively.

The presence of an aromatic structure may explain the broadening of the second peak. Indeed, the plasma polymerization of AN implies a high monomer fragmentations and recombinations resulting in a branched and crosslinked PPAN, containing still nitrile groups and some triazine structures due to the cyclization of AN, combined with some crosslinking (confirmed by the imine groups contributions).

3.2 ISS

ISS spectra are presented in Figures 4 and 5, and the results obtained from the spectra are presented in the Table 3. The values are the atomic ratios O/C and N/C as measured using surface peak area ratios. To use these values in a quantitative way for each spectrum, one should know the influence of different factors varying with the energy (cross section, transmission factor, efficiency of the channeltron...). before taking into account the possibility of shadowing (because of the sensitivity to the topmost surface). This is quite difficult and ISS results are therefore mostly used as here in the comparative way.

Table 3 : Atomic ratios and presence of contamination
(stars) for untreated and plasma treated carbon fibres.

Sample	O/C	N/C	Contamination
untreated	1.19 (±0.2)	-	*
P1 (APN$_A$NB)	0.83 (±0.1)	0.83 (±0.1)	**
P2 (APN$_A$N)	0.39 (±0.07)	1.75 (±0.25)	
P3 (APN$_{Ar}$)	0.68 (±0.03)	1.7 (±0.1)	
P4 (PN$_A$N)	0.46 (±0.1)	1.8 (±0.05)	
P5 (AP)	0.61 (±0.07)	1.88 (±0.05)	
PAN	-	2.4 (±0.1)	

The values displayed in Table 3 are obtained as the average of four atomic ratios. Error bars are presented under the form of an absolute maximum error. The presence of N in the topmost layer is not a surprise even if the presence of contaminants could have masked some atoms. The importance of contamination is underlined by stars in the third column of Table 3 (probably Na or Mg; F in addition for P1). Untreated carbon fibre and P1 are amongst the most contaminated.

Similarities to XPS results can be observed : (1) oxidation of untreated fibres, less N and (slightly) more O amount for P_1 (PPAN layer removed after drenching in acetone bath); (2) no difference between P_2 and P_4 about the activation and (3) some differences for neutralization (oxidation for neutralization in Ar and air). It is interesting to underline this similarity although the surface sensitivities are different. This indicates the homogeneity of the treatment within the sampling depth of those techniques.

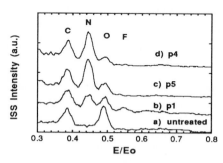

Figure 4 : ISS spectra of a) untreated, b) P_1, c) P_5 and d) P_4 samples.

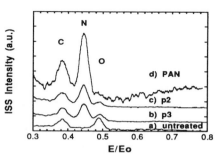

Figure 5 : ISS spectra of a) untreated, b) P_3, c) P_2 and d) PAN samples.

3.3 Static SIMS

Negative spectra in the range 0 to 50 amu are presented in Figure 6 for the different fibres and for PAN powder. Some obvious changes in the functionalities can be seen ; in particular, the importance of the nitrile (CN^-) peak at mass 26 is clear ; peaks are also present at mass 16 (O^-), 17 (OH^-), 38 (C_2N^-) and 42 (CNO^-). Other peaks at mass 19, 35 and 32 are related to F, Cl and S contaminants. However, the presence of these contaminants does not compromise our analysis. We also measured positive ion SIMS spectra on sample P_4, where the two prominent peaks are at mass 23 (Na^+) and 39 (K^+). The high intensity of these peaks and the nature of the assumed functionalisation imply that negative spectra should be first performed.

After a calibration, the spectra were transformed under the form of an histogram (and then one intensity value is associated to one mass only). To make the comparison meaningful, the intensity values for each mass are normalized to the total intensity (except intensity for mass 1 which is too dependant on the spectrometer settings). These values are presented in the Table 4. Reproducibility was tested by taking three spectra for each sample. The result is not very positive as found by Hearn and Briggs (1991). It can be seen by the error values (expressed as a percentage of the mean value). The differences of an order of magnitude seen in this Table are deemed significative.

Table 4 : Relative intensities for negative peaks (in %)

M		untreated	P_1	P_2	P_3	P_4	P_5	PAN
12	C^-	17 (15%)	6.7 (3%)	7 (9%)	4.8 (54%)	4.1 (8%)	7.2 (33%)	7.8 (15%)
13	CH^-	32.(12%)	13.4 (8%)	9 (17%)	5.9 (52%)	4.6 (3%)	14 (25%)	9.9 (16%)
16	O^-	13 (60%)	10.6 (5%)	2.9 (32%)	2.5 (60%)	1.5 (5%)	6.9 (12%)	2.1 (58%)
17	OH^-	5 (52%)	5.3 (16%)	0.9 (2%)	0.9 (47%)	0.5 (7%)	3.4 (11%)	0.5 (61%)
19	F^-	0.6 (37%)	0.2 (27%)	/	/	/	0.3 (36%)	0.5 (48%)
24	C_2^-	13.5 (7%)	8.8 (6%)	5.8 (61%)	4.4 (34%)	3.7 (5%)	6.2 (18%)	4.4 (25%)
25	C_2H^-	10 (40%)	12 (1%)	6 (82%)	4.6 (23%)	3.7 (11%)	8.1 (6%)	4.8 (7%)
26	CN^-	2 (60%)	36 (10%)	64 (16%)	72 (15%)	76 (1%)	46 (8%)	63.9 (5%)
32	S^-	0.4 (20%)	0.2 (29%)	/	/	/	0.1 (31%)	0.2 (33%)
35	Cl^-	0.9 (45%)	0.3 (22%)	/	/	/	0.1 (30%)	0.5 (64%)
36	C_3^-	0.4 (46%)	0.4 (1%)	0.3 (50%)	0.3 (30%)	0.2 (2%)	0.5 (24%)	0.2 (24%)
37	Cl^-/C_3H^-	0.4 (31%)	0.2 (13%)	/	0.1 (19%)	0.1 (21%)	0.3 (45%)	0.2 (51%)
38	C_2N^-	/	0.4 (7%)	0.6 (23%)	0.8 (31%)	0.9 (12%)	0.6 (32%)	0.4 (10%)
42	CNO^-	/	0.2 (47%)	0.2 (75%)	0.3 (32%)	0.3 (29%)	0.4 (61%)	/
50	C_3N^-	/	0.3 (66%)	0.5 (94%)	0.8 (21%)	0.9 (8%)	0.7 (43%)	2 (50%)

As we have just observed, the neutralization seems to have an effect on the PPAN structure. We suppose that under AN monomer, there is a rearrangement of the surface and that under other atmospheres, the active surface is oxidized.

Due to the form of presentation of these results (ratio to the total intensity), the values presented for the characteristic peaks in the case of the P_1 sample are clearly closer to those of the untreated carbon fibre (as the spectrum in itself) : more relative intensity for C^-, CH^-, O^-, OH^-, C_2^-,C_2H^- and less for very characteristic peaks as 26 (CN^-), 38 (C_2N^-) and 50 (C_3N^-) which are not detected in the case of untreated fibres.

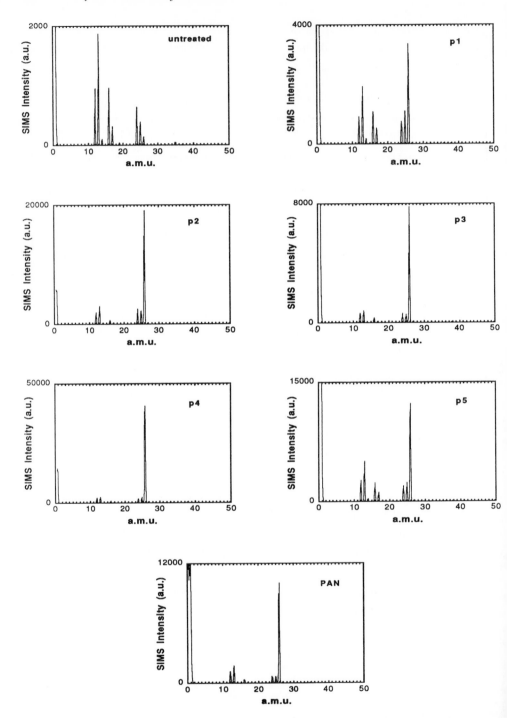

Figure 6 : SSIMS (-) spectra of untreated carbon fibres, plasma treated carbon fibres (P1 to P5) and reference PAN.

As for the contamination, it is clear that untreated carbon fibre, P_1 and PAN samples are contaminated [see Cl⁻ (35 and part of 37), F⁻ (19) and S⁻ (32)]. In the case of the P_5 sample, contamination is less present but could not be excluded. We suppose that the low sensitivity of our XPS spectrometer does not allow us to observe these contaminations in the XPS results.

No real difference is observed between P_2 and P_4. As underlined before, activation effects are probably masked by following processes effects.

The presence of CN⁻ (26), C_2N^- (38) and C_3N^- (50) indicates a polymerization of AN (PPAN) but this layer is different from PAN reference by CNO⁻ (42) (Table 4). The intensity values presented in Table 5 are observed for the same data treatment, but come from a 100-0 amu spectrum (not shown). The relative intensities for peaks at masses 62 (C_4N^-), 74 (C_5N^-), 86 (C_6N^-), and 98 (C_7N^-) exhibit interesting differences between P_2, P_3, P_4, P_5 samples and PAN. In fact, we notice that the structure of PPAN is more complicated as can be expected, because of the high degree of fragmentation undergone by the monomer in the plasma. Indeed, these peaks, more present in the case of treated fibres are not expected in the case of PAN structure. Nevertheless, the P_2 sample is the only one to show, as PAN, lower relative intensities for all these peaks. There was not such a significant difference for the peaks at masses 26 (CN⁻) and 38 (C_2N^-) and 50 (C_3N^-). Thus, we can assume that we need activation step (even if it is useful effect probably masked) and neutralization step (under AN atmosphere) to appraoch the real PAN structure.

Table 5 : Relative intensities of nitrogenated peaks at higher masses ($\times 10^{-3}$)

Mass		P_2	P_3	P_4	P_5	PAN
62	C_4N^-	0.45	1.06	0.66	0.8	0.09
74	C_5N^-	0.82	2.23	1.27	1.26	0.19
86	C_6N^-	0.17	0.47	0.26	0.3	0.13
98	C_7N^-	0.21	0.46	0.32	0.4	0.05

4. CONCLUSIONS

From our first work comparing the abilities of XPS, SSIMS and ISS for the analysis of plasma treated carbon fibre, it was deduced that XPS still remains the most information-rich technique, mainly because of its broad use and the ease of data interpretation. ISS can serve as an important complement to XPS due to its different depth resolution. But Static SIMS, by providing extreme sensitivity to trace elements or minor surface functionalization modifications, is the performant complement to XPS in order to study the structure of plasma polymers. The nature of PlasmaPAN layer is better characterized by XPS but information about its structure is brought by Static SIMS.

Except the fibre drenched in an acetone bath, XPS and ISS can not help to identify the changes occurred during the processes before (activation) and after (neutralization) the plasma polymerization. Nevertheless, by Static SIMS, differences between these latter processes could be detected. The neutralization step under monomer atmosphere leads to a PPAN structure close to PAN reference one. Unfortunately, the changes due to the activation step seem to be masked by the plasma polymerization and neutralization processes. The use of IR and SIMS techniques (in the dynamic mode with mass profiles) could be useful to characterize the interface bringing more information about plasma activity.

5. REFERENCES

Hearn M J and Briggs D 1991 Surface and Interface Analysis 17 421

Ishitani A 1982 Carbon 19 269

Kozlowski C and Sherwood P M A 1986 Carbon 24 357

Lefebvre C and Verbist J 1991 Proceedings of ICCM-8, Composites Tsai and
        Spinger editors 4 p 39

Proctor A and Sherwood P M A 1982 Surface and Interface Analysis 4 212

Raynaud M and Riga J 1991 J. Electron Spectroscopy and Related Phenomena 53 251

Takahagi T and Shimada I 1986 J. Pol. Sc. A Poly. Chem. 24 3101

Xie Y and Sherwood P M A 1990 Chem. Mater. 2 293

Photoacoustic Fourier transform IR spectroscopic study of polymer–dentin interaction

I. Stangel and E. Ostro, Faculty of Dentistry, McGill University, Montréal, Québec, H3A 1A4, CANADA

and

A. Domingue, E. Sacher and L. Bertrand, Département de Génie Physique, Ecole Polytechnique, Montréal, Québec, H3C 3A7, CANADA

ABSTRACT: Photoacoustic Fourier transform IR spectroscopy has been used to study the interaction of dentin adhesive system components with human dentin and its individual phases. Spectroscopic evidence indicates that the quantitative variations in untreated dentin spectra, at the same positions at which the system components absorb, preclude the use of dentin itself in this study. Treatment of the components, which give reproducible spectra when untreated, indicates that both calcium hydroxylapatite and collagen are capable of interacting.

1. INTRODUCTION

Historically, various materials have been used as tissue replacements in teeth. Lacking adhesion, they have been retained by being mechanically locked in geometric shapes cut into the body of the tooth. This procedure can be rigorous, requires the removal of otherwise healthy tooth tissue and often results in the weakening of the tooth structure. Clearly, materials that demonstrate successful, long-term bonding to tooth substrates would simplify operative procedures and minimize tooth structure removal. Additionally, such a development would facilitate the displacement of dental amalgam from many routine procedures, an aim of the dental profession for a number of years.

The problem of bonding to enamel, the highly mineralized outer layer of tooth crowns, has been solved by taking advantage of its homogeneous, crystalline structure: acid pretreatment preferentially removes inter-rod material, permitting tag formation to occur via the penetration of subsequently applied low viscosity resin. Such micromechanical bonding is substantial and has been clinically demonstrated to be highly successful. However, bonding to dentin, the vital, internal substance of teeth, is far more difficult, given its heterogeneity. Human dentin contains about 69% (w/w) polycrystalline calcium hydroxylapatite $[Ca_{10}(PO_4)_6(OH)_2]$, 13% water and the rest, 18%, consists of proteins, mainly in the form of collagen bundles (Asmussen and Munksgaard 1985, Nakabayashi 1984, 1985). Morphologically, tubules containing cell processes are present within the body of dentin, so the dentin exists as peri- and inter-tubular phases. Although the collagen contains about 20 amino acids, it is composed mainly of aspartate (~375 residues asp/1000) and partly phosphorylated serine (~420 residues ser(P) + ser/1000) according to Veis et al (1972).

Adhesion to the collagen could occur through reaction with the carboxylate group of asp or the aliphatic alcohol of the unphosphorylated component of the ser. Adhesion to the calcium hydroxylapatite could occur through complexation with the Ca (or its replacement ion).

Whether adhesion occurs through the inorganic or the organic component, or both, it should be capable of providing a strong, stable, nonhydrolyzable interface, such as that obtained with enamel. While mechanisms to accomplish this have been proposed (Asmussen and Munksgaard 1985, Nakabayashi 1984, 1985), there is little direct evidence to demonstrate specific chemical changes at the dentin surface which these mechanisms require. A major difficulty involves the chemical analysis of these surfaces.

Surface analytical methods useful for determining the chemical states of opaque materials fall broadly into two categories: electron and optical spectroscopies. Of the various electron spectroscopies available, X-ray photoelectron spectroscopy (XPS) (Ghosh 1983) involves a single electronic transition and is the easiest to understand. The optical technique which best lends itself to the study of chemical structure is infrared (IR) spectroscopy, which measures characteristic vibrations of the groups present in the molecule.

IR spectroscopy may be adapted for surface studies, particularly of opaque materials, through the use of the photoacoustic technique (Rosencwaig 1980). It is based on the measurement of the pressure of a gas above the sample, pressure changes being due to the release of heat by the sample on absorbing IR radiation. It is the technique explored in this study. A major difference between photoacoustic and transmission spectroscopies is that, in contrast to the latter, photoacoustic spectroscopy is a null technique: no absorption, no signal. Minor peaks and variations are more easily seen by a null technique.

There are several other advantages to the use of this technique. For one, the instrument used is a modern, computer-controlled, Fourier transform spectrometer, with ease of data manipulation and spectral addition; the latter is extremely important for the comparison of component spectra with that of the dentin. Another advantage is the possible use of phase separation (Bertrand 1988). While not discussed in this article, phase separation permits the separation of surface and volume signals; the depth probed by the surface signal is a function of sample thickness, thermal properties and the modulation frequency used; it can approach 150 A of the surface under conditions obtainable with the spectrometer used (Bordeleau et al 1987). Moreover, the ability to separate phases permits us to overcome photoacoustic artifacts found with heterogeneous samples such as dentin (Bertrand, unpublished).

Here we use phase Fourier transform infrared photoacoustic spectroscopy (PHAS-FTIR/PAS) to follow the specific changes which occur in the chemical state of the surface of the dentin substrate and its components when applying agents proposed for adhesion promotion. It is our purpose to show both the usefulness and limitations of this technique in this area.

2. EXPERIMENTAL

For the purpose of the study, baseline data were obtained using synthesized calcium hydroxylapatite, collagen, and both demineralized and intact human dentin. The calcium hydroxylapatite powder was obtained from Professor H. Margolis, Department of Physical Chemistry, Forsyth Dental Center, Boston, MA. The collagen was obtained from Sigma Chemicals as Type I, retrieved from bovine tendon. Demineralization of dentin was carried out by placing dentin fragments in 0.25N HCl for 48 hours; the demineralized material was kept in aqueous solution until just prior to spectral analysis, when it was filtered onto a fine steel mesh and dried. Dentin samples were obtained from a freshly extracted human molar, sectioned so as to obtain a dentin disk parallel to the occlusal surface; it was treated with 1,2-ethylene diamine-

N,N,N',N'-tetracetic acid (EDTA) to remove the smear layer, washed and fractured into several sample fragments.

The photoacoustic cell was specially designed (Rousset et al 1983). It uses He as the carrier gas and detects pressure changes with a miniature Knowles Electronics BL1785 microphone. S/N was kept high by keeping the total gas volume small, ~15 mm^3. Spectra were obtained on a BOMEM DA-3 FTIR spectrometer, in the range 500-4000 cm^{-1}. Two hundred spectra were coadded at a resolution of 4 cm^{-1} and an interferometer mirror speed of 0.1 cm/sec, corresponding to photoacoustic frequencies between 100 and 800 Hz.

3. RESULTS

3.1 Dentin and Its Components

The PHAS-FTIR/PAS spectra of calcium hydroxylapatite and demineralized dentin are found in Figures 1a and 1b, respectively; that for the apatite is in excellent agreement with previously published vibrational spectra (Bertoluzza et al 1981, Engel and Klee 1972, González-Díaz and Hidalgo 1976, González-Díaz and Santos 1977). The addition of the spectrum of water, Figure 1c, to them gives the spectrum in Figure 1d. It is in excellent qualitative agreement with the PHAS-FTIR/PAS spectrum of dentin, several examples of which are seen in Figure 2.

In that second figure are found spectra from different areas of the same exposed plane of dentin. While the spectra are qualitatively identical, there are quantitative variations in peak heights and ratios. These variations are certainly due, in part, to differences in the relative concentrations of the components. They are also due, in part, to positioning problems since repositioning the same areas led to similar quantitative variations.

3.2 Treated Dentin

Dentin samples were treated with several reagents used in preparing teeth for bonding. These include:

1.   1,2-ethylenediamine-N,N,N',N'-tetracetic acid (EDTA) and

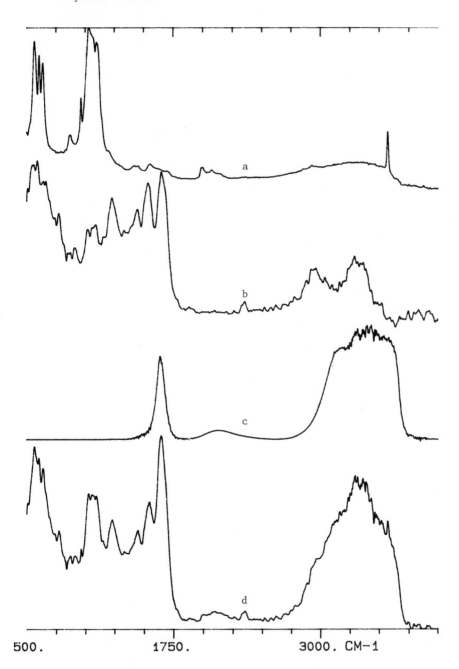

Fig 1. PHAS-FTIR/PAS spectra of (a) calcium hydroxylapatite, (b) demineralized dentin, (c) water and (d) their sum.

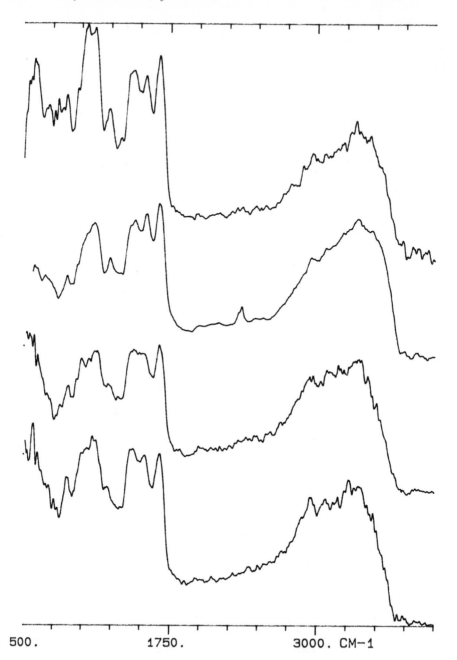

Fig 2. PHAS-FTIR/PAS spectra of dentin obtained at different areas of the same exposed plane.

2.   1,2-di(2-aminoethoxy)ethane-N,N,N',N'-tetracetic acid (EGTA), both
      used to modify the cut dentin;

3.   phosphorylated ester/hydroxyethyl methacrylate (Prisma primer,
      commercial - L.D. Caulk, Milford, DE, USA);

4.   EDTA, followed by hydroxyethyl methacrylate/gluteraldehyde (GLUMA
      primer, commercial-Bayer, Dormagen, Germany);

5.   EGTA, followed by GLUMA primer.

Because the vibrational peaks of these treatments (not shown) fall at
positions at which the dentin also manifests peaks (e.g., 1300, 1400, 1600,
1700, and 3000 cm^{-1}), and the dentin, itself, suffers from the previously
mentioned quantitative variations at these same positions, it is difficult
to draw conclusions as to the effects of these treatments, even using
subtractive techniques.

3.3 Treated Calcium Hydroxylapatite

The reagents used to treat the apatite were similar to those used to treat
the dentin in the previous section, including:

1.   Prisma primer;

2.   EGTA.

PHAS-FTIR/PAS spectra are seen, subsequent to a deionized water wash, in
Figure 3. A comparison with that of untreated apatite, in Figure 3a, clearly
shows new peaks at 1300, 1400 and 1700 cm^{-1} for the Prisma primer (Figure
3b) and at 1400 and 1600 cm^{-1}, for the EGTA (Figure 3c). Interestingly,
spectra subsequent to a tap water wash show the removal of much of the EGTA
while the Prisma primer is unaffected; this is a clear demonstration of the
complexation equilibrium of the EGTA with the Ca ion at the dentin surface
and the cations (Ca, Fe, Mg) present in the tap water.

3.4 Treated Collagen

Collagen was treated with Prisma primer and washed in deionized water. While
the spectra for the untreated (Figure 4a) and treated (Figure 4b) collagen
appear similar, subtraction of the first from the second gives the spectrum
in Figure 4c. The peak at 1700 cm^{-1} is characteristic of Prisma primer, as
seen for similarly treated apatite in Figure 3b. The peak at 3300 cm^{-1} in
Figure 4c lies some 75 cm^{-1} higher than similar peaks in Figures 4a and 4b,
indicating that it is real.

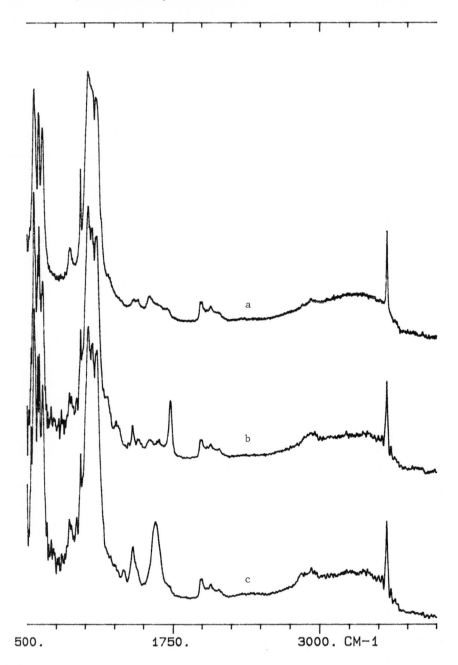

Fig 3. PHAS-FTIR/PAS spectra of (a) untreated calcium hydroxylapatite and calcium hydroxylapatite treated with (b) Prisma primer and (c) EGTA.

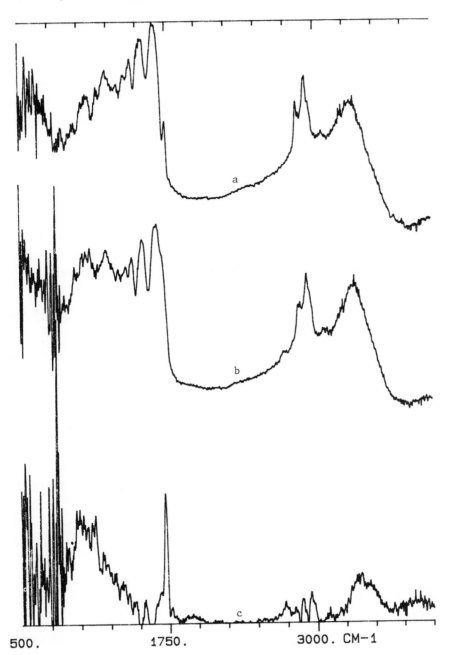

Fig 4. PHAS-FTIR/PAS spectra of (a) untreated collagen, (b) collagen subsequent to treatment with Prisma primer and (c) their difference.

4. DISCUSSION

The quantitative variations in the PHAS-FTIR/PAS spectra of dentin, found even when the same area is repositioned, makes the use of dentin as a substrate problematical. While such quantitative variations are absent when the pure components are used, making surface reactions more visible, the use of components obscures any synergistic effects which may occur in dentin. Nonetheless, given these quantitative variations, it may be necessary to use the components in place of the actual dentin, at least initially.

The problem of intercomponent synergism must be addressed and, since it cannot be addressed through the use of a dentin substrate, a dentin model must be created. We believe three major problems must be overcome:

1.   a variation in component concentration across a dentin-surface;
2.   a repositioning difficulty, and, perhaps underlying this,
3.   porosity variations (i.e., tubular concentrations) which differ with position and, at any position, with depth.

    This argues for a model dentin substrate made up of:

1.   dentin components and, if necessary,
2.   a binder which does not absorb in the IR.

These should be thoroughly mixed in a given weight ratio and compressed under given conditions of pressure temperature and time; the resultant pellet should be a reasonable representative of dentin.

Finally, given the initial use of component substrates, the spectra obtained on calcium hydroxylapatite clearly show that PHAS-FTIR/PAS is a viable technique for the study of dentin interactions: the photoacoustic technique reproduces spectra obtained by transmission IR and Raman spectroscopies, and demonstrates both complex formation and how it is affected by water wash. Indeed, the question is raised as to whether the EGTA retained on the dentin surface after water washing blocks reaction with primer. Similarly, the use of spectral subtraction in the case of collagen treatment, shows it to be useful in delineating obscured spectra. The use of a reliable dentin model will permit data handling by computer, so that both specific and synergistic reactions may be easily studied.

5.   ACKNOWLEDGMENTS

The authors wish to thank the following agencies for funding these studies: the Natural Sciences and Engineering Research Council of Canada and the Fonds pour la Formation de Chercheurs et l'Aide à la Recherche du Québec.

6.   REFERENCES

Asmussen E and Munksgaard EC 1985 Posterior Composite Resin Dental Restorative Materials ed F Vanherle and D Smith (St Paul MN: 3M Company) p 217
Bertoluzza A, Fagnano C, Fawcett V, Long DA and Taylor LH 1981 J Raman Spectrosc. 11 10
Bertrand L 1988 Appl. Spectrosc. 11 10
Bertrand L and Domingue A 1991 Proceedings of the 7th International Topical Meeting on Photoacoustic and Photothermal Phenomena ed D Bicanic (New York: Springer Verlag) in course of publication
Bordeleau A, Bertrand L and Sacher E 1987 Spectrochim. Acta 43A 1189
Engel G and Klee WE 1972 J Solid State Chem. 5 28
Ghosh PK 1983 Introduction to Photoelectron Spectroscopy (New York: Wiley-Interscience)
González-Díaz PF and Hidalgo A 1976 Spectrochim. Acta 32A 631
González-Díaz PF and Santos M 1977 J Solid State Chem. 22 193
Nakabayashi N 1984 CRC Crit Rev. Biocompat. 1 25
Nakabayashi N 1985 Int. Dent. J 35 145
Rosencwaig A 1980 Photoacoustics and Photoacoustic Spectroscopy (New York: Wiley)
Rousset G, Monchalin JP and Bertrand L 1983 J Phys. (Paris) 44 Colloque C6 165
Veis A, Spector AR and Zamoscianyk H 1972 Biochem. Biophys. Acta 257 404

7.   APPENDIX

7.1   The Suppression of Photoacoustic Artifacts by PHAS-FTIR/PAS

As noted in the Introduction, PHAS-FTIR/PAS overcomes artifacts inherent in photoacoustic spectroscopy. While how this is accomplished has no bearing on the present paper, a short description and references are given here. PHAS-FTIR/PAS uses complex transforms to obtain the in-phase and quadrature components of the photoacoustic signal. The proper treatment of these components (Bertrand 1988, Bertrand and Domingue 1991) leads to the correction of artifacts such as those due to sample dilatation and light scattering. In the case of signal saturation with increasing absorbance, the fact that the in-phase component is decreasing in the absorbance range in which the quadrature component is saturating permits signal saturation to be put off to significantly higher absorbances than possible with normal PAS.

SECTION II

METALS ON POLYMERS

Diffusion and interface formation at polymer–metal interfaces

F. Faupel

Institut für Metallphysik der Universität Göttingen, Abteilung Metallkunde, Hospitalstraße 3-7, 3400 Göttingen, FR Germany

ABSTRACT: There is evidence now from various investigation in polyimides that metals of low reactivity may diffuse into polymers. Measurements by a radiotracer technique show that Arrhenius plots for single-atom diffusion of Cu and Ag in PMDA-ODA exhibit a convex curvature and a discontinuity at the glass transition. This points to free-volume controlled diffusion. For reactive metals thermally activated bond breaking appears to be dominant. They form relatively uniform interfaces, whereas noble metals tend to cluster at the surface and, at low evaporation rates and high temperatures also inside the polymer.

1. INTRODUCTION

While metals and polymers have been the subject of extensive research throughout the last decades our present knowledge of the interfaces between both materials is still rather limited. On the other hand, metal-polymer layered structures are nowadays widely used in microelectronics and many other fields. In microelectronics mainly polyimides are employed, e.g. for advanced packaging and on the chip level (Tummala and Rymaszewski 1989). Utilizing their unique combination of properties, such as thermal stability, chemical resistance, ease of planarization, and low dielectric constant, it was possible to increase the density and transmission speed of packaging structures substantially. Heat treatment during processing of these structures may result in crack formation and delamination induced by thermal stress (Chen et al. 1988). Therefore, adhesion between metal and polymer is a major concern (Ho and Faupel 1988). Adhesion crucially depends on the extent of chemical interaction of the metal with the polymer, which also determines the interfacial morphology. In this process the diffusion behaviour of the metal appears to play an important role (Ho et al. 1991). Moreover, diffusion of the metal, typically copper or a Cu-Al alloy, into the bulk of the polymer can create serious problems in the dielectric and insulating characteristics of the latter, thus limiting the utility of directly deposited Cu.

While adhesion promoters such as chromium also act as barriers for Cu diffusion, it is desirable to keep the brittle adhesion layer as thin as possible for reasons of mechanical stability (Faupel et al. 1989, Faupel 1990).

The present paper reviews our degree of understanding of diffusion and its relation to interface formation and, furthermore, aims to give perspectives, of how open questions could be addressed. Until now investigations of polymer-metal interfaces were mostly restricted to polyimides, noble metals, aluminium, and adhesion promoters like chromium, because of their importance to microelectronics. Marked differences in interface formation were observed depending on whether the metal was deposited onto a fully cured polyimide or a polyamic acid precursor was spin coated (Kowalczyk et al. 1988) or vapour deposited (Grunze et al. 1988) onto the metal and cured subsequently to form the polyimide. At interfaces formed by curing the polyamic acid on the metal surface chemical reactions can occur between the metal and the polyamic acid. In spin coating the polar solvent may plasticize the matrix and promote dispersion of compound particles (mostly oxides) into the polyimide. This process involves mechanisms completely different from metal atom diffusion into fully cured polymers. We shall therefore confine ourselves to the latter case.

2. DIFFUSION

Valuable information on the diffusion behaviour of various metals in polyimides, in particular regarding the correlation with the reactivity of the metal have been obtained from ultraviolet photoemission spectroscopy (UPS) and x-ray photoemission spectroscopy (XPS) studies (Hahn et al. 1983, Chou and Tang 1984, Bartha et al. 1985, White et al. 1987, Ohuchi and Freilich 1988, Ho et al. 1991). Although the details of the underlaying reaction mechanisms are still controversial (for recent reviews see Ho et al. 1991 and Kowalczyk 1990) the following qualitative conclusions can be drawn: Cr and Ti interact strongly with the polymer, whereas the reactivity of Ag and Cu is relatively weak and that of Al and Ni is somewhere in between. For Cu the core level intensity turned out to decrease substantially upon heat treatment, indicating that copper diffused into the polymer. The signals of reactive metals like Cr and Ti proved to be thermally stable. Furthermore, a thin buffer layer of these materials blocks permeation of Cu into polyimide.

Recently, diffusion of a number of metals in polymers has been measured by backscattering techniques. Tromp et al. (1985) investigated interdiffusion at the copper-polyimide interface employing high resolution medium energy ion scattering (MEIS). In this novel technique the energy scale of the backscattered ions is also a depth scale because of the inelastic energy loss of the particles on both their ingoing and outgoing path through the sample. In comparison to conventional Rutherford backscattering spectrometry (RBS) depth resolution is largely improved by use of medium ion energies (50 - 200 keV vs. 1 - 2 MeV for RBS) and an electrostatic analyzer. MEIS spectra of Cu evaporated on to pyromellitic dianhydride-oxydianiline (PMDA-ODA) exhibited a very large broadening at high deposition temperatures, pointing to strong copper diffusion into the polyimide. However, the depth distributions observed did not resemble simple diffusion profiles. In particular, the occurrence of a maximum concentration below the surface indicated the existence of Cu clusters buried in the polymer (Tromp. et al. 1985). The formation of copper and silver clusters in the bulk of polyimides, and thus the ability of these metals to diffuse into polymers has also been confirmed by means of cross sectional transmission electron microscopy. These experiments will be discussed in section 3.

Shanker and MacDonald (1987) utilized Rutherford backscattering to monitor diffusion of Cu and Ag into polyimide at elevated temperatures. No mobility could be detected for Ni and Mo up to 375 °C. Das and Morris (1989) also found indications of cluster formation of ion-implanted Cu in Kapton (PMDA-ODA) films by means of RBS. Based on the same technique Cu diffusion was demonstrated in parylene, too (Yang et al. 1991). Paik and Ruoff (1989) reported substantial diffusion of Cr into polyimide. A diffusion constant as high as 2×10^{-19} m^2/s at 200 °C was found by these authors from RBS. However, this value exceeds the diffusivity of Cu as determined in the direct diffusion measurements discussed below and is also in sharp contrast to the aforementioned results. In polymers due to the low energy loss in these materials, the depth resolution of RBS is limited to some ten nm. Therefore, results from this indirect technique should be treated with caution, at least as far as measurement of small diffusion constants of the order of 10^{-19} m^2/s in polymers are concerned.

There has long been a lack of direct measurements of metal diffusion in polymers owing to the experimental difficulties concomitant with the low mobility of the metals. Even diffusion constants of Cu, which is relatively mobile, are many orders of magnitude smaller than those of common gases, water, and typical organic solvents (Frisch and Stern, 1983). Lately, a radiotracer technique in conjunction with ion beam sputtering for depth profiling has proven to provide the required depth resolution and sensitivity at least for the faster diffusing noble metals (Faupel et al. 1989). In this technique a thin layer of radioactive tracer is deposited onto the sample surface and is allowed to diffuse into the sample during annealing. The tracer penetration profile is obtained by means of serial sectioning and counting the activity of each section. In case of Fickian diffusion the thin film solution of Fick's second law:

$$c(x,t) = \frac{S_o}{\sqrt{\pi Dt}} \exp\left(-\frac{x^2}{4Dt}\right) \tag{1}$$

holds for sufficiently thin tracer layers. Then the diffusion constant D of the tracer can be determined from a the slope $-1/4Dt$ of a semilogarithmic plot of the activity, which is directly proportional to the concentration c, versus the penetration depth x squared. Here t is the annealing time and S_o is the original coverage at $x = t = 0$. It should be pointed out that the determination of diffusivities based on penetration profiles is much more direct than the evaluation of permeation experiments, which are often employed for the measurement of diffusion coefficients of gases and organic solvents in polymers.

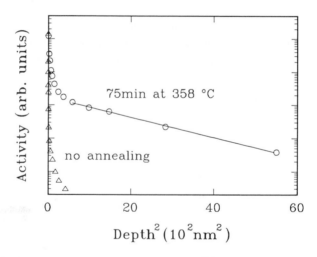

Fig. 1. Semilogarithmic penetration profile of [110m]Ag into PMDA-ODA. The tracer was deposited at room temperature at about 4 ML/min. A profile of a sample which was not annealed is displayed for comparison.

For depth profiling the surface of the rotating sample was eroded by means of an intense beam of low-energy argon ions (Spit et al. 1989, Faupel et al. 1992). The sputtered-off material was collected on a foil which was advanced quickly after each section like a film in a camera without interrupting the sputtering process. Neutralization of the ions substantially enhanced the sputtering rate. The rate was determined from the total sputter time and the height of a surface step, which was created by covering a small area of the sample during sputtering. The step height was measured with a step scanner or an ellipsometer with a typical accuracy of ± 10 nm. Since the square of the step height enters into the evaluation of D this can be the dominant source of error. For very short profiles the uncertainty may exceed 30 %.

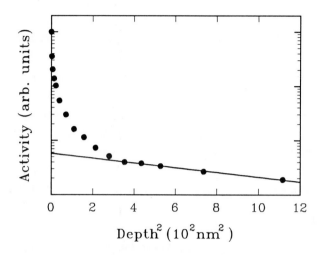

Fig. 2. Penetration profile of ^{67}Cu into PMDA-ODA after annealing at 252 °C for 69 min. The tracer was flash evaporated at room temperature.

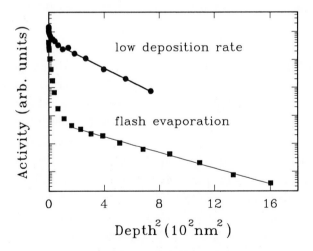

Fig. 3. Effect of deposition rate on penetration profiles of ^{110m}Ag into PMDA-ODA. The tracers were deposited continuously at a rate of about 0.1 ML/min during annealing at 588 K and flash evaporated at the annealing temperature of 601 K, respectively.

Meanwhile diffusion of ^{67}Cu and ^{110m}Ag in spin coated polyimide films of pyromellitic dianhydride-oxydianiline (PMDA-ODA) (Faupel et al. 1989, 1990) and of ^{67}Cu in biphenyl dianhydride-phenylene diamine (BPDA-PDA) (Vieregge and Gupta 1991) has been studied over an extended temperature range. In these experiments typically 0.25 - 10 monolayers (ML) of the tracer were evaporated at room temperature or at the annealing temperature at different rates. A typical penetration profile of silver, which was deposited at a medium rate onto PMDA-ODA at room temperature prior to annealing is shown in Fig. 1, together with a profile of an unannealed sample. The latter essentially represents the resolution function of the presents IBS technique. Since it is well known that ion bombardment may cause substantial broadening of diffusion profiles, and polymers were expected to be particularly prone to sputtering artifacts, the effect of the sputtering process on the penetration profiles was investigated carefully (Faupel et al. 1992, Willecke and Faupel 1992). In order to minimize broadening due to ion bombardment the use of ion energies far below 1 keV turned out to be essential. Typical energies were in the range of 100 - 200 eV. (About one order of magnitude higher energies are generally employed in secondary ion mass spectrometry, SIMS). The resolution function and the penetration profile shown in Fig. 1 were obtained at 180 eV. One notes that the width of the resolution function is very small in comparison to the diffusional broadening. It can be neglected in the evaluation of diffusion constants from the slope of the linear range (see below). This view was further corroborated by the fact that variation of the diffusion time at constant temperature did not affect the resulting diffusivities (Willecke and Faupel 1992).

The effect of annealing in Fig. 1 gives direct evidence of silver diffusion into the polymer. However, most of the activity is detected very close to the surface and only a small fraction of Ag diffused into the bulk by a uniform mechanism, as reflected in the linear range of the profile. This behaviour was also found for flash evaporated copper (Fig. 2.). Apparently, only a few atoms are highly mobile, while the mobility of the others is strongly reduced. For silver in PMDA-ODA the shape of the diffusion profiles was investigated in detail (Willecke and Faupel 1992). While in first measurements (Faupel et al. 1990), which involved only a rather limited range of relatively high deposition rates, no significant effect could be resolved, the ratio of mobile to less mobile atoms turned out to increase markedly at very low deposition rates and high temperatures. A penetration profile where the tracer was evaporated continuously and at a very low rate during annealing is depicted in Fig. 3. Here the linear range of the highly mobile atoms dominates. (For an accurate evaluation equation (1) has to be replaced by the continuous source solution in this case). In contrast, flash evaporation of silver at an even higher temperature results in pronounced immobilization of almost all silver atoms. (Note the logarithmic activity scale in Fig. 3.).

The shape of the penetration profiles and the effect of the deposition rate is consistent with the aforementioned tendency of noble metals to form clusters. Within this framework only a few isolated atoms are able to diffuse into the bulk at high deposition rates and low temperatures, before being trapped and thus being at least partly immobilized by neighbouring metal atoms. The number of atoms which escape into the bulk increases with decreasing evaporation rate and increasing temperature. In order to corroborate this picture experiments will be performed where a continuous inactive film of silver is deposited onto the polymer prior to the deposition of the tracer. Under these conditions no diffusion of the tracer into the polymer should occur.

Diffusivities, which were attributed to diffusion of isolated atoms, were determined from the linear ranges of the penetration profiles. Obviously, most accurate diffusion constants can be obtained if the tracer is evaporated at very low rates. The results for the diffusion of Cu and Ag in PMDA-ODA are summarized in an Arrhenius plot (ln D versus 1/T) in Fig. 4. One sees that the much smaller Cu atoms diffuse considerable faster than the Ag atoms. The Arrhenius plots for both penetrants exhibit a pronounced nonlinearity.

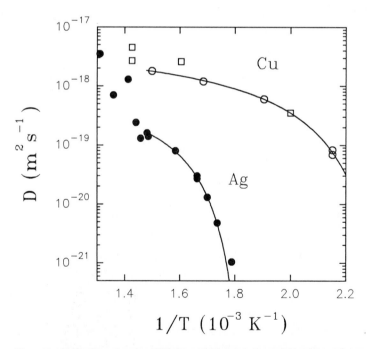

Fig. 4. Arrhenius plot for diffusion of isolated Cu and Ag atoms in the polyimide PMDA-ODA. ln D is displayed as function of the inverse absolute temperature. Solid lines represent fits to the data based on equation (5). Open squares are from Vieregge and Gupte (1991).

The interpretation of the diffusion behaviour requires information on the molecular structure of the polymer. PMDA-ODA is an aromatic polyether with an extremely rigid backbone. The chain can freely rotate about the phenyl ether linkage (Fig 5). The length of the repeat unit and its molecular weight are 1.8 nm and 382, respectively. The chain molecular weight is of the order of 1×10^5. The molecular weights and diameters of the metal atoms are 108 and 0.288 nm for silver and 64 and 0.256 nm for copper. Diffraction profiles of PMDA-ODA have yielded an average molecular chain separation distance of about 0.5 nm and an intramolecular periodicity of 1.55 nm, reflecting the projection of the 1.8 monomer unit onto the chain axis. Based on diffraction results and on scattering measurements it has been proposed that PMDA-ODA forms ordered aggregates. In these aggregates of 2 - 2.5 nm thickness - which cannot be considered crystalline owing to the absence of well-defined intermolecular reflections - the local alignment of chains is ordered in a smectic fashion along the chain axis for several monomeric units. A coherence length of some 5 - 6 nm was determined (Isoda et al. 1981, Russell et al. 1983, Takahashi et al. 1984, Russell and Brown 1986). Consequently, the morphology is diffuse rather than being made up of two well defined crystalline and amorphous phases. Its development depends on the initial imidization temperature. In addition to the segmental ordering of PMDA-ODA a molecular orientation parallel to the surface of the film was observed, which is most pronounced for thin films in the μm range (Russell 1984). Unoriented films are obtained by removing the film from the substrate prior to imidization.

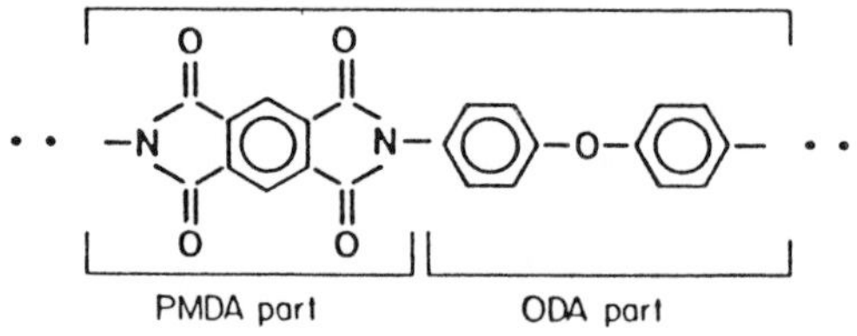

Fig. 5. Molecular structures of the monomeric unit of PMDA-ODA.

In real semicrystalline polymers the diffusion and solubility of penetrants in the crystalline phase can usually be neglected. Here these quantities should at least be substantially reduced in the more ordered regions. Presumably, the ordered aggregates serve as tortuous impediments to diffusion and their quantitative effect should dependent on the degree of ordering. Hence, it is very important for the discussion of diffusion, whether the structure of the polymer changes upon annealing. This was investigated by Takahashi et al. (1984) in the temperature interval 250 - 400 °C. In oriented polyimide films the molecular order improved somewhat by annealing at temperatures above 350 °C, but no major changes were observed. Unoriented films, however, exhibited some crystalline features at 400 °C. Above 400 °C more pronounced ordering should occur. For diffusion samples which were not equilibrated at the highest temperature involved this could lead to structural changes during the diffusion process and thus to diffusivities decreasing with time. In this light no major time dependence of the diffusivities is to be expected up to the maximum curing temperature of 400 °C in PMDA-ODA. In general, measurements of the diffusivity as function of time could be a sensitive tool for detecting structural changes in the polymer during heat treatment. Detailed investigations of relaxation effects are in progress (Willecke and Faupel 1992).

A glass transition, although not known precisely, occurs above 400 °C in spin coated PMDA-ODA (Russell et al. 1983). The glass transition temperature T_g depends on the preparation conditions of the polyimide. Lower values down to 360 °C have also been reported (Du Pont technical information: "Kapton, Summary of properties"). No measurable effect at T_g is seen in differential scanning calorimetry (Yang et al. 1985). Apparently, the glass transition is not very pronounced. This is also reflected in the thermal expansion coefficient. It changes gradually from room temperature to 400 °C rather than being constant below T_g in the vicinity of the glass transition and exhibiting an almost step like change at T_g to a much higher value.

Another structural feature that might effect diffusion and the formation of metal clusters inside the polymer is the presence of voids arising from water and solvent driven off during imidization. Russell (1984a) performing small angle X-ray scattering observed void sizes ranging from 5 - 15 nm with a volume fraction of about 7×10^{-4}. Two distinct averages below 5 nm and above 10 nm were found. The technique was not sensitive to voids < 2 nm. Since PMDA-ODA was spin coated and imidized on silicon wafers tensile stress arising from the large mismatch in thermal expansion was induced into the films. However, due to the small Youngs modulus of polymers, the stress level is relatively low and decreases as the temperature approaches the maximum curing temperature of 400 °C (Chen et al. 1988). Therefore, no significant influence on diffusion should be expected.

Until now, no models are available for diffusion of metal atoms in polymers. However, many theories have been developed describing diffusion of small penetrants such as gas, liquid, and organic solvent molecules (for a review see e.g. Frisch and Stern 1983). These theories can be classified as either molecular or free-volume models. Molecular models involve specified motions of the penetrant molecule and of the polymer chains relative to each other and take into account the intermolecular forces. It is assumed that the penetrant molecules are localized in holes of suitable size. Occasionally they acquire sufficient thermal energy, from a collision with a surrounding polymer segment, to jump into a neighbouring hole large enough to accommodate these molecules. Consequently, the diffusion coefficient D exhibits an Arrhenius-type of behaviour. The most comprehensive theory of this type appears to have been proposed by Pace and Datyner (1980). Although in molecular theories the activation energy can be temperature dependent they are largely based on experimental results of diffusion of small gas molecules, which appears to be decoupled almost completely from the polymer dynamics. Arrhenius plots are linear over the whole temperature range and, in particular, through the glass transition. Diffusion of larger molecules, on the other hand, is often characterized by a curved Arrhenius plot, as seen in Fig. 4 for Cu and Ag, and by a discontinuity near the T_g. Such a behaviour can be explained quite successfully in terms of free-volume models, despite the fact they are based on an over-simplified view of the molecular process. We note that though molecular theories relate some of their parameters to detailed molecular properties of the polymer-penetrant system they still contain disposable parameters.

In free-volume models, which were originally developed for simple liquids, the specific free volume $\hat{V}_F$ is given by the difference of the total specific volume $\hat{V}$ and the specific occupied free volume $\hat{V}_o$, which is usually assumed to be the total free volume at 0 K. As the temperature is raised the material expands homogeneously due to the increasing amplitude of the anharmonic vibrations with temperature. In addition, holes are formed, which are distributed discontinuously throughout the material. The first type of volume is called interstitial free volume $\hat{V}_{FI}$. It is distributed uniformly among the molecules. $\hat{V}_F$ can then be expressed as:

$$\hat{V}_F = \hat{V}_{FI} + \hat{V}_{FH} \tag{2}$$

With respect to the hole free volume $\hat{V}_{FH}$ it is assumed that it can be redistributed with no increase in energy. Hence it is available for molecular transport. Vrentas and Duda (1978) developed a free-volume theory for diffusion in the limit of zero penetrant concentration. Following Cohen and Turnbull (1959) these authors used the equation

$$D = D_o \exp\left(-\frac{\gamma \bar{V}^*}{\bar{V}_{FH}}\right) \exp\left(-\frac{E^*}{kT}\right) \tag{3}$$

for the diffusivity of a one-component simple liquid. Here $\bar{V}^*$ is the critical local hole volume required to jump into a new position. The bar indicates that the volumes are now taken per molecule. γ is an overlap factor (typically between 0.5 and 1) which takes into account that the same free volume is available to more than one molecule. E^* is the critical energy a molecule must obtain to overcome the attractive forces holding it to its neighbour. The preexponential factor D_0 is a much weaker function of temperature than the exponential terms. In a liquid the hole free volume is very small generally for $T_g < T < T_g + 100\ °C$ and the diffusion process is free-volume controlled. Hence, it is possible to absorb the E^* term into the preexponential factor. For a binary system Vrentas and Duda further assumed that the average hole free volume per molecule is given by the total hole free volume of the system divided by the number of solute molecules plus the number of polymer jumping units. The jumping unit of a polymer is that small portion of a polymer chain, which can jump from one position to another if sufficient hole free volume is available. The penetrant tracer diffusion coefficient D_1 is then given by

$$D = D_1 = D_{01} \exp\left(-\frac{\gamma \xi \hat{V}_2^*}{\hat{V}_{FH2}}\right) \qquad (4)$$

$\xi = \hat{V}_1^* M_1 / \hat{V}_2^* M_j$ is the coupling parameter between polymer and penetrant diffusion and M_j the molecular weight of the jumping unit. (Note that the $\hat{V}_i^*$ are now taken per gram of material). Full coupling corresponds to $\xi = 1$. Within this framework $\hat{V}_1^*$ should be a purely geometrical constant and should not depend on the polymer. Equation (5) can be written in the form of a Williams-Landel-Ferry equation (Ferry, 1980)

$$\ln\left(\frac{D(T)}{D(T_g)}\right) = \frac{A\xi\ (T - T_g)}{B + T - T_g} \qquad (5)$$

Vrentas and Duda extended their model to glassy polymers. They assumed that the material is in an equilibrium structure above T_g and in a non-equilibrium structure, which remains invariant during the diffusion process, below the glass transition temperature. The authors further introduced the assumption that the thermal expansion coefficient can be adequately approximated by an average value above and below T_g in the temperature range of the diffusion studies. The rapid change in the expansion coefficient, generally observed in the vicinity of the glass transition, was idealized as a step change at T_g. Within this framework equation (5) still holds for the glassy state if B/λ is substituted for B. The parameter λ $(0 \leq \lambda \leq 1)$ describes the character of the change in volume contraction which can be attributed to the glass transition. If the equilibrium liquid structure is also realized below T_g then $\lambda = 1$. If the structure is frozen in at T_g one has λ

= 0 and the hole free volume below the glass transition is given by the hole free volume at T_g. It is evident from equation (5) that a plot of ln D versus 1/T will not be linear above and below T_g if $\lambda > 0$. The nonlinearity becomes more pronounced (and thus easier to detect experimentally) as the size of the solvent increases. Moreover, a step change in the Arrhenius plot at T_g is predicted for $\lambda < 1$. The ratio of the apparent activation energies of diffusion $E_D \equiv - k \, \partial \ln D / \partial (1/T)$ below and above T_g in the vicinity of T_g is equal to λ:

$$E_D(T_g^-)/E_D(T_g^+) = \lambda \qquad (6)$$

Equation (6) allows to determine λ. Very recently, Ehlich and Sillescu (1990), in a systematic study involving several polymers and solvents of different sizes, demonstrated that solute diffusion in polymers can be well described within the free-volume approach above and below the glass transition temperature. However, $\hat{V}_1^*$ proved to be matrix dependent, indicating that solute-matrix coupling cannot be understood in a purely geometrical sense.

The salient features of the present results for Cu and Ag diffusion (Fig. 4), in particular the size effect, the convex curvature, and the discontinuity near the glass transition temperature, are predicted by the free-volume theory of Vrentas and Duda. For Cu the break in the Arrhenius plot at T_g cannot be established definitely due to the scattering of the data in the interesting temperature range. However, exclusion of only one of eight data points largely reduces the scattering and results in the expected discontinuity. Certainly, it is desirable to measure additional diffusion constants near T_g for confirmation.

The solid lines in Fig. 4 represent fits to the data in the glassy range based on equation (5). One notes that the experimental points are well described. Resulting fit parameters are of the right order of magnitude, too. Apparently, the considerable temperature dependence of the thermal expansion coefficient can still be represented by an average value in the limited temperature interval of the measurements. The E^* term (equation (3)) could also play a role since the chemical interaction between noble metals and the polyimide, though weak in comparison with metals like Cr or Ti, might not be negligible. Incorporation of the E^* term did not improve the fit significantly. While more quantitative conclusions will be postponed until further data are available, one can state that diffusion of Cu and Ag seems to be controlled largely by the availability of free volume.

In contrast, for metals like chromium and titanium, which are bound to the polymer chains much stronger than noble metals, we expect diffusion to be governed by thermally activated bond breaking. This should result in a very low mobility in accordance with the diffusion barrier properties of Cr and Ti their thermal stability as observed in the spectroscopic studies alluded to above. It would be interesting to measure the diffusivity of such metals and to find out how much their penetration profile extends into the bulk of the polymer. Cross-sectional transmission electron microscopy does not provide sufficient sensitivity to address this question. First attempts to improve the depth resolution of the present technique by further reduction of the ion energy and deconvolution of the resolution function appear to be encouraging (Willecke and Faupel 1992). The experiments, which have to be carried out under UHV conditions since the reactive metals are prone to oxidation, should at least yield an upper limit of the diffusion coefficients.

The aforementioned measurements of Vieregge and Gupta (1991) have shown that Cu diffusion in BPDA-PDA proceeds substantially slower than in PMDA-ODA. The chemical reactivity of both polymers is expected to be much the same. However, BPDA-PDA forms a more compact polymeric structure (Best et al. 1991). Apparently, the diffusion of weakly interacting metals, such as Cu and Ag can serve as an effective probe to examine the molecular morphology. Particularly, the pronounced discontinuity in the diffusion coefficient of Ag at the glass transition, which does not show up in DSC, demonstrates that Ag as a spherical particle with a large mass is very sensitive to changes in polymer properties.

3. INTERFACE FORMATION

The importance of metal diffusion to the formation of metal-polymer interfaces has already been pointed out. Most information concerning the interfacial microstructure and its dependence on the type of metal and the preparation conditions has been obtained from cross-sectional transmission electron microscopy (TEM). The first systematic study of this kind was carried out by LeGoues et al. (1988) in the Cu/PMDA-ODA system. After evaporation of Cu at different rates and temperatures samples were thinned to electron transparency in a direction perpendicular to the interface. Cu was seen to form nearly spherical agglomerates when evaporated at very low rates. The size of these particles and their average distance from the interface markedly increased with increasing deposition temperature. At 300 °C cluster sizes in the 10 nm range were observed. However, clusters in the bulk of the polymer were only detected at very low rates of the order of one monolayer per min. At higher rates and low deposition temperatures Cu forms a uniform film at the surface. Subsequent annealing results in island formation only on top of the polyimide. At high evaporation rates even high-temperature deposition does not produce Cu clusters inside the polymer film.

In the light of our investigation of Ag diffusion in PMDA-ODA, specifically with respect to the dependence of the penetration profiles on the deposition conditions, (Faupel 1990a, Willecke and Faupel 1992) we expected silver to form clusters inside the polymer, too. Since the mobility of Ag is substantially lower than that of Cu (Fig. 3) Ag aggregation was studied at relatively high temperatures. Samples were either thinned to electron transparency or electron transparent sections were cut perpendicular to the interface by means of a microtome. Formation of spherical Ag particles was indeed observed in the bulk of PMDA-ODA during evaporation at low rates (Foitzik and Faupel 1991). An example is depicted in Fig. 6. Like in case of Cu, Ag deposition at room temperature resulted in a continuous film (Fig. 7) and subsequent annealing did not lead to cluster formation inside the polymer. Apparently, for both Ag and Cu sizable diffusion may take place at the metal-polymer interface, resulting in a spread-out nonuniform structure. On the other hand, Cr, which exhibits a very low mobility, always seem to form sharp interfaces. The behaviour of Al and Ni appears to be somewhere in between (LeGoues et al. 1988). A strong correlation between the interfacial microstructure and the diffusivity of the metal, which in turn is largely determined by its chemical reactivity, is obvious.

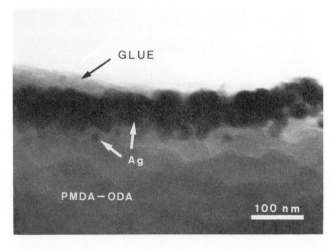

Fig. 6. Cross-sectional TEM micrograph showing the morphology of the silver-poly-imide interface formed after metal evaporation at 360 °C and a deposition rate of about eight monolayers per min. The sample was held at the deposition temperature for 40 min. Small Ag clusters are resolved in the bulk of the polymer near the interface (left).

The formation of big metal globules deep in the bulk of a glassy polymer, as observed for Cu and Ag in PMDA-ODA is quite remarkable. It is well known that metals with low reactivity tend to cluster in islands on polymer surfaces due to their high cohesive energy (Wetzel et al. 1985). However, until now, subsurface particle formation has only been reported for polymers above the softening temperature, where metal aggregates sank completely into polymer substrate (Kovacs and Vincett 1982). The driving force for this process is again related to the high cohesive energy of the metal, which gives rise to a surface energy σ_M exceeding that of the polymer σ_P by one to two orders of magnitude. If, in addition to the surface energy of the metal is lager than the sum of the interfacial energy σ_{MP} and the polymer surface energy

$$\sigma_M > \sigma_P + \sigma_{MP} \tag{7}$$

then there is a driving force for a metal cluster to be imbedded completely in the bulk of the polymer. Below T_g the strain energy for the cluster in the polymer matrix has to be taken into account, too. The importance of surface energy considerations is also suggested by the fact that pinning of copper particles at the silicon substrate was observed (LeGoues et al. 1988). The aforementioned voids with sizes comparable to those of the metal clusters in the bulk could also play a role, e.g. in reducing the strain energy for cluster formation. Void filling would additionally eliminate surface energy of the voids.

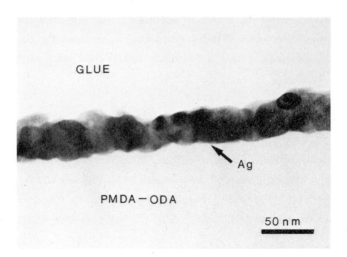

Fig. 7. Cross-sectional TEM micrograph showing the morphology of the silver-poly-imide interface formed after metal evaporation at room temperature. The deposition rate was much the same as in Fig. 6. No clusters can be seen inside the polymer.

The kinetics of how relatively big particles can form in a glassy polymer are not yet fully understood. Perrin et al. (1985) observed motion of gold particles at $T > T_g$ in gold-fluorocarbon-polymer composite films. The onset of mobility above T_g points to the important role of polymer self-diffusion in this process. Recently, diffusion of copper particles has been reported even through the bulk of non-porous amorphous alumina substrates in certain gas atmospheres at 200 °C (Gai et al. 1990). On the other hand, formation of big clusters inside the polymer only occurs at very low deposition rates and elevated temperatures where a large portion of the metal atoms is able to diffuse into the bulk, as shown by the radiotracer studies (Fig. 3). This suggests that the metal particles inside the polymer do not form by motion of surface clusters but rather by diffusion and agglomeration of single atoms or small associates.

Recently, Monte Carlo calculations were carried out to elucidate the effect of metal agglomeration on the diffusion (Silverman 1991) and cluster formation (LeGoues et al. 1988). In these calculations the shape of the diffusion profiles and the cluster size and distribution could be reproduced quantitatively based on metal and polymer diffusion.

In order to get deeper inside into the underlying mechanisms of metal particle formation inside glassy polymers interfaces which were prepared at different deposition and annealing conditions are being studied by means of cross-sectional transmission electron microscopy and scanning electron microscopy. We expect to clarify the contribution of metal diffusion by comparing the maximum depth of cluster formation for Cu and Ag under identical annealing conditions. If polymer diffusion dominates this depth should be more or less equal. Otherwise a marked difference reflecting the much higher diffusivity of copper should result.

ACKNOWLEDGMENTS

The author would like to thank many co-workers, in particular P. S. Ho and D. Gupta from the IBM Th. J. Watson Research Center as well as R. Willecke and A. Foitzik from University of Göttingen. The author has also benefited from discussions with Th. Hehenkamp from University of Göttingen. Part of this work has been supported by VW-Stiftung.

REFERENCES

Bartha J W, Hahn P O, LeGoues F K and Ho P S 1985 J. Vac. Sci. Technol. A3 1390
Best M E, Moylan C R and Ree M 1991 J. Polym. Sci. Polym. Phys. 29 87
Chen S T, Yang C H, Faupel F and Ho P S 1988 J. Appl. Phys. 64 6695
Chou N J and Tang C H 1984 J. Vac. Sci. Technol. A2 751

Das J H and Morris J E 1989 J. Appl.Phys. 66 5816

Ehlich D and Sillescu H 1990 Macromolecules 23 1600

Faupel F 1990 Haftung bei Verbundwerkstoffen und Werkstoffverbunden ed W Brock-
mann (Oberursel: Deutsche Gesellschaft f. Materialkunde) p. 53

Faupel F 1990a Advanced Materials 2 266

Faupel F, Gupta D, Silverman B D and Ho P S 1989 Appl. Phys. Lett. 55 357

Faupel F, Gupta D, Silverman B D and Ho P S 1990 Advanced Materials and Processes
eds H E Exner and V Schumacher (Oberursel: Deutsche Gesellschaft f. Materialkunde)
p 887

Faupel F, Hüppe P W, Rätzke K, Willecke R and Hehenkamp Th 1992 submitted to J.
Vac. Sci. Technol. A

Faupel F, Yang C H, Chen S T and Ho P S 1989 J. Appl. Phys. 65 1911

Ferry J D 1980 Viscoelastic Properties of Polymers, 3rd ed (New York:Wilely)

Foitzik A and Faupel F 1991 Mater. Res. Soc. Symp. Proc. 203 59

Frisch H L and Stern S A 1983 CRC Rev. Solid State and Mater. Sci 11 123

Fujita H 1961 Fortschr. Hochpolym. Forsch. 3 1

Gai P L, Smith B C and Owen G 1990 Nature 348 430

Grunze M, Unertel W N, Gnanarajan S and French J 1988 Mater. Res. Soc. Symp, Proc.
108 189

Hahn P O, Rubloff G W and Ho P S 1983 J. Vac. Sci. Technol. A2 756

Ho P S and Faupel F 1988 Appl. Phys. Lett 53 1602

Ho P S, Haight R, White R C, Silverman B D and Faupel F 1991 Fundamentals of
Adhesion ed L. H. Lee (New York: Plenum Publishing Corporation) p 383

Isoda S, Shimada H, Kochi M and Kambe H 1981 J. Polym. Sci. Polym. Phys. 19 1293

Kowalczyk S P 1990 in Metallization of Polymers ed E Sacher, J Pireaux and S P
Kowalczyk ACS Symposium Series 440 (Washington DC: American Chem. Soc.) p 10

Kowalczyk S P, Kim Y H, Walker G F and Kim J 1988 Appl. Phys. Lett. 52 375

Kovacs G J and Vincett P S 1982 J. Colloid. Interface. Sci. 90 335

LeGoues F K, Silverman B D and Ho P S 1988 J. Vac. Sci. Technol. A6 2200

Ohuchi F S and Freilich S C 1988 J. Vac. Sci. Technol. A6 1004

Pace R J and Datyner A 1980 J. Polym. Sci. Polym. Eng. Sci. 20 51

Pace R J and Datyner A 1980a J. Polym. Sci. Polym. Phys. 18 1103

Paik KW and Ruoff A L 1989 Mat. Res. Soc. Symp. Proc. 153 143

Paul C W 1983 J. Polym. Sci. Polym. Phys. 21 425

Pawlowski W P, Jacobson M I, Teixeira M E and Sakorafos K G 1990 Proc. MRS Fall
Meeting '89 Boston Massachusetts

Perrin J, Despax B and Kay E 1985 Phys. Rev. B. 32 719

Russell T P 1984 J. Polym. Sci. Polym. Phys. 22 1105   1984a Polym. Engin. Sci. 24
345

Russell T P and Brown H R 1986 IBM Research Report RJ 5032

Russell TP, Gugger H and Swalen J D 1983 J. Polym. Sci. Polym. Phys. 21 1745

Shanker K and MacDonald J R 1987 J. Vac. Sci. Technol. A5 2894

Silverman B D 1991 Macromolecules 24 2467

Spit F H M, Gupta D and Tu K N 1989 Phys. Rev. B39 1255

Takahashi N, Yoon D Y and Parrish W 1984 Macromol. 17 2583

Tromp R M, LeGoues F K and Ho P S 1987 J. Vac. Sci. Technol. A3 782

Tummala R R and Rymaszewski E J 1989 Microelectronics Packaging Handbook (New

York: Van Nostrand Reinhold)

Vieregge K and Gupta D 1991 private communication

Vrentas J S and Duda J L 1978 J. Appl. Polym. Sci. 22 2325

Vrentas J S, Duda J L and Ling H-C 1985 J. Polym. Sci. Polym. Phys. 23 275

Wetzel J T, Smith D A and Appleby-Moughham G 1985 Mat. Res. Soc. Symp. Proc. 40 271

Willecke R. and Faupel F 1992 to be published

White R C, Haight R, Silverman, B D and Ho P S 1987 Appl. Phys. Lett. 51 48

Yang D K, Koros W J Hopfenberg H. B and Stannett V T 1985 J. Appl.Polym. Sci. 30 1035

Yang G, Dabral S, You L, Bakhru H, McDonald J F and Lu T M 1991 Mat. Res. Soc. Symp. Proc. 203

Theoretical investigations of the aluminum/polythiophene interface

S. Stafström and M. Boman
Department of Physics, IFM, Linköping University, S-581 83 Linköping, Sweden

R. Lazzaroni and J.-L Brédas
Service de Chimie des Matériaux Nouveaux, Université de Mons, Place du Parc, 20
B-7000 Mons, Belgium

ABSTRACT: The interactions at the interface between aluminum and polythiophene are studied theoretically. Aluminum is found to interact strongly with the polythiophene chain. The fundamental interaction unit corresponds to an aluminum dimer (Al_2) bound to a single thiophene ring. This type of interaction modifies strongly the geometrical structure of the polymer chain and induces localization of the π-electron system. The calculated charge transfer between aluminum and the thiophene system is found to be in good agreement with the charge transfer deduced from experimental core level spectra of the aluminum/polythiophene interface. Higher concentrations of aluminum atoms on a polythiophene chain results in the formation of additional Al_2-thiophene complexes. Within the first monolayer, clustering of aluminum on polythiophene is shown to be energetically unfavorable.

1. INTRODUCTION

During the past fifteen years, a lot of interest has focused on unsaturated π-conjugated polymers. Primarily, this is due to the ability to dope these polymers to a very high electrical conductivity. The quasi one-dimensional nature of the conjugated polymers has also been shown to give rise to other interesting features, for instance: nonlinear excitations, nonlinear optical properties, and semiconducting properties. The use of semiconducting conjugated polymers as an electroactive material in microelectronic devices is a new and rapidly growing research field. Burroughes *et al.*, 1988, reported the first examples of high performance Schottky diodes, MIS (metal-insulator-semiconductor) diodes, and MISFET (MIS-field-effect-transistors) structures involving conjugated polymers. The all-organic, high-mobility transistor invented by Garnier *et al.* (1990) is an excellent example of how novel organic materials can be exploited to produce components with almost the same characteristics as those made out of conventional, inorganic semiconductors but with new properties such as flexibility. Furthermore, polymer light emitting diodes (LEDs) show very promising characteristics (Burroughes *et al.*, 1990 and Braun and Heeger, 1991). LEDs fabricated with oriented semiconducting polymers represent a particularly interesting type of device, because they have the possibility of emitting polarized light. This is a unique concept which is made possible only by the existence of highly anisotropic emitters such as oriented conjugated polymers. A first step towards the achievement of polarized light emitting diodes has been successfully accomplished: polarized photo-luminescence has been demonstrated from highly oriented conjugated polymer blends (Hagler *et al.*, 1991).

In the context of electronic devices such as Schottky or MIS diodes, the interactions at the metal-polymer interface are important, not only as concerns the mechanical and thermal stability of the interface but also in relation with the electronic properties of the device. Some of the problems of prime interest for theoretical work in this field are directly related to metal/polymer interfaces. For instance, it is of great importance to investigate to which extent the delocalized π-electron system, which is responsible for the semi-conducting properties of the conjugated polymer, is affected by chemical reactions at the metal-polymer interface. This type of studies are also related to the demand for high mechanical stability at the interface, a feature which requires a significant amount of chemical bonding between the metal and the polymer.   Several different π-conjugated polymers have been studied in the context of microelectronic devices. Since processable (soluble) polymers are preferable, the most frequently studied conjugated polymers in use nowadays are *trans*-polyacetylene, poly(3-alkylthiophenes) and polyparaphenylene vinylene. In this article, we discuss polythiophene (Fig. 1), which has a π-electronic system identical to that of poly(3-alkylthiophenes) (P3AT) but has no alkyl side chains (whose role is to ensure solubility properties). For the studies reported here, these side chains do not matter (see below) and their consideration would only complicate the theoretical work.

The solubility of P3AT makes the polymer processable. This property is particularly important for making high quality electronic devices (Burroughes *et al.*, 1988). Field-effect-transistors (FET) (Paloheimo *et al.*, 1990) as well as Schottky diodes (Tomozawa *et al.*, 1987 and Gustavsson *et al.*, 1991) are fabricated from thin films of poly(3-hexylthiophene). The electronic properties of P3AT and polythiophene are, however, practically identical since they are dominated by the π-conjugated system of the thiophene backbone to which the saturated alkyl side chains do not couple. Furthermore, it has been shown from X-ray photoelectron spectroscopy (XPS) that the modifications of poly(3-octylthiophene) (P3OT) due to deposition of aluminum take place within the thiophene units (Lazzaroni *et al.*, 1991). This shows also that the octyl side chains are unimportant as far as the interaction chemistry is concerned. Since the unit cell of polythiophene contains a smaller number of atoms than P3OT, polythiophene is more tractable in the theoretical studies.

Reports of XPS core level studies of aluminum deposited on poly(3-octylthiophene) (P3OT) (Lazzaroni *et al.*, 1991), show that aluminum interacts strongly with the thiophene units. A second component in the S(2p) core level spectrum appears as the amount of Al deposited onto the P3OT surface increases. This new feature lies at approximately 1.6 eV lower binding energy than the main component, which indicates that a substantial amount of electron density is transferred to the sulfur atom of the thiophene unit. The absolute value of the new S(2p) binding energy is similar to those found in the literature for pure metal sulfides. The Al(2p) core level peak is located at 74.0 eV at low coverage, which is 1.3 eV below its position in the bulk metal. The electron-deficient Al atoms giving rise to this peak are those donating electrons to the thiophene system. The carbon content of P3OT is dominated by the saturated carbons of the octyl chains, which makes it very difficult to observe chemical shifts in the C(1s) peak due to charging of carbon atoms with in the polymer backbone.

The extended nature of the lowest unoccupied molecular orbitals of the polymer introduces the possibility of delocalization of the electronic charge donated to the polymer chain, i.e., surface doping of the polymer. However, XPS core level data show two different S(2p) peaks, one for sulfur interacting with Al and one peak corresponding to unaffected sulfur. These data indicate that the interaction between Al and P3OT is localized to, approximately, one thiophene unit. This feature of the interaction between Al and P3OT is, however, fundamentally different from what is observed in similar studies on doped polythiophene (Lazzaroni *et al.*, 1990). The doping induced charge distribution is approximately equally spread over the thiophene units, which gives rise to a single S(2p) peak.

2. METHODOLOGY

The molecular orbital (MO) calculations presented in this article are performed at the *ab initio* Hartree-Fock level with both 6-31G and 6-31G** basis sets, using the Gaussian 90 program package (see references) within a direct SCF approach. For comparison with the *ab initio* results, we also perform calculations using the semiempirical Modified Neglect of Diatomic Overlap, or MNDO, method (Dewar and Thiel, 1977). The MNDO method is also applied to systems that are too large to treat at the *ab initio* level. Complete optimizations (i.e., without any symmetry constrains) of the geometrical structure and calculations of the total energy and the electronic structure are performed with both *ab initio* and MNDO methods.

In order to investigate the interaction chemistry at the aluminum/polythiophene interface, we perform calculations on two different thiophene systems: α-trithienylene (α-3T) (see Fig. 1) and α-sexithienylene (α-6T). The α-6T oligomer is particularly interesting since thin films build up from this molecule are used as semiconducting material in an all organic, thin film transistor (Garnier *et al.*, 1990). Initially, we studied α-6T at the MNDO level and let two Al atoms interact with the thiophene system. Since an Al atom has an odd number of electrons, these two atoms form a closed shell dimer. The metal/polymer interactions are such that the dimer prefers to bind to a single thiophene unit. The same interaction unit is also studied at the *ab initio* and MNDO levels for a system consisting of two aluminum atoms and the α-3T molecule (see Fig. 1). In this way, detailed comparisons as concerns the geometry and the electronic structure can be made between the results of the MNDO and *ab initio* calculations and between MNDO results on systems of different size, i.e., α-3T and α-6T.

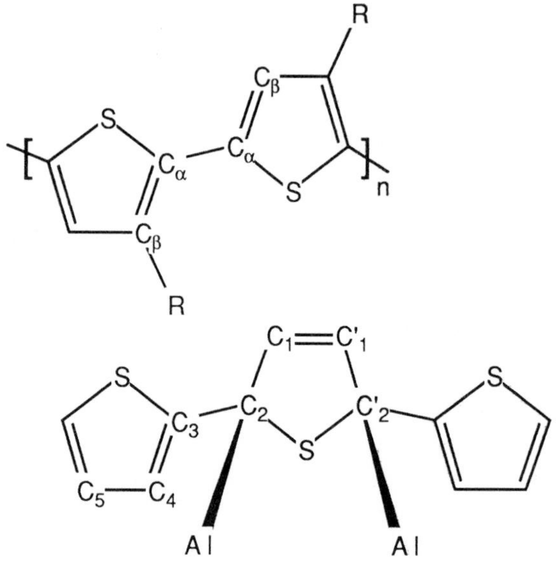

Fig. 1. Schematic structure of polythiophene (above) and Al_2/α-3T (below).

Even though the Al_2/α-3T can be thought to be a highly simplified model for the aluminum/polythiophene interface, it turns out that it is able to graps the essence of the chemical interactions at the interface. This is supported by results from surface sensitive XPS studies of Al deposited on P3OT (Lazzaroni *et al.*, 1991). The positions of the various core levels exhibit very small changes in going from submonolayer deposition of Al to multilayer coverage, which shows that the interaction at the interface is dominated by the deposition of the first monolayer or submonolayer of aluminum. Furthermore, we show below that the features of the Al_2/α-3T system are maintained in larger model systems of the aluminum/polythiophene interface (Al_4/α-6T).

3. RESULTS AND DISCUSSION

In order to discuss the interaction chemistry at the aluminum/polythiophene (or Al/P3AT) interface, we first examine the α-6T system interacting with two aluminum atoms. We considered five different input geometries for this system. In four of these, the aluminum atoms are placed on the same side of the thiophene chain and separated by 1, 2, 3, and 4 thiophene units, respectively (see Fig. 2). A system with the two aluminum atoms placed on opposite sides of the thiophene chain is also studied. Calculations with these five input geometries are performed at the MNDO level. In each case, the optimizations include all geometrical parameters related to the orientation of the aluminum atoms relative to α-6T as well as all geometrical parameters of the α-6T system itself.

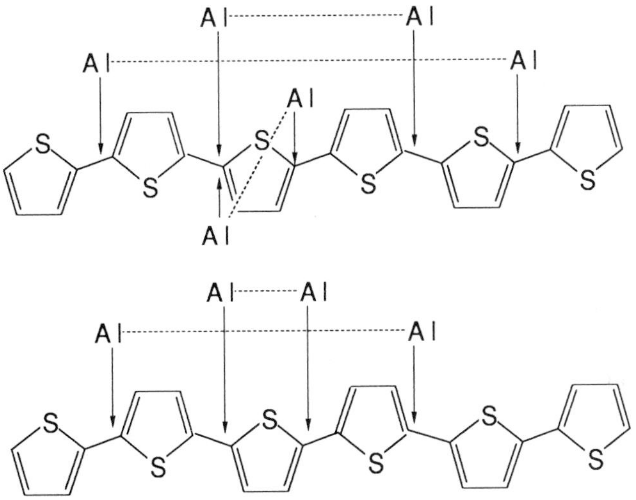

Fig. 2. Schematic description of the five different input geometries discussed in the text.

The geometry optimization technique works in such a way that it searches for configurations for which the force on each atom is zero (given that this configuration corresponds to a state of minimum energy). The technique, as such, can not distinguish a local minimum from the true ground state of the system, since in both cases, the force on every atom is zero. Therefore, if the input geometry is close to a local energy minimum, the final optimized geometry will be that corresponding to this minimum. Indeed, this situation occurs for the input geometries we have chosen. We obtain five different optimized structures, one for each of the five input geometries described above. The four optimized structures with the aluminum atoms placed on the same side of the thiophene chain all have roughly the same separation between the aluminum atoms as in the starting geometries (see Fig. 2). This shows that there are local energy minima along the polymer chain for two Al atoms to interact with poly-thiophene. The interaction site for the aluminum atom to the thiophene system is in all cases at carbon C_α positions (see Fig. 1). The lowest total energy of the five different cases is obtained for the configuration in which the two aluminum atoms bind to the two C_α of the same thiophene unit, which thus appear to correspond to the ground state. This is due to the fact that the unpaired electrons of each Al atom pair off most efficiently, with a state of lowest energy as a result, if they are close to each other.

The difference in heat of formation (or total energy) between the ground state configuration and the other four cases described above is: 12.82 kcal/mole (separation of 4 thiophene units), 7.67 kcal/mole (3 units), 2.81 kcal/mole (2 units), and 0.15 kcal/mole (the two Al atoms on opposite sides of the α-6T system). In the case where the Al atoms are forced to lie on opposite sides of the thiophene chain we obtain the minimum energy configuration in the case when the two Al atoms also attach to C_α atoms on the same thiophene ring. As a matter of fact, the total energy of this arrangement is nearly identical to what above is referred to as the ground state configuration. The geometrical modifications and the electronic properties of these two systems are also very similar. We will therefore restrict the following presentation to only one of these cases, namely, the case for which the aluminum atoms attach to the same side of the thiophene system (see Fig. 1). A more detailed presentation of the electronic structure and symmetry requirements related to the different arrangements will be given elsewhere (Stafström *et al.*, 1991).

In order to investigate in more detail the ground state configuration obtained in the MNDO optimization, we study this case at the *ab initio* level. A complete geometry optimization is performed on the α-3T system, with and without two aluminum atoms present. The basis set in these calculations is the standard 6-31G basis set. The *ab initio* optimization results in an interaction picture which is nearly identical to that obtained from MNDO optimizations, namely, the two aluminum atoms bound to the two C_α's on the central thiophene unit. Table 1 lists the optimized bond lengths around this unit for three different optimizations: *ab initio* calculations on the α-3T system (6-31G basis set), MNDO calculations on the α-3T system, and MNDO calculations on the α-6T system.

All three cases exhibit qualitatively the same type of geometrical modifications of the thiophene system due to the interaction with aluminum. The changes in the bond lengths of the central thiophene unit are such that the five-membered ring transforms from an aromatic structure to a quinoid like structure. The geometrical modification are confined to one thiophene ring and its C_α-C_α linkage with the neighboring rings. This localized nature of the interaction is further confirmed by the similarities between the MNDO optimizations on the α-3T and α-6T systems.

	6-31G				MNDO				
	α-3T	Al$_2$/α-3T	Δ	α-3T	Al$_2$/α-3T	Δ	α-6T	Al$_2$/α-6T	Δ
C$_2$-S	1.808	1.949	0.141	1.688	1.770	0.082	1.688	1.769	0.081
C$_1$-C'$_1$	1.434	1.335	-0.099	1.443	1.354	-0.089	1.441	1.356	-0.085
C$_1$-C$_2$	1.349	1.492	0.143	1.386	1.505	0.119	1.388	1.502	0.114
C$_2$-C$_3$	1.445	1.477	0.032	1.449	1.491	0.042	1.446	1.489	0.043
C$_3$-C$_4$	1.350	1.351	0.001	1.385	1.385	-0.003	1.388	1.387	-0.001
C$_4$-C$_5$	1.439	1.439	0.000	1.448	1.449	0.001	1.441	1.445	0.004

Table 1. Optimized bond lengths for thiophene chains with and without Al$_2$ present, Δ denotes the difference in bond length between the two cases. The atom labelling in Table 1 refers to that shown in Fig. 1. (units: Ångström)

The Al atoms are located out of the plane of the α-3T molecule and bound to the C$_\alpha$'s of the central thiophene unit. The *ab initio* optimized Al-C$_\alpha$ (Al-C$_1$ and Al-C'$_1$ in Fig. 1) distance is 2.083 Å (1.896Å in the MNDO optimization), which is well below the sum of the Van der Waals radii of these atoms. The short Al-C$_\alpha$ bond length is indicative of the formation of a semi covalent bond between these atoms. In order to create the necessary space for the electrons forming this bond, the local geometry around the C$_\alpha$ has to change considerably. In addition to the modifications in the bond lengths presented above, there is also a substantial change in the steric conformation around this bond.

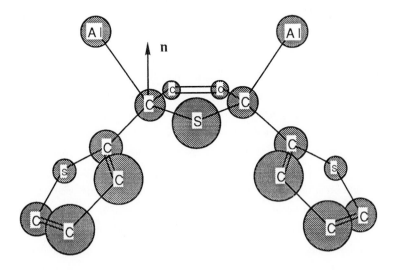

Fig. 3. Optimized structure of the Al$_2$/α-3T complex.

The optimized structure of the Al_2/α-3T complex is shown in Fig. 3. In comparison with the almost coplanar α-3T oligomer, the torsions which occur in the Al_2/α-3T system constitute a significant geometrical modification. The C_α-C_α bonds of the Al_2/α-3T system stick out of the plane of the central thiophene unit, in opposite direction to the Al-C_α bond. This bonding pattern corresponds to an sp^3 hybridization around the C_α atom of the central thiophene unit. In the *ab initio* optimized structure the C_α-C_α bond points at $112°$ towards the normal of the thiophene plane (**n** in Fig. 3) whereas the Al-C_α bond is directed in the opposite direction, at $48°$ towards the normal of the thiophene plane. Nearly identical values are obtained in the case when the two aluminum atoms bind to opposite sides of the thiophene system and in the MNDO calculations. In this context it should be noted that our geometry optimizations are performed on single chains. Interchain interactions will certainly affect the values of the torsion. However, the tendency of making the thiophene chain strongly nonplanar is present and must be considered in future studies of these type of systems.

The non-coplanarity of neighboring thiophene rings induced by the bonding of Al to the system reduces the π-electron overlap between these rings. This is very clearly observed in the longer thiophene system. Without Al present, the thiophene units of α-6T are close to co-planar and include π-states that are delocalized over all six rings of the system. The sp^3 hybridization around the C_α atoms in the presence of the Al dimer prevents the delocalization of the π-electrons. This is an important result as concerns the electronic properties of the polythiophene chains in contact with the aluminum surface. Clearly, the charge transport along these chains becomes strongly reduced due to the localization of the π-system. Whether this will effect the properties of an operating device such as a Schottky diode or an LED is an important question that we hope to return to in the near future.

The geometrical modifications of the thiophene system in contact with Al consists of a change from aromatic to quinoid structure (see Table 1). This change in the geometry is related to the fact that electronic charge is transferred to the α-3T system. The total electron transfer is a result of two different processes. In one of these processes, electrons from the aluminum dimer are transferred to the lowest unoccupied molecular orbital (LUMO) of α-3T, which in Al_2/α-3T becomes the highest occupied molecular orbital (HOMO). Both the LUMO of α-3T and the HOMO of Al_2/α-3T are π-type orbitals. The doubly occupied HOMO of Al_2/α-3T is localized to the central thiophene unit and is anti-bonding in the C_α-C_β and C_α-S bonds and bonding in the C_α-C_β bond. The population of this orbital is consistent with the geometrical changes observed for Al_2/α-3T relative to α-3T, namely, an increase in the C_α-C_β and C_α-S bond lengths and a decrease in the C_β-C_β bond length (see Table 1).

As usual in organometallic complexes, a charge transfer from the metal to an unoccupied orbital of the organic molecule is accompanied by a back-donation of charge from the occupied part of the spectrum of the molecule to the metal. In this case this occurs via a hybridization of σ- and π-orbitals of the thiophene system with Al 3s and 3p orbitals. A more detailed description of the electronic structure of the Al/thiophene systems will be given elsewhere (Stafström *et al.*, 1991)

The chemical shifts in the core level spectrum of P3OT upon aluminum deposition show that Al interacts strongly with P3OT at the interface (Lazzaroni *et al.*, 1991). As presented in Sec. 1 above, the chemical shifts of the Al (2p) and S (2p) core levels indicate that electronic charge is transferred from Al to the sulfur atoms of the thiophene system. This observation can be compared, qualitatively, with the theoretically calculated Mulliken atomic charges. The changes in the Mulliken atomic charges of α-3T due to the interaction with aluminum are shown in Table 2. Each Al atom is observed to lose approximately half an electron to the α-3T system which accommodates this additional charge mainly on the C_α atoms (C_2 and C'_2) and on the central sulfur atom. As concerns the sulfur atom, this result is in agreement with the large shift in the S(2p) core level observed experimentally (Lazzaroni *et al.*, 1991). No chemical shift is observed in the case of carbon. However, the experimental studies are performed on P3OT for which the contribution to the C 1s core level peak is completely dominated by the contributions from the carbons of the side chains. It is not possible, therefore, to observe any chemical shift due to the change in the atomic charge of a single carbon atom of the P3OT unit cell.

	6-31G (α-3T)	6-31G** (α-3T)	MNDO (α-3T)	MNDO (α-6T)
Al	+0.549	+0.403	+0.454	+0.461
S	-0.388	-0.259	-0.324	-0.318
C_1	+0.038	+0.019	-0.031	-0.033
C_2	-0.431	-0.304	-0.157	-0.169
C_3	+0.111	+0.055	+0.010	+0.003
C_4	+0.001	+0.009	-0.046	-0.043
C_5	+0.001	-0.005	+0.012	+0.020

Table 2. Changes in Mulliken atomic charges due to interaction with aluminum (units: $|q_e|$).

It is striking that practically no change in the Mulliken charges occurs outside the thiophene unit which interacts with the aluminum dimer. This result is in agreement with the XPS core level data of the Al/P3OT interface, which show two different S(2p) peaks. Both the theoretical and experimental results therefore indicate that the region around the Al/polythiophene interface contains two different sulfur atoms only, one type which is affected by Al and another type which remains unaffected.

For the system with the two aluminum atoms placed on opposite sides of the thiophene backbone, we observe an identical charge distribution as the one shown in Table 2. The two bonding patterns, which correspond to the ground state of the system, are therefore both in agreement with experimental data. There is no bonding pattern of Al_2 to the thiophene system other than the Al dimer formation, for which we observe a major charge transfer to the sulfur atom. This result is a further evidence that the theoretically obtained ground state correctly represents the bonding pattern of the Al/P3OT interface.

We continue our studies of the aluminum/polythiophene interface by adding more aluminum atoms to the α-6T system. Again, this system is too large to study using *ab initio* techniques with extended basis sets. However, the results presented above clearly indicate that the MNDO method provides an interaction picture which is in excellent qualitative agreement with the results of the *ab initio* calculations. There are no reasons to believe that this agreement will disappear when the system is extended further.

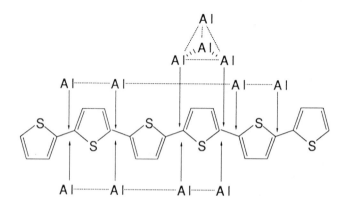

Fig. 4. Schematic picture of three different Al_4-6T systems (see text for details).

The optimized MNDO geometries of three different Al_4/α-6T systems are calculated, these systems are: two Al dimers in direct contact with the thiophene system separated by 2 and 1 thiophene rings, respectively, and a cluster of four aluminum atoms in which only two of the Al atoms are in direct contact with the thiophene system and the other two are located on top of this aluminum dimer (see Fig. 4). We obtain different local minima for all of these input configurations. The global minimum, however, corresponds to the case where the aluminum dimers are separated by one thiophene unit. As was shown above (see Tables 1 and 2), the modifications of the thiophene system hardly extend outside one aluminum dimer/thiophene unit. The interactions between two such units are, therefore, quite small and the energy difference between the two different separations of the aluminum dimers is very small. The case of the aluminum cluster has, however, a substantially larger energy (43.0 kcal/mole higher energy than the ground state energy). This result shows that the system gains energy by letting the deposited aluminum atoms interact with the thiophene system directly instead of forming aluminum clusters.

The geometrical modifications and the charge transfer associated with each Al dimer-thiophene unit in the ground state Al_4/α-6T system are identical to those of a single dimer in the Al_2/α-6T system (see Tables 1 and 2). Therefore, the agreement between the XPS data, which indicate the existence of two different types of sulfur atoms, and the calculated charge transfer for the model system is extended to be valid also in the case of a higher Al coverage. Furthermore, the results of the case where the aluminum atoms are put on top of each other show that the same interaction picture yields for this type of aluminum tetramer as for a single dimer, i.e., the same amount of charge is transferred to the thiophene system in both cases. The charge is accommodated on the thiophene system in precisely the same way in both cases. Therefore, we have been able to show that the same interaction chemistry yields for the cases of a single dimer, an Al monolayer and an Al multilayer system. This result is in perfect agreement with experimental data of the Al/P3OT interface (see above).

4. SUMMARY AND OUTLOOK

The results presented above for the aluminum-polythiophene system are the first to relate the effects of metallization of conjugated, semiconducting polymers, to the electronic properties of the polymer. The results show that aluminum interacts strongly with polythiophene and forms aluminum dimers on the polymer chain. In the ground state, the aluminum dimer is bound to a single thiophene ring. The interaction between Al and the thiophene system results in a transfer of electronic charge from aluminum to the sulfur and the α-carbon atoms of this thiophene ring. No modifications of the charge density of the thiophene system outside the interaction unit is observed. This type of charge transfer is in agreement with the measured chemical shifts of the S (2p) core level upon aluminum deposition. For no other interaction unit than the Al_2-thiophene unit do we observe any major increase in the charge on the sulfur atom. Therefore, from combined theoretical and experimental studies we have been able to obtain a detailed knowledge of the interaction chemistry at the aluminum/polythiophene interface. Furthermore, we have been able to show that the interactions between aluminum and the thiophene system is independent on the amount of Al interacting with the thiophene system. This result is in perfect agreement with experimental XPS spectra. It also shows that the Al/α-3T model system for the aluminum/polythiophene interface is relevant.

The geometrical modifications of the polymer chains which are in direct contact with the aluminum atoms are such that the delocalization of the π-electron system is blocked. This raises the question of how the semiconducting properties of the polymer are affected by the interfacial interactions with aluminum and if there are other metals that affect the polymer less and therefore should be more suited for device applications. These type of questions certainly have to be studied in the future in order to find the best suited materials for the electronic devices including semiconducting conjugated polymers. The application of a metal/polymer interface in an LED device (see Sec. 1) is also related to the result presented above. In the studies of the charge injection process from the metallic contact into the polymer, from which the photons are emitted, one must take into account the bonding pattern at the metal/polymer interface. In particular for the system discussed above, the effects of the changes in the π-electron system due to interaction with aluminum are important for the charge transport across the interface.

ACKNOWLEDGEMENTS

We gratefully acknowledge the National Supercomputing Center (NSC) in Linköping, Sweden for the use of the computer time.

REFERENCES

Braun, D. and Heeger, A. J., 1991, *Appl. Phys. Lett.,* **58**, 1982.

Burroughes, J. H., Jones, C. A., and Friend, R. H., 1988, *Nature*, **335**, 137.

Burroughes, J. H., Bradley, D. D. C., Brown, A. R., Marks, R. N., Mackay, K., Friend, R. H., Burns, P. L., and Holmes, A. B., 1990, *Nature*, **347**, 539.

Dewar, M.J.S. and Thiel W., 1977, *J. Am. Chem. Soc.*, **99**; 4899 and 4907.

Garnier, F., Horowitz, G., Peng, X., and Fichou, D., 1990, *Adv. Mater.*, **2**, 592.

Gaussian 90, Revision F, Frich, M. J., Head-Gordon, M., Trucks, G. W., Foresman, J. B., Schlegel, H. B., Raghavachari, K., Robb, M., Binkley, J. S., Gonzalez, C., Defrees, D. J., Fox, D. J., Whiteside, R. A., Seeger, R., Melius, C. F., Baker, J., Martin, R. L., C. M., Kahn, Stewart, J. J. P., Topiol, S., and Pople, 1990, *Gaussian, Inc., Pittsburgh PA*.

Gustavsson, G., Sundberg, M., Inganäs, O., and Svensson, C., 1991, *J. Mol. Electonics*, in press.

Hagler, T. W., Pakbaz, K., Voss, K. F., and Heeger, A. J.., 1991, *Phys. Rev. B*, in press.

Lazzaroni, R., Lögdlund, M., Stafström, S., Salaneck, W. R., and Brédas, J.-L., 1990, *J. Chem. Phys.*, **93**, 4433

Lazzaroni, R., Brédas, Dannetun, P., Lögdlund, M., Uvdal, K., and Salaneck, W. R., 1991,*Synth. Met.*, **41-43**, 3323.

Paloheimo, J., Kuvalinen, P., Stubb, H., Vuorimaa, E., and Yli-Lahti, P., 1990, *Applied Phys. Lett.*, **56**, 1157.

Stafström, S., *et al.*, 1991, to be published.

Tomozawa, H., Braun, D., Phillips, S., Heeger, A. J., and Kroemer, H.,1987, *Synth. Met.*, **22**:,63.

Chemical and electronic structure of the early stages of interface formation between aluminium and α,ω-diphenyltetradecaheptaene

P. Dannetun, M. Lögdlund, C. Fredriksson, M. Boman, S. Stafström, and W. R. Salaneck

Department of Physics, IFM, Linköping University, S-581 83 Linköping, Sweden,

  and

B. E. Kohler and C. Spangler

Department of Chemistry, University of California, Riverside, CA 92521, USA

ABSTRACT: A model molecular system for *trans*-polyacetylene, consisting of a diphenylpolyene with 14 carbon atoms (7 C=C double bonds) in the polyene chain, has been used to study the early stages of the interface formation with aluminum, using X-ray and ultraviolet photoelectron spectroscopy. The spectra are interpreted with the help of the results of both MNDO (geometry optimization) and VEH (electronic structure) quantum chemical calculations. It has been found that upon physical vapor deposition, aluminum reacts preferentially with the polyene chain, rather than with the phenyl end-units, and does so in a pair-wise fashion. The π-electronic structure at the interface is disrupted.

1. INTRODUCTION

Polymers as *electronic materials* are finding an ever increasing number of places in proposed applications in industry (Duke, 1987; Clark, 1991). Recently, an all organic transistor (Garnier, *et al*, 1990) and light emitting diodes (Burroughes, 1990; *et al*, Braun and Heeger, 1991), as well as several other devices involving polymer-metal interfaces have been studied. These organic-polymer-based devices each have unique and advantageous properties. The study of the electronic and chemical structure of polymer interfaces has received relatively little attention, however, compared with the surfaces and interfaces of metals and inorganic semiconductors (see, for example, Sacher, *et al*, 1990).

A most interesting class polymer materials today, from a point of view of electronic properties and future applications, are the polyconjugated polymers, which can be *doped* to a state of very high electrical conductivity, so called "conducting polymers", referred to as "conjugated polymers" in their undoped, insulating state. The interconnection between the chemical (geometrical) and electronic structure of conjugated polymers, as well as the relationship to surface properties, are important issues in characterizing the metal-polymer interface. It has been shown that for well prepared samples, the surface electronic structure of certain ideal conjugated polymer materials can be equal to that in the bulk (Salaneck, *et al*, 1990). In the initial stages of metal-polymer interface formation, however, certain chemical and electronic structural changes can, and do, occur (Lazzaroni, *et al*, 1990a, 1991).

Historically, a variety of sophisticated surface-sensitive spectroscopic methods have been used to study clean polymer surfaces (Clark and Feast, 1978). The surfaces, as well as the initial stages of metal-interface formation on conventional semiconductor surfaces, have also been relatively heavily studied (see, for example, Margaritondo, 1988). Reports of analogous studies of metal-on-polymer interfaces, however, are, in comparison, still relative few (Burkstrand, 1981; and see discussions in Sacher, *et al*, 1990). This later fact is particularly true in the case of conjugated polymers. Today, because it is still not yet possible to prepare surfaces of all conjugated polymers with the same high degree of well-defined chemical and electronic structure as can be done for conventional inorganic semiconductor materials, very little attention has been paid to conjugated polymer-metal interfaces.

The spectroscopic tools of choice for the investigation of surface chemical and electronic structure are ultraviolet and X-ray photoelectron spectroscopy, UPS and XPS, respectively. Within photoelectron spectroscopy (both UPS and XPS) is contained a maximum of both chemical and electronic information, as well as a high degree of surface sensitivity, such that, for example, monolayers of metals on polymers can easily be studied. The method is essentially non-destructive for most organic systems.

2. MODEL MOLECULAR SYSTEMS FOR INTERFACES

High quality, ultra-thin films of vapor-deposited organic molecular solids can easily be prepared in ultra-high vacuum (UHV), with surfaces which are chemically clean and molecularly well defined (see, for example, Salaneck, 1981). Such films, as used in the study reported herein, are usually amorphous or very polycrystalline, but very homogeneous and pure, by nature of the preparation method. *Such vapor-deposited molecular solid films, of well chosen model molecules, can serve as spectroscopically ideal starting points for studies of the initial stages of metal-on-polymer interface formation.* Taking the model system approach one step further, such model molecules may then be, in turn, vapor-deposited on to well prepared and characterized metal surfaces in order to study the initial stages of polymer-on-metal interface formation.

Studies of the initial stages of both metal-on-polymer and polymer-on-metal interface formation are of importance in the context of how the interfaces may affect actual device applications. In other words, when a few isolated atoms of an active metal are applied in some way to a polymer surface, they will react in a very different way than when an isolated polymer chain is applied in some way to the clean surface of an otherwise three dimensional metal substrate, with or without the presence of an oxide (see, for example, Salaneck, *et al* 1988). In the future, knowledge obtained from such studies of model molecular systems might then be used to interpret the results of similar studies on real polymer surfaces, or even to predict in advance the behaviour of real polymer interfaces.

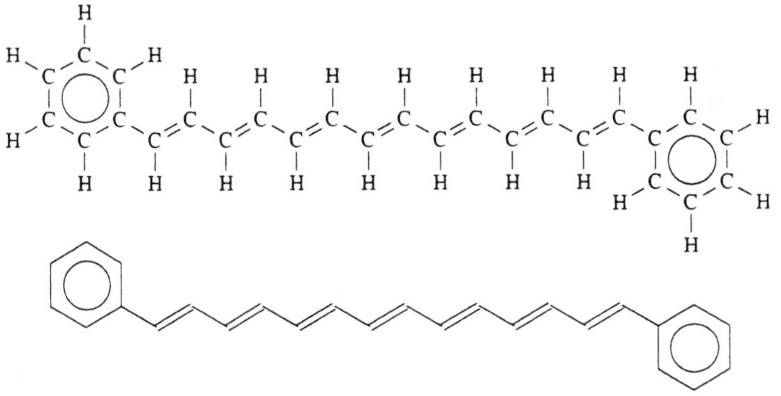

Fig. 1: The molecular structure of DP7 is shown in full (above) and abbreviated (below) forms.

In this contribution are reported some initial results of a study of the surface of vapor-deposited α, ω-diphenyltetradecaheptaene, and the interaction of the surface of this molecular solid with physically vapor deposited (PVD) aluminum. The α, ω-diphenyltetradecaheptaene molecule can be considered as a model molecule for *trans*-polyacetylene, or *trans*-$(CH)_x$. The name α, ω-diphenyltetradecaheptaene is represented by the short term "DP7" for obvious reasons. The chemical structure of DP7 is shown in Fig. 1. Some aspects of the electronic structure of some of the diphenylpolyenes have been studied by optical absorption spectroscopy by Kohler and coworkers (Hudson, *et al*, 1982; Horwitz, *et al*, 1987). The original UPS of the diphenylpolyenes, with 4 or less C=C double bonds (i.e., DP4 in the present nomenclature), was studied in the gas phase by Hudson and coworkers (1976). The UPS gas phase data was analysed by Duke and coworkers using the CNDO/S2 model (Yip, *et al*, 1976). In these shorter molecules, the polyene chain segments are too short to represent *trans*-$(CH)_x$. In DP7, the polyene chain length is long enough to support a soliton-antisoliton excitation, as in *trans*-$(CH)_x$ (Brédas and Heeger, 1989). The vapor pressure of a large polyene like DP7, however, is too small for this molecule to be studied in the gas phase.

3. EXPERIMENTAL DETAILS

High quality, ultra-thin films of DP7 were prepared *in situ* in an ultra high vacuum (UHV) photoelectron spectrometer of our own design and construction. The UHV system consists of separate sample preparation and analysis chambers, pumped with a special combination of ion- turbomolecular- and cryogenic-pumping, such that the ultimate base pressure in the system is considerably below 1×10^{-10} Torr. XPS is carried out using unfiltered MgK_{α} radiation, and with an electron energy analyzer resolution such that the $Au(4f_{7/2})$ line has a FWHM of 0.9 eV. UPS was carried out with a He-resonance lamp, filtered through a 2 m grazing-incidence uv-monochromator. The energy resolution is 0.2 eV in the XPS spectra and the He II spectra, but 0.1 eV in the He I spectra.

Thin films of DP7 were made by physical vapor deposition, PVD, using a filament-heated borosilicate glass crucible for vapor deposition of DP7, through a cryogenic shield, held at 20 K, onto sputter-cleaned gold substrates, which were held at 200 K during deposition. The gold substrates themselves consisted of a few thousand Ångströms of gold, vapor-deposited upon optically flat Si(110) wafers. Thicknesses of the deposited films were estimated by observing the attenuation of the $Au(4f_{7/2})$ XPS line, and calculating the effective thickness using the well-tested data of Clark and Thomas (1977) for the inelastic mean-free-path of electrons in typical organic polymers. The resultant films were in the range of 100 Å to 200 Å, and were free from oxygen (often a persistent problem with "real" polymer surfaces), as determined by XPS. Oxygen free surfaces are of particular importance when the interaction with aluminum is to be studied. UPS spectra were subsequently taken at room temperature, at which the DP7 films were stable in UHV indefinitely. Optical absorption spectra (not shown) of similar films prepared on quartz substrates were used to verify that no damage was done to the DP7 molecules during handling of the molecular source in air, loading into UHV, or from the vapor deposition process itself. It can not excluded, however, that a self-distillation process, which is automatically built into the PVD process, may be partially responsible for the molecular purity and the high quality of the resultant thin films.

After the films of DP7 were studied by UPS and XPS, aluminum was vapor-deposited by PVD from a liquid-nitrogen-shielded hot-filament source, onto the films at temperatures near room temperature. The surface temperature is estimated to be above room temperature, however, due to i.r. heating from the hot-filament. Effective coverage of the surface was estimated by the build-up of intensity of the Al(2p) XPS intensity relative to the decreasing C(1s) intensity. Since it is well known, however, that Al forms islands as well as diffuses easily on essentially all semiconductor surfaces (depending upon deposition parameters, substrate temperature, and the specific semiconductor), it is almost impossible to give more than an effective-thickness value. The spectra reported here are taken for Al deposition "in the first stages of interface formation", or roughly in the "monolayers" range.

4. THEORETICAL CALCULATIONS

The geometries reported here were obtained through geometry optimization calculations using the semiempirical MNDO method of Dewar and Thiel (1977). The MNDO method is known to yield reliable predictions for the geometry of large organic molecules. It is important to note that the results of the MNDO modelling indicate that the geometry of DP7 is planar, which is consistent with the results of optical absorption measurements on DP7 (Kohler, 1991), and the known crystal structure of the shorter chain diphenylpolyenes (Drenth and Weibenga, 1955). The planar geometry was also used in a previous analysis of gas phase UPS spectra of the shorter linear diphenylpolyenes (Yip, *et al*, 1976).

The electronic structure of DP7 was calculated using the VEH method (Nicolas and Durand, 1979, 1980, and André, *et al*, 1979). The VEH method has been shown to yield usable estimates of the single particle energy states of organic molecules and polymers. The UPS valence band spectra are compared with the calculated density-of-valence-states, DOVS, obtained from the VEH method in the usual manner, and convoluted with a Gaussian broadening factor in order to match the experimental resolution.

5. ELECTRONIC STRUCTURE OF DP7

The electronic structure of DP7 will be discussed in detail eslewhere (Lögdlund, *et al*, 1992). Here, however, certain points must be presented to facilitate the discussion of the PVD of aluminum. The valence electronic structure of DP7, obtained with He II (40.8 eV) photons, is shown in Fig. 2, and compared with the DOVS generated from the output of the VEH calculations on the DP7 molecule. Note that the overall agreement is very good. Cross-section effects are not included, however, in the calculated DOVS, and hence the intensities agree most well for lower binding energies, where the molecular states involved are derived mostly from C(2p) atomic orbitals, and as such the photoionization cross section is the same for all of the states. Towards higher binding energies, however, where C(2s)-derived states occur, the photoionization cross section for uv light decreases markedly, and the *intensities* of the calculated DOVS are in less good agreement. The binding energies, the major structural features, and the total band width, however, are all in good agreement. The effects of the photoionization cross sections on the intensity of UPS spectra of *trans*-polyacetylene have been discussed recently by Rasmusson and coworkers (1991). Of particular interest in this study is the fine-structure near lowest

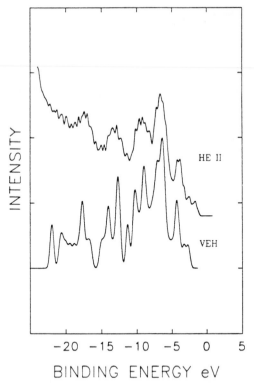

Fig. 2: Valence band energy spectra of DP7; UPS (above) and the VEH DOVS (below).

binding energies, i.e., for $-5 < E_B < 0$ eV in the UPS spectrum of Fig. 2. According to the VEH wave functions, the first two peaks, down from the Fermi level (0 eV in the figure) correspond to π-states essentially localized (80 %) on the polyene chain (integrated intensity on the 2 benzene rings is less than 20 %). The larger peak at about -4 eV is dominated by a doubly-degenerate π-state almost totally localized on the benzene rings. The spatial character of these wave functions from the VEH calculations is identical to those reported in previous studies of the shorter diphenylpolyenes (Yip, *et al*, 1976). The essentially *localized* character of the wave functions of these 3 molecular states (2 non-degenerate and 1 degenerate) will be useful in the discussions of the interaction with Al, below.

In Fig. 3 is shown a portion of the C(1s) XPS spectrum for *trans*-(CH)$_x$ (Keane, *et al*, 1991), DP7 and our own data on condensed benzene (Clark, *et al*, 1976). Because of space considerations, the main C(1s) peaks are not shown, but only the relatively weak satellite features which appear on the high binding energy side of the main C(1s) XPS peak. These so-called shake-up (s.u.) peaks correspond to real many-electron effects, which arise from excitation of (mainly) $\pi-\pi^*$ valence electronic transitions simultaneous with the sudden photoionization of the C(1s) core state during the absorption of an X-ray photon. The s.u. spectra represent the electronic excitations of the molecular ion system, i.e., the spectrum of 1-electron 2-hole states generated in connection with the core photoionization process (Wendin, 1981). The s.u. spectra of a variety of organic systems have been analysed in detail (Freund and Bigelow, 1987). In certain cases, however, the s.u. features can often be used as a finger print of certain molecular constituents, since the s.u. features of a variety of organic systems are known or can be recorded for comparison, as is done in this study. The intensities of s.u. features in general are very low, compared with the main C(1s) peak, especially in pure σ-bonded hydrocarbons. When π-electrons are

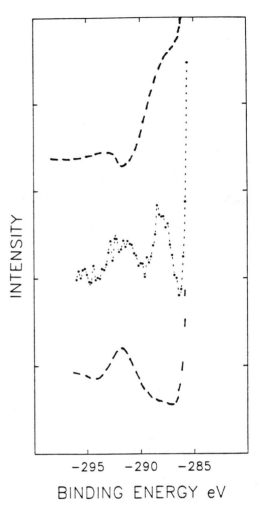

Fig. 3: The s.u. portion of the C(1s) spectra are shown for *trans*-(CH)$_x$ (top), DP7, and benzene (bottom)

present, however, the intensities of the s.u. feature can be significant, up to almost 10 % of the main C(1s) peak in certain conjugated systems (see, for example, Clark, *et al*, 1976; Riga, *et al*, 1977). It is clear from Fig. 3 that the s.u. spectrum of DP7 appears to be a approximately superposition of the s.u. spectra of benzene and *trans*-(CH)$_x$, which will be discussed elsewhere (Lögdlund, *et al*, 1992). Note that for a shorter polyene chain, as in DP7, the major s.u. peak for *trans*-(CH)$_x$ at about 288 eV moves toward higher binding energy relative to the main C(1s), and appears more intense, as was shown in general for π-systems of varying localization by Riga and coworkers (1977).

6. A MODEL FOR THE Al-(CH)ₓ INTERFACE

In Fig. 4 is shown the He I (21.2 eV photons) UPS spectra obtained during the first stages of interface formation with PVD aluminum. Notice that the first two features which appear between 1 and 3 eV are sharper than in Fig. 2 (because the resolution is 0.1 eV) and that they are strongly affected by the Al. The strong feature at about 4 eV is essentially unaffected by the aluminum. The two affected peaks are associated with molecular wave functions highly localized on the polyene chain, while the unaffected peak is dominated by photoelectrons from molecular wave functions almost totally localized on the phenyl end groups.

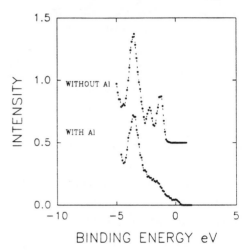

Fig. 4: Low binding energy portion of the high resolution He I spectra of DP7 without and with Al.

The C(1s) XPS main peak (not shown) occurs at 284.2 eV. In Fig. 5 is shown the s.u. portion of the C(1s) spectrum, obtained at the same time as the UPS spectra of Fig. 4, during the first stages of Al-interface formation. Notice that the s.u. feature near 288 eV, which is associated with the polyene chain, is strongly affected by the presence of the Al, while the higher energy feature near 292 eV, associated mostly with the phenyl end groups, is essentially unaffected. The binding energy of Al(2p) in bulk aluminum is 72.7 eV. During deposition of aluminum, however, the Al(2p) binding energy first appears at about 74 eV and shifts only to about 73 eV for the largest coverage obtained, indicative of the initial "monolayer coverage range" of thickness of the aluminum overlayers.

The results presented in the previous two paragraphs indicate that during the initial stages of interface formation, under the conditions described in § 3, the Al atoms reacts preferentially with the polyene chain, and not significantly with the phenyl end groups. As is well known, Al atoms are mobile on most surfaces at room temperature, and may therefore preferentially seek out the polyene chain portion of the molecule to react.

Theoretical modelling of the interaction of Al with DP7, however, reveals an interesting feature of the interaction. Using MNDO modelling, it is found that two Al atoms are required to react with the DP7 molecule to form a lower total energy (bound state) state. The Al interacts with the polyene chain, as indicated in Fig. 6, pair-wise. This pair-wise interaction is conceivably due to the fact that the Al atom contains an odd number of valence electrons. The interaction of only one Al atom with the DP7 molecule would result in an open shell system. If two Al atoms interact, however, a closed shell system is formed (see Stafström, these proceedings). The high surface mobility of the Al atoms at room temperature presumably allows the Al atoms to interact with the DP7 molecule in this fashion, after which the usual island formation starts to occur. In the Fig. 6 is also indicated the type of geometrical changes which the modelling predicts occur upon interaction with two Al atoms. The π-conjugation along the chain is broken, and the remaining part of the chain in between the Al atoms twists out of the plane of the molecule, which is otherwise planar, and even the phenyl end groups twist out of the plane by a large amount. The C-atoms, to which the Al atoms are attached, become sp³ hybridized. According to the MNDO calculations,

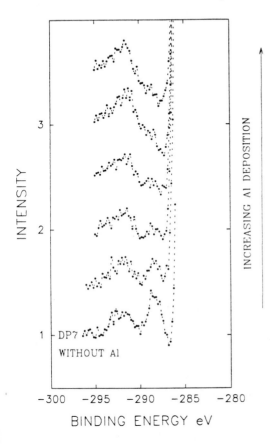

Fig. 5: The s.u. portion of the C(1s) spectra of DP7 during PVD of Al are shown, with no coverage at the bottom, and increasing coverage as indicated.

the Al atoms do not separate too much from one another along the polyene chain, otherwise the system approaches the case of a single Al atom interacting with the polyene chain, which is energetically undesirable.

The pair-wise interaction of Al with a conjugated system is not unique. In the case of Al on polythiophene, an affect on the the sulfur atoms is observed experimentally (Lazzaroni, *et al*, 1991). *Ab initio* level calculations of the interaction of Al with polythiophene, however, indicate that,for a single Al atom, only reaction with the carbon atoms in the thiophene ring is expected. When two Al atoms are allowed to interact with the polythiophene, however,

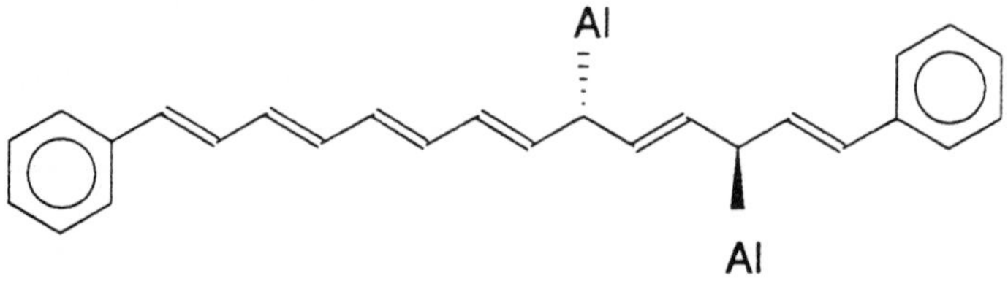

Fig. 6: An example is shown one of several possible pair-wise bonding configurations of two Al atoms to the polyene portion of the DP7 molecule .

they react with the carbon atoms in the thiophene ring in such a way that the ring twists, resulting in a significant change in the effective charge on the sulfur atom, in exact agreement with the experimental results (see Stafström, these proceedings).

7. SUMMARY

The initial stages of interface formation of aluminum on a model molecule for *trans*-polyacetylene, namely α, ω-diphenyltetradecaheptaene, or DP7, have been studied using surface-sensitive X-ray and ultraviolet photoelectron spectroscopy.   Because of space limitations, the details of the electronic structure of α, ω-diphenyltetradecaheptaene are reported separately.

During the initial stages of interface formation, the aluminum atoms preferentially interact with the polyene chain portion of the molecule, verifying the choice of DP7 as a model for the study of the interaction of aluminum with *trans*-$(CH)_x$. The preferential interaction with aluminum shows up clearly in the UPS valence band spectra, as well as in the XPS C(1s) s.u. spectra. Theoretical modelling of the interactions indicates that the Al atoms prefer to interact pair-wise with the polyene chain. The pair-wise interaction is similar to the situation of aluminum on polythiophene, where the effect is even more distinct. These results indicate that the π-system of the conjugated polymer is disrupted at the interface when a metal has been vapor-deposited upon it. These results also indicate the importance of including the interfacial electronic structure in considerations of charge injection and transport over, or near, the polymer-metal interface.

8. ACKNOWLEDGEMENTS

The authors acknowledge support for work on conjugated polymers by grants from the Swedish Board for Technical Development (STU), the Swedish Natural Sciences Research Council (NFR), and the Neste Corporation, Finland.

9. REFERENCES

André, J. M., Burke, L. A., Delhalle, J., Nicolas, G. and Durand, Ph., 1979, *Internat. J. Quant. Chem. Symp.* **13**, 283.

Braun, D. and Heeger, A. J., 1991, *Appl. Phys. Lett.*, **58**.

Brédas, J. -L., and Heeger, A. J., 1989, *Chem. Phys. Lett.*, **154**, 56.

Burroughes, J. H., Bradley, D. D. C., Brown, A. R., Marks, R. N., Mackay, K., Friend, R. H., Burns, P. L., and Holmes, A. B., 1990, *Nature*, **347**, 539.

Clark, D. T., Adams, D. B., Dilks, A., Peeling, J., and Thomas, H. R., 1976, *J. Electr. Spec.*, **8**, 51.

Clark, D. T., and Thomas, H. R., 1977, *J. Polym. Sci., Polym. Chem. Ed.*, **15**, 2843.

Clark, D. T., 1991, in *Science and Applications of Conducting Polymers*, Salaneck, W. R., Clark, D. T., and Samuelsen, E., Ed's, (*Adam Hilger*-IOP, Bristol, UK).

Dewar, M. J. S., and Thiel, W., 1977, *J. Am. Chem. Soc.* **99**, 4899, and 4907.

Drenth, W., and Weibenga, E. H., 1955, *Acta Crystallog.* **8**, 755.

Duke, C. B., 1987, *Synth. Met.*, **21**, 5.

Freund, H. -J., and Bigelow, R. W., 1987, *Physica Scripta*, **T17**, 50.

Garnier, F., Horowitz, G., Peng, X., and Fichou, D., 1990, *Adv. Mat.* **2**, 592.

Horwitz, J. S., Itoh, T., Kohler, B. E., and Spangler, C. W., 1987, *J. Chem. Phys.*, **87**, 2433.

Hudson, B. S., Ridyard, J. N. A., and Diamond, J., 1976, *J. Am Chem. Soc.*, **98**, 1126.

Hudson, B., Kohler, B. E., and Schulten, K., 1982, in *Excited States*, Lim, E. C., Ed. (Academic, New York).

Keane, M., Naves de Brito, A., Correia, N., Svensson, S, Karlsson, L., Lunell, S., Salaneck, W. R., Lögdlund, M., Swansson, D. B., and MacDiarmid, A. G., 1991, *Phys. Rev. B.*, in press.

Kohler, B. E., and Pescatore Jr, J. A., 1991, in *Conjugated Polymeric Materials: Opportunities in Electronics, Optoelectronics and Molecular Electronics*, Brédas, J. -L., and Chance, R. R., Ed's (Kluwer, Dordrecht), in press.

The metal/polyaniline interface: an x-ray photoelectron spectroscopy study

R. Lazzaroni,[1*] C. Grégoire,[2] M. Chtaïb,[2] J.J. Pireaux.[2]

[1]Service de Chimie des Matériaux Nouveaux, Département des Matériaux et Procédés, Université de Mons-Hainaut, 21 Avenue Maistriau, B-7000 Mons (Belgium)

[2]Laboratoire Interdisciplinaire de Spectroscopie Electronique, Facultés Universitaires Notre-Dame de la Paix, 61 Rue de Bruxelles, B-5000 Namur (Belgium)

ABSTRACT: In this work, we investigate the interface between a metal and a conjugated polymer, polyaniline, to determine the nature of the interactions taking place between the two species. We use X-ray Photoelectron Spectroscopy (XPS) to follow the adsorption of the first metal layers (up to 10 Å) onto the surface of the pristine polymer. In the first part of the paper, polyaniline is maintained in the base (non protonated) form and we consider the deposition of Aluminum for two redox states, emeraldine and leucoemeraldine. The second part of the work deals with the adsorption of copper onto emeraldine films in the base form and in the protonated form (hydrochloride salt).

1. INTRODUCTION

Over the last decade, solid/polymer interfaces have been widely studied, as some basic properties of these materials, such as adhesion and chemical stability, appear to be related the nature of the interfaces. Recently, conjugated polymers, i.e., polymers with a network of conjugated π electrons along the chains, have been shown to be promising compounds for the design of new semiconducting devices (Burroughes et al 1988, Renkuan et al 1991), due to their outstanding electronic properties. The properties of these devices strongly depend on the interplay between the layers, which in turn is related to the chemical nature and structure of the interfaces involved (Garnier et al 1990). Therefore, fundamental studies of the interactions taking place between the surface of a conjugated polymer and organic or inorganic adsorbates appear to be of particular interest.

*: Chargé de Recherches du Fonds National de la Recherche Scientifique.

Our work deals with the structure and the chemical and electronic properties of metal/conjugated polymer interfaces. Here, we focus on polyaniline, whose structure is shown in Figure 1. A common feature of conjugated polymers is that their electrical conductivity can be tuned over 10 orders of magnitude by a redox process, from the insulating to the semiconducting up to the metallic regime. In the case of polyaniline, the oxidation state is determined by the ratio of imine nitrogen atoms and quinoid rings to amine nitrogen atoms and aromatic rings. The conductivity of polyaniline therefore depends directly on the value of y (see Figure 1).

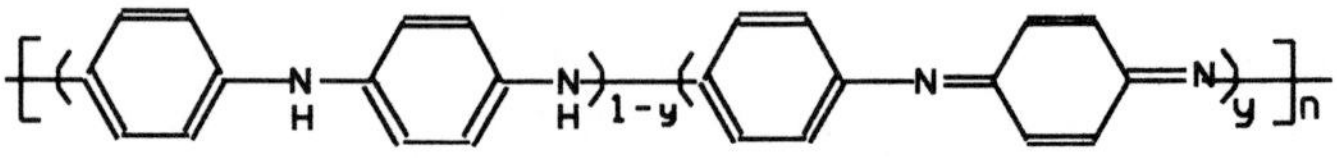

Figure 1: The structure of polyaniline (0 < y < 1).

Polyaniline stands out as a conjugated polymer, due to the fact that one can also use an acid/base equilibrium (protonation of the imine nitrogen) to modify reversibly the electrical properties of the system, without changing the number of electrons on the chain. The highest conductivities (350 S/cm) are obtained for the fully-protonated form of the intermediate (y = 0.5) oxidation state, emeraldine (Monkman and Adams 1991).

The first part of the paper consists of an X-ray Photoelectron Spectroscopy (XPS) study of the first stages of Aluminum deposition on the surface of polyaniline in two different oxidation states: i) the fully-reduced form (y = 1), called leucoemeraldine, and ii) emeraldine (y = 0.5). Both polymers are in the base form (non protonated). In the second part of the work, we investigate the copper/emeraldine system. In order to determine the influence of the acid/base equilibrium on the properties of the interface, we compare the results obtained on the base form and on the protonated polymer as copper is gradually deposited onto the surface.

2. EXPERIMENTAL

Thin polyaniline films are prepared electrochemically from aniline (0.1 M) dissolved in 1M hydrochloric acid. Polymerization is carried out potentiostatically at + 0.8 V vs. the standard calomel electrode, in a two-compartment cell. The anode, i.e., the electrode where the polymer film is grown, typically consists of an Indium Tin Oxide (ITO) glass slide and the counterelectrode is made of a platinum wire. When the synthesis is completed, the electrochemical potential is set to the value corresponding to the desired oxidation state. Here we used - 0.2 V to reduce the polymer to leucoemeraldine; emeraldine is obtained at + 0.4 V. Finally, the samples are washed with 1M HCl and dried in air. Additional treatment with 1M NH4OH and rinsing with water leads to the corresponding base forms.

The spectrometer used for these experiments is a Surface Science Instruments ESCA-206 apparatus equipped with a monochromatized AlKα source. XPS spectra are collected on a 600 μm-diameter spot with a resolution of 1.0 eV. When checking the polyaniline base films before evaporation, a small oxygen contamination ($\approx$ 1 oxygen atom per 5 rings) is detected. Aluminum and copper are evaporated from Knudsen cells inside a separate chamber (P < 5×10^{-10} Torr). The deposition rate, which is monitored with a quartz microbalance, was 0.2 Å/min for Al and 4 Å/min in the case of Cu, and the pressure during the evaporation remained below 1 10^{-9} Torr; the amount of deposited metal will be expressed in Å.

3. RESULTS AND DISCUSSION

3.1. The Al/ polyaniline interface

a: Emeraldine. The C1s and N1s core level spectra of polyaniline in the emeraldine form (imine/amine ratio = 1) are shown in Figure 2 (a,c). The C1s spectrum consists of a slightly asymmetric peak containing the contribution of all carbons. The experimental resolution is not sufficient for the C-N component to be clearly observed.

The N1s spectrum exhibits a more complex structure: in addition to the imine and amine components, located at 398.2 and 399.2 eV, respectively, a broad band appears with a low intensity at 402.0 eV.

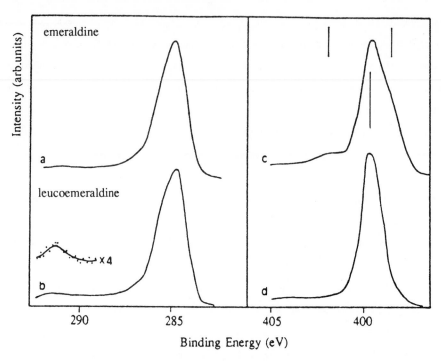

Figure 2: C1s (left) and N1s (right) spectra of pristine emeraldine (a,c) and leucoemeraldine (b,d).

The study of polyanilines in different oxidation states, i.e., with different imine/amine ratios, clearly indicates that this last band always lies at the same position and that its intensity is related to the relative amount of imine nitrogens in the polymer (Snauwaert et al 1990). This feature has been assigned to a shake-up satellite related to the imine group. Shake-up satellites are common in aromatic compounds, and correspond to π-π^* electronic transitions in the presence of the core hole. As such, they are very sensitive to the electron density on the molecule. For instance, the energy separation of the shake-up from the main peak and its relative intensity have been shown to be related to the degree of electron delocalization in acenes and polyphenyls (Riga et al 1977, 1981). The shake-up structure has also been used in the study of the temperature-induced conformational modifications of polyalkylthiophenes (Salaneck et al 1988, 1990).

The assignment of the shake-up in the N1s spectrum of emeraldine has been confirmed by recent INDO CI calculations (Sjögren and Stafström 1992). In the case of an all-imine model molecule for polyaniline, the most intense satellites are calculated to lie in the 3-5 eV region from the main peak. The calculated shake-up pattern of the corresponding all-amine compound, which can be compared to leucoemeraldine (see below), shows only weak structures at higher binding energies.

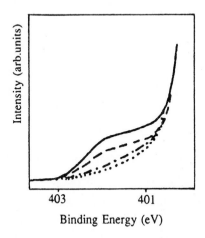

Figure 3: Evolution of the imine N1s shake-up satellite as a function of Al coverage in emeraldine (———: pristine; - - -: 0.5Å; - - - --; 1Å; ⋯⋯: 3Å coverage).

Upon Al deposition, the imine-related shake-up vanishes gradually (Fig. 3), indicating that the electronic density on the imine nitrogen sites is strongly affected. This suggests that aluminum preferentially interacts with the imine-quinoid-imine part of the chain. This is in agreement with recent theoretical calculations which indicate that imine-quinoid-imine groups along the polyaniline chain are preferentially charged, e.g., upon protonation (dos Santos and Brédas 1989a,b). It is noteworthy that the nitrogen atoms which have reacted with Al do not give rise to a new feature in the N1s spectrum, e.g., as a low binding energy peak. This might be due to the fact that the extra charge density brought in by the aluminum delocalizes over the N=quinoid=N segment. As a consequence, the shift of the carbon and nitrogen atoms would remain too small to be detected. Accordingly, the binding energy of the C1s main peak is constant at 284.7 eV throughout the deposition. However, a low-intensity feature appears at 6.5 eV from this peak upon metallization. This position corresponds to a shake-up transition involving the phenyl ring, which is observed in leucoemeraldine and which is not present in pristine emeraldine (Snauwaert et al 1990). This suggests that the polymer is driven towards a more reduced state upon Al deposition.

For initial exposures (0.5-1 Å), Aluminum appears to be in the oxidized form, with a rather broad Al2p peak centered around 75 eV. This is a clear indication of the charge transfer towards the organic molecules. This peak probably contains a minor contribution from the Al-O species due to the small oxygen contamination. Note, however, that Al deposition does not seem to affect the position and the line shape of the corresponding O1s peak. This behavior is also observed at the Al/polyimide interface (Ho et al 1985); it is explained by assuming that the oxygen atom acts simply as a bridge for the electron transfer from aluminum to the carbon backbone.

At higher coverage (2-4 Å), a sharp band corresponding to unaffected Al atoms starts to grow at 73.2 eV. The energy difference relative to the bulk metal (72.6 eV) reflects the size effect of the Al clusters. Note that the absolute intensity of the oxidized component also increases with Al deposition.

b: Leucoemeraldine. Comparison of the spectra of pristine leucoemeraldine (the all-amine form) with those of emeraldine (Fig. 2b,d) shows that the C1s levels are very similar. However, in leucoemeraldine, a shake-up is located at 6.5 eV from the main line. This band, which is analogous to that found in benzene, aniline, and their oligomers (Riga et al 1981), is assigned to a π-π^* electronic transition within the phenyl ring.

The N1s spectra are quite different; in leucoemeraldine, the main peak (399.3 eV) is sharp due to the presence of only one type of nitrogen: the amine species. Accordingly, the shake-up related to the imine group is not present. Again, this evolution quantitatively agrees with the results of theoretical calculations (Sjögren and Stafström 1992).

Deposition of the first Al atoms (< 0.5 Å) has no influence on the N1s spectrum, probably because of the small contribution of the reacted sites relative to the overall intensity. Further deposition leads to a gradual shift in the peak position, from 399.4 eV to 398.6 eV for 3Å-coverage (Fig. 4).

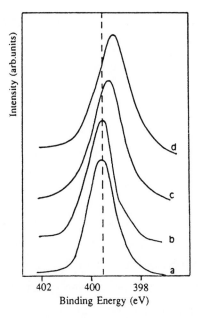

Figure 4: N1s spectra of the Al/leucoemeraldine interface
(**a**: pristine; **b**: 0.5Å; **c**: 1Å; **d**: 3Å coverage).
The dashed line corresponds to the position of the peak in the pristine polymer.

This evolution is consistent with a charge transfer from aluminum atoms towards the amine nitrogen. One could expect that, due to the presence of unaffected nitrogen sites and the presence of reacted nitrogen species, the N1s spectrum would display two distinct components, the former being located at the same position as in the pristine polymer and the latter at lower binding energy. In leucoemeraldine, instead, the whole peak is shifted. Along with this shift, the peak width at half maximum increases from 1.3 eV in the pristine system to 1.7 eV for 3 Å-coverage. In our opinion, this broadening is a consequence of the existence of a second component too close to the first one to be resolved. As a matter of fact, the upper spectrum (Fig. 4d) can be fitted by the sum of two contributions of approximately the same intensity. The first one corresponds to unaffected nitrogen atoms (399.3 eV), and the second one, which represents the Al-amine complex, is shifted 1 eV downwards. At low coverage, the second component is detected as a shoulder on the low energy side of the peak. In emeraldine, the interaction of Al with the imine group did not induce any significant shift, since the extra electron density can be efficiently delocalized over the neighboring quinoid ring. The fact that a shift is indeed observed in leucoemeraldine indicates that charge delocalization from the amine atom in the Al-N complex to the phenyl ring does not occur to a large extent. This is further confirmed by the fact that the C1s main peak and shake-up satellite are unaffected by the deposition process.

3.2. The Cu/polyaniline interface.

In the case of copper, we focussed on a single oxidation state, emeraldine, and we varied the degree of protonation of the imine nitrogen atoms.

a: base form. No significant changes are observed in the lineshapes of the C1s and N1s spectra upon Cu deposition for coverages ranging from 0.5 Å to 10 Å. This indicates that the interaction between copper and emeraldine base is rather small. The only effect appearing on the spectra is a shift of the whole C1s line towards higher binding energies. While it is located at 284.5 eV in the pristine polymer, it gradually moves to 284.9 eV for a 10 Å coverage. This shift seems to indicate that the electron density on the phenyl rings has slightly decreased, and this may be interpreted as the result of a small charge transfer towards the Cu atoms. The absence of a sizable C1s shake-up satellite in emeraldine base prevents any detailed investigation of the modifications which would be likely to occur in the π electronic structure of the aromatic system involved in the charge transfer. Therefore, the observed shift of the whole line is the only experimental feature suggesting that a charge transfer does exist.

The analysis of the copper core levels does not provide a clear confirmation of this phenomenon. For 0.5 Å coverage, the $Cu2p_{3/2}$ peak appears as a single line (FWHM = 1.65 eV) located at 932.0 eV. This value is smaller than the binding energy of pure copper (932.5 eV); in principle this is consistent with the direction of the charge transfer. However, the position of the $Cu2p_{3/2}$ line remains unchanged throughout the deposition, even for coverages where unaffected metallic-like Cu atoms certainly represent the major contribution to the signal. For 10 Å Cu deposition, this line is found to lie at 931.8 eV; this small shift is probably due to the growth of copper clusters on the surface. It thus appears that copper atoms or small copper clusters on the emeraldine surface show a binding energy of 932.0 eV.

Finally, we observe that the amount of copper present at the upper surface, as determined by the intensity by the XPS signal, is not strictly proportional to the amount which has been deposited, as measured with the quartz microbalance inside the evaporation chamber. This suggests that copper can easily diffuse towards the bulk of the polymer, that is at least 50 Å away from the upper surface, if we consider 50 Å as a reasonnable value for the probe depth of photoelectrons with a 550 eV kinetic energy. It is worthwhile to notice that copper diffusion from the interface to the bulk of nitrogen-containing polymers has already been observed for polyacrylonitrile systems (Wu 1988).

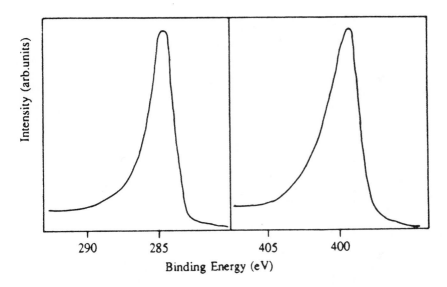

Figure 5: C1s (left) and N1s (right) spectra of HCl-protonated emeraldine.

b : salt form. The C1s and N1s spectra of HCl-protonated emeraldine are shown in figure 5. The protonation of the imine nitrogen is followed by a structural rearrangement of the chain, which can be modeled as a semiquinone radical cation (polaron) lattice (Glarum and Marshall 1987) or a dication (bipolaron) lattice (Stafström et al 1987). As a consequence, the electron density on the nitrogen sites is strongly modified. Positively-polarized nitrogen atoms now appear on the XPS spectrum at higher binding energies (around 400-401 eV), giving rise to the observed tailing of the lineshape. Simultaneously, the contribution of neutral imine nitrogen at 398.2 eV has vanished. A slight broadening is also observed on the high energy side of the C1s peak. Again, it can be related to the delocalization of the positive charge along the conjugated system.

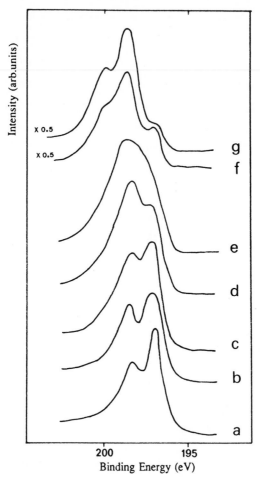

Figure 6: Cl2p spectra of HCl-protonated emeraldine as a function of Cu coverage (**a**: pristine; **b**: 0.1 Å; **c**: 0.5 Å; **d**: 1.0 Å; **e**: 2.0 Å; **f**: 5.0 Å; **g**: 10.0 Å).

In these samples, the Cl2p$_{3/2}$ line of HCl-protonated emeraldine appears at 197.0 eV. This value clearly corresponds to chloride ions. The Cl2p lineshape changes dramatically upon Cu deposition, as can be seen from figure 6. The doublet corresponding to the original species gradually disappears while a strong increase in the intensity of the signal is observed around 198.5 eV. Finally, at 10 Å Cu coverage, a well-defined doublet is located at 198.5 eV whereas the shoulder on the low energy side indicates that a small part of the original chloride is still present in the polymer. The new chlorine species is most probably a chloride ion, associated to a chemical species other that the protonated polymer chain (the observed binding energy is too small to correspond to covalently bonded chlorine atoms). It could result from the formation of copper(I) chloride, CuCl. The $Cl2p_{3/2}$ binding energy in this compound is 198.2 eV (Kishi and Ikeda 1973), which is reasonably close to the value observed here. The slight discrepancy may arise from the fact that we probably probe small clusters while the binding energy found in the literature corresponds to the bulk material.

The binding energy we measure for $Cu2p_{3/2}$ (932.4 eV) is consistent with the formation of a copper(I) chloride-like species, since pure CuCl is expected to give rise to a peak at 932.5 eV. However, it should be noted that metallic copper would also appear in this region. Therfore, the position of the $Cu2p_{3/2}$ line can not be considered as a clear-cut confirmation of the presence of CuCl on the surface of the polymer.

The amount of chlorine relative to carbon and nitrogen appears to increase as the Cu deposit gets thicker. At the final stage (10 Å coverage), the chlorine/carbon and chlorine/nitrogen ratios are more than twice as large as the value for the pristine polymer. Since the whole experiment was carried out in UHV, contamination on the surface by chlorine is unlikely. One possible explanation is that the surface of the polymer becomes enriched in chlorine by migration of chloride from the bulk to form CuCl.

If we assume that part of the copper atoms adsorbed onto the surface are oxidized in Cu(I), one surface species must obviously be reduced. The oxygen impurity is likely to be the most reactive site. However, its concentration (1 oxygen per 3 rings) is too small to account for the amount of CuCl which would be formed. Moreover, the position and lineshape of the O1s signal remain unchanged troughout the deposition. Similarly, the C1s and N1s peaks do not appear to be significantly modified by the adsorption of copper. On the basis of these results, we can not propose a reasonable mechanism for the reaction taking place at the surface. This phenomenon clearly deserves further investigation.

4. SYNOPSIS

Aluminum, which is known to be very reactive, interacts strongly with the surface of polyaniline. In the case of emeraldine base, the Al atoms form a complex with the N=quinoid ring=N segments of the chain. The formation of the complex strongly perturbs the electronic structure of the π system, as indicated by the vanishing of the shake-up satellite typical of the quinone diimine group. When such a group is not present, e.g. in leucoemeraldine base, the Al atoms bond to the amine nitrogen sites, without interacting significantly with the phenyl rings.

We find that copper interacts only weakly with the surface of emeraldine base; no large changes are observed in the core lines of the polymer. It also appears that copper atoms can readily diffuse from the surface towards the bulk of the samples. In contrast, adsorption of copper strongly modifies the surface of HCl-protonated emeraldine; a new chlorine species is formed, whose structure is probably close to Cu(I) chloride.

REFERENCES

Burroughes J H, Jones C A and Friend R H 1988 Nature 335 137
dos Santos M C and Brédas J L 1989a Phys. Rev. Lett. 62 2499
dos Santos M C and Brédas J L 1989b Phys. Rev. B 40 11997
Glarum S H and Marshall J H 1987 J. Electrochem. Soc. 134 2160
Garnier F, Horowitz G, Peng X and Fichou D 1990 Adv. Mater. 2 592
Ho P S, Hahn P O, Bartha J W, Rubloff G W, LeGoues F K and Silverman D B 1985 J. Vac. Sci. Technol. A 3 739
Kishi K and Ikeda S 1973 Bull. Chem. Soc. Jpn. 46 342
Monkman A P and Adams P 1991 Synth. Met. 41 627
Renkuan Y, Shucheng Y, Hong Y, Ruolian J, Huizuo Q and Decheng G 1991 Synth. Met. 41 727
Riga J, Pireaux J J, Caudano R and Verbist J 1977 Phys. Scr. 16 346
Riga J, Pireaux J J, Boutique J P, Caudano R, Verbist J and Gobillon Y 1981 Synth. Met. 4 99
Salaneck W R, Inganäs O, Thémans B, Nilsson J O, Sjögren B, Österholm J O, Brédas J L and Svensson S 1988 J. Chem. Phys. 89 4613
Salaneck W R, Lazzaroni R, Sato N, Lögdlund M, Sjögren B, Keane M P, Svensson S, Naves de Brito A and Correia N 1990 Conjugated Polymers: Opportunities in Electronics, Optoelectronics, and Molecular Electronics, eds J L Brédas and R R Chance, NATO-ARW Series E 182 (Dordrecht: Kluwer) pp 101-16
Sjögren B and Stafström S 1992 to be published
Snauwaert P, Lazzaroni R, Riga J, Verbist J and Gonbeau D 1990 J. Chem. Phys. 92 2187
Stafström S, Brédas J L, Epstein A J, Woo R S, Tanner D B, Huang W S and MacDiarmid A G 1987 Phys. Rev. Lett. 59 238
Wu C R 1988 Synth. Met. 26 21

Aluminium-simple model polymer interface formation: an x-ray photoemission study

Ch. Grégoire, Ph. Noël, R. Caudano and J.J. Pireaux
Facultés Universitaires Notre-Dame de la Paix, Laboratoire Interdisciplinaire de Spectroscopie Electronique (LISE), 61, Rue de Bruxelles, 5000 Namur Belgium.

ABSTRACT : X-Ray photoelectron Spectroscopy (XPS) has been used to study the interface formation between in-situ evaporated aluminium and two polymers containing different functionalities (Polycaprolactone and poly 2,2 dimethyl phenyl oxide). In the case of polycaprolactone, the chemical attack first takes place on the carboxyl groups. In a second step, the ester (C-O) entities are affected. Both mechanisms lead to the formation of an Al-O-C complex layer. The formation of metallic clusters was observed. The analysis of the shake-up satellite of poly 2,2 dimethyl phenyl oxide revealed that Al atoms strongly affect the electronic structure of the ring before reacting with the ether groups. Organometallic Al-O-C complexes were also observed.

1. INTRODUCTION

Since the late seventies, many studies have been performed on metallized polymers due to the technological progress in the field of microelectronics and packaging applications. Known for its thermal and mechanical properties [1,2], polyimide was extensively studied using a variety of techniques such as XPS [3-6], HREELS [7,8] and IR spectrometry [9, 10].

This polymer is a complex material containing many different functionalities : Nitrogen, Phenyl rings, Carboxylic (C=O) and ether (C-O) groups. It is not straightforward to understand the first stages of the metal/polyimide interface formation : which chemical bonds do react first with the incoming metal atoms? Does further deposited metallic atoms stick on the so-created intermediate layer?

Nowadays, a fundamental study of the metal/polymer interface formation is needed. We used X-Ray Photoelectron Spectroscopy (XPS) in order to collect atomic, chemical and microscopic information to answer these questions.

This paper is part of a work [11] aiming at determining the nature of the interaction occuring between Aluminium and organic surfaces containing more and more electronegative sites. Focussing on hydrocarbon compounds like polyethylene and polystyrene, we found no indication of a significant interaction between Aluminium and polyethylene. The metallization of polystyrene was quite different because of the presence of phenyl rings, giving rise to a shake-up satellite in the C1s core level spectrum. The shake-up peak, which involves $\pi \rightarrow \pi*$ transitions in the filled and empty electronic bands of the material, is expected to be an ideal active spectator of a bonding formation, if any, with the evaporated metal atoms. The formation of a weakly-bonded complex of Al with polystyrene was observed.

It is now well established that metal atoms adsorbed on polymer surfaces interact preferentially with strongly electronegative atoms [12-14], leading for instance to the formation of a metal-oxygen-polymer complex at the interface. The purpose of this work is to follow our fundamental study by analysing the metallization of polymers containing oxygen species and mixed oxygen-phenyl entities, in order to confirm or infirm that oxygen atoms are the preferential sites for Aluminium deposition.

2. EXPERIMENTAL SECTION

2.1. Analytical tools

The polymer surface and polymer/metal interface reported here were investigated using X-Ray Photoelectron Spectroscopy (XPS). Measurements were carried out in an X-probe 206 ESCA spectrometer (from Surface Science Instruments).

The instrument is equiped with three UHV chambers, dedicated to the following tasks : sample preparation, metallization and analysis (all these chambers have a base pressure of 10^{-9}-10^{-10} Torr). A sample entry lock allowing sample transfer from atmosphere into the system is also available.

The spectra were recorded using a monochromatized AlK_α source (1486.6 eV) and the X-Ray spot size was set to 600 µm diameter in order to achieve a good data collection rate. The binding energies observed were referred to the aliphatic carbon (CH) located at 284.5 eV.

2.2. Materials

The polymer samples (Fig. 1) were thin films prepared onto gold-covered silicon wafers. The preparation of these thin films was made by dissolution of the solid polymer (powder or thick film) in an appropriate solvant. We chose toluene for the polycaprolactone sheet and chloroform in the case of poly 2,2 dimethyl phenyl oxide powder on the basis of 1 mg of material in 1 ml of solvant. The solutions were then spin-coated onto the substrates.

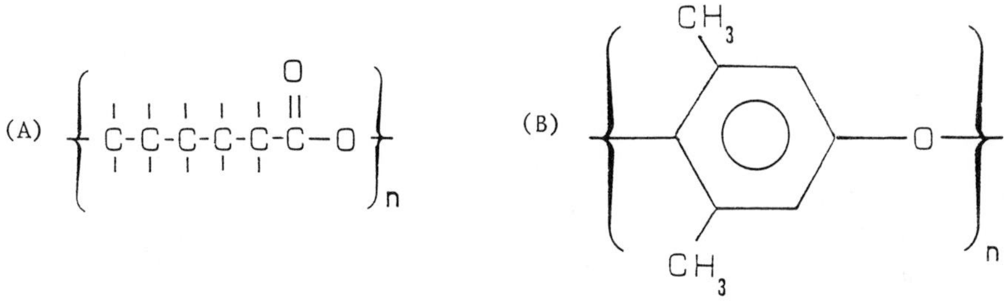

Figure 1 : Molecular structure of the unit cell of polycaprolactone (A) and poly 2,2 dimethyl phenyl oxide (B).

High purity Aluminium (Johnson Matthey, Puratronic Grade) was used for the evaporation.

2.3. Metallization set-up

Aluminium was evaporated in situ in an extremely well controlled way from a Knudsen effusion cell, previously outgassed for a few days. This technique provides very low and reproducible evaporation rates (less than 1Å/min) without any variation of the pressure (10^{-10} Torr) in the chamber. The metal coverage is determined using a quartz microbalance and is expressed hereafter in Å.

3. RESULTS AND DISCUSSION

3.1. XPS analysis of the clean polycaprolactone surface

The C1s and O1s core level spectra of polycaprolactone are presented in Fig.2. The C1s line exhibits three well resolved peaks at 284.5 eV (1), 286.1 eV (2) and 288.6 eV (3). We can assign these components as follow : peak (1) is due to aliphatic carbon (CH type), peak (2) is the contribution of the ester functionnality (C-O) and peak (3) at higher binding energy is attributed to the carbon atoms involved in carboxylic groups (C=O) [15]. The intensity ratio between the three components corresponds to the stoichiometry of the polymer (4 aliphatic carbons per 1 C-O and 1 C=O) except for a slight depletion of the carboxylic contribution.

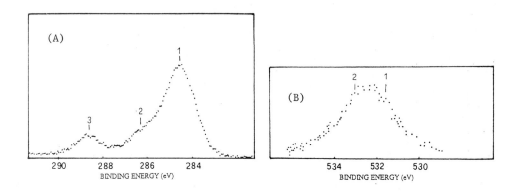

Figure 2 :   C1s (A) and O1s (B) core levels of polycaprolactone. The assignment of the different peaks is discussed in the text.

The oxygen region shows one broad band (FWHM ≃ 2.5 eV); this feature is built from the contributions of both types of oxygen atoms. It can be fitted with two peaks of equal intensity located at 531.6 eV (C=O) and 533.0 eV (C-O). Again, this is in good agreement with the molecular structure of the polymer.

3.2.   The Aluminium/polycaprolactone interface formation

Starting with a pristine polycaprolactone surface, Aluminium was evaporated stepwise.

In Fig. 3 A-B-C, we show the XPS data for C1s, O1s and Al2p core levels to follow the evolution of Al/polymer interface for different coverage. The spectra have been normalized to constant intensity in order to emphasize the chemical changes occuring at different metallization steps.

Upon Al deposition, the binding energy of aliphatic carbon is found to increase from 284.5 eV in the clean polymer to 285.0 eV for 0.5 Å coverage. Further Al evaporation does not appear to induce another shift of this peak (see Fig.3A). Simultaneously, we observe a dramatic decrease of the intensity of the carboxyl emission at 289.1 eV. This component has completely disappeared at 2.1 Å Al coverage. Up to 0.5 Å, the position and intensity of the contribution of the ester (C-O) carbons remain unaffected (286.3 eV). As the Al deposition grows further, this signal slowly decreases. As a consequence, the relative intensity of the main line at 285.0 eV becomes more and more important. This appears clearly for coverages higher than 2.0 Å.

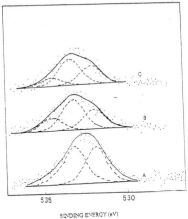

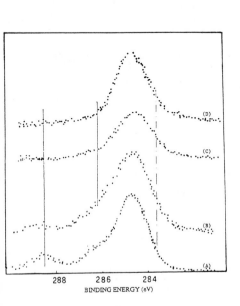

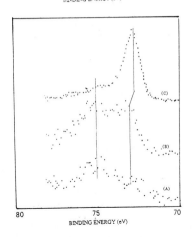

Figure 3 : A: Evolution of the C1s core level of polycaprolactone as a function of Aluminium coverage. (A) 0.0 Å, (B) 0.5 Å, (C) 2.0 Å, (D) 15.0 Å.
B : Evolution of the O1s core level of polycaprolactone as a function of Aluminium coverage. (A) 0.0 Å, (B) 0.5 Å, (C) 2.0 Å
C : Evolution of the Al2p signal. (A) 0.5 Å, (B) 3 Å, (C) 15 Å.

A broadening towards the lower binding energies is also observed, which can be related to the appearance of a fourth component located at 283.7 eV. The intensity of this peak remains small throughout the deposition : it roughly represents 5 % of the total peak area at 15 Å coverage. This can be explained by the formation of an Al-O-C like complex. Indeed, if additional carbon, in the form of the peak at 283.7 eV, appears on the surface, the fractional concentration of the 'untreacted" species should have declined. Within fthe experimental error, the intensity of the main C1s line at 285.0 eV remains unchanged at low coverage, whereas the contribution of the C=O species declined from 13.1 to 6.4 atomic percent. This decrease should be completely accounted for the appearence of the peak at 283.7 eV. The low binding energy of this peak is thus indicative for an electron-rich environment typical of oxides and carbides of metal. This type of complex has been found at the interface between Al and other polyesters, such as Polyethyleneterephthalate (PET) [16]. The formation of metal-O-C species appears to be a general process, as it is also observed upon Ti [17] and Cu [13] deposition. The evolution of the lineshape therefore indicates a charge transfer between Al atoms and C=O entities, confirming the decrease of the intensity of this component.

The slow decrease of the C-O intensity indicates that the ester carbon is also affected by the metallization. This suggests that the Aluminium does not interact only with the C=O groups. This can be understood in the following way : Al first reacts with C=O groups. Once these functionalities have all reacted, the Aluminium atoms start to attack the C-O entities.

As expected, at higher coverage (> 5 Å), the total intensity of the C1s emission decreases, due essentially to the screening effect by the metallic layers.

The most important modification observed in the O1s core level spectrum upon Al deposition is the decrease in the intensity of the C=O contribution at 531.7 eV (Fig.3B). This is consistent with the evolution of the corresponding component in the C1s spectrum. An additional weak component appears at a higher binding energy (534.0 eV). The origin of this feature is not clearly understood at this stage of our work; it may arise from the contamination of the Al surface by residual water molecules to form hydroxyl species (which typically appears at 533-534 eV, whereas the oxide shows up at a lower binding energy : 531.5 eV). The oxygen atom of the Al-O-C complex has been proposed to appear roughly at the same position that the C=O oxygen [16]. Note that our data do not provide a clear-cut confirmation of that assignment. Studying the interaction of Al with the surface of a polymer containing either C=O or C-O groups may be helpful to clarify this point.

As can be seen from Fig.3C, the Al2p photoelectron peak presents two components. Although the intensity of the signal is very low, we observe that at low coverage, the component at higher binding energies (75.0 eV) predominates, whereas the intensity of the first component increases mostly after $\simeq$ 3 Å Al deposition. The component appearing at higher binding energy is clearly due to the interaction of Al atoms with oxygen in the polymer. It is to be noticed that this peak is very broad (> 2 eV) which suggests that it contains several contributions of slightly different chemical nature, such as the above mentioned hydroxyl species. The low energy component, which corresponds to unaffected Al atoms is found to shift towards lower binding energies as the Al coverage is increased (from 73.5 eV at 0.3 Å coverage to 72.8 eV for a 15 Å thick deposit). This shift suggests that Al clusters form and grow gradually on the surface. This final value is close to the position of Al2p in the bulk metal.

As a conclusion, these results show that the first stage of Al deposition on polycaprolactone consists in a reaction between Al and the oxygen of C=O groups, leading to the formation of an Al-O-C complex. When the C=O sites have all reacted, the main interaction takes place via the oxygen of the ester group, as demonstrated by the evolution of the C1s spectrum. At higher coverages, we observe the growth of metallic-like clusters.

3.3. XPS analysis of clean poly 2,2 dimethyl phenyl oxide surface

The C1s and O1s core level spectra poly 2,2 dimethyl phenyl oxide are presented in Fig. 4. We can observe two main components in the C1s line. Peak (1) at 285.0 eV is the fingerprint of the C-C species (including both aliphatic and aromatic parts) while peak (2) located at higher binding energies (286.3 eV) is assigned to the C-O ether in the polymer. The two components of the main peak are found to be in the expected intensity ratio (3:1:1) confirming that the structure of the polymer is maintained at the surface.

The assignment of the different peaks is discussed in the text.
As it is often observed in aromatic systems, one can also distinguish a low intensity shake-up peak centered 6.7 eV up from the main C1s line. This energy separation is to be compared to the value found in polyparaphenylene [18].

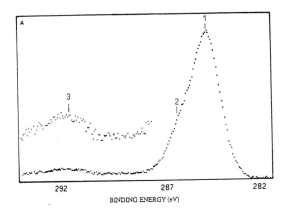

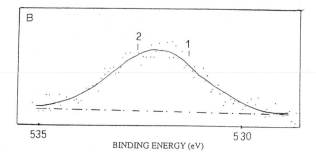

Figure 4 : C1s (A) and O1s (B) core levels of poly2,2 dimethyl phenyl oxide.

Surprisingly, the O1s spectrum shows two peaks whereas a single line would be expected on the basis of the molecular structure of poly 2,2 dimethyl phenyl oxide. The most intense component is located at 533.2 eV; this value is consistent with the presence of the ether group. The existence of the second component, which makes up roughly 25 % of the total O1s intensity, cannot be explained on the basis of the molecular structure nor can it arises from the procedure used to prepare the samples. Hence, we are led to relate this feature to a contaminant whose nature could not be determined. In order to minimize this problem, we follow only the evolution of C1s and Al2p signals during the metallization.

3.4.   The Aluminium/poly 2,2 dimethyl phenyl oxide interface formation

For an Al coverage lower than 1 Å, only a subtle change can be observed in the C1s lineshape : the intensity of the shake-up satellite is decreased by the presence of Aluminium (see Fig.5A). In the pristine polymer, this satellite represents 6 % of the total C1s intensity. After the deposition of 0.5 Å, its intensity is roughly 1 % of the total area. It is well known that the shake-up satellite of the phenyl ring is very sensitive to the electronic structure of the aromatic system [18], e.g., its position and intensity are strongly affected by the degree of delocalization of the π electrons. Therefore, the effect we observe here clearly suggests that the π electronic system of the phenyl rings is perturbed by the presence of Aluminium atoms. In the first stages of the interface formation, the Al atoms form a π-type complex with the phenyl rings.

For coverages higher than 2 Å, the intensity of the component corresponding to the C-O sites gradually decreases. Simultaneously, a new feature appears at lower binding energy (1 eV down from the main line). Again, this evolution can be related to the formation of a Al-O-C organometallic species. Upon the charge transfer from Al to the molecules, the carbon sites which were positively polarized due to the presence of the oxygen atoms become slightly negatively charged. The shift of the corresponding line to lower binding energies accounts for the evolution we observe in the C1s lineshape.

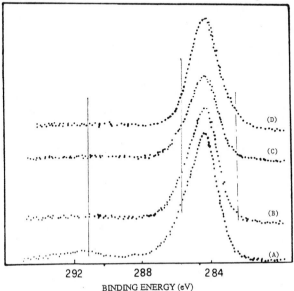

BINDING ENERGY (eV)

Figure 5 :   A: Evolution of the C1s core level of polycaprolactone as a function of Aluminium coverage. (A) 0.1 Å, (B) 0.3 Å, (C) 0.5 Å, (D) 3.0 Å.

The Al2p core level spectrum contains two contributions. Similar to the metallization of polycaprolactone, we can observe for low coverages, a more intense increase of the component at 75.0 eV, than the peak of unaffected Al atoms, located at 72.7 eV. This peak is found to be very broad and contains both the Al interaction with the phenyl rings and the oxygen present in the polymer. Indeed, as mentioned in the discussion concerning the C1s core level, Aluminium first reacts with the phenyl rings before exercising an influence on the ether (C-O) groups. In both cases, there is a charge transfer from Al to the interacting entity, so that its contribution in the Al2p spectrum appears at higher binding energies. The interaction with the C-O groups leads to the formation of an Al-O-C complex as in the case of polycaprolactone.

For higher Al deposition, the component located at 72.7 eV and corresponding to metallic Al is increasing and predominates the Al2p spectrum.

From these results, we conclude that Aluminium atoms <u>first</u> react with the phenyl rings of the polymer, leading to an Al-O-(polymer) complex. In a second step, the ether functionalities participate to the interaction. Finally, a metallic film grows on these "organometallic" species.

4. CONCLUSION

The analysis of the first step of Aluminium/polymer interface formation was realized in order to determine the preferential adsorption sites. The metallized polymers were composed of different functionalities (C=O, C-O and phenyl rings).

We found in the case of polycaprolactone, that Al atoms saturates the carboxylic entities before interacting with the second oxygen of the esters groups.

The intensity evolution of the shake-up satellite of poly 2,2 dimethyl phenyl oxide shows that phenyl rings are affected at rather low coverages. For higher Al deposition, Al atoms react with the oxygen of the ether entities.

In both cases, the interaction of Al atoms with the oxygen contained in the polymer leads to the formation of Al-O-polymer complex.

5. REFERENCES

1. See, for example, *Polyimides : Synthesis, Characterization and Applications*, edited by K.L. Mittal (Plenum, New York, 1984), Vol. 1.
2. D.L. Allara, F.M. Fowkes, J. Noolandi, G.W. Rubloff and M. Tirrell, Mater. Sci. Eng. 83, 213, (1986).
3. H.J. Leary and D.S. Campbell, Surf. Interface Anal. 1, 75 (1979).
4. P.O. Hahn, G.W. Rubloff and P.S. Ho, J. Vac. Sci. Technol. A2, 756 (1984).
5. J.R. Salem, F.O. Sequeda, J. Duran, W.Y. Lee and R.M. Yang, J. Vac. Sci. Technol. A4, 369 (1986).
6. M. Grunze and R.N. Lamb, Chem. Phys. Lett. 113, 283 (1987).
7. J.J. Pireaux, C. Grégoire, P.A. Thiry, R. Caudano and T.C. Clarke, J. Vac. Sci. Technol. A5, 598 (1987).
8. N.J. DiNardo, J.E. Demuth and T.C. Clarke, J. Vac. Sci. Technol A4, 1050 (1986).
9. R.A. Dine-Hart and W.W. Wright, J. Appl. Polym. Sci., 11, 609 (1969).

10. J. Zurakowska_Orszah, T. Chreptowicz, A. Orzeszko and Kaminski, Eur. Polym. J. , 15, 409 (1979).

11. C. Grégoire, P. Noel, R. Caudano, J.J. Pireaux, to be published.

12. J.M. Burkstrand, Phys. Rev. B20 (1979) 4853.

13. M. Chtaib, J. Ghijsen, J.J. pireaux, R. Caudano, R.L. Johnson, E. Orti, J.L. Brédas, Phys. Rev. B., in press.

14. Y. Depuydt, P. Bertrand and P. Lutgen, Surf. Interf. Anal., 12 (1988) 486.

15. L.J. Atanasoka, Steven G. Anderson, H.M. Meyer, J.H. Weaver, J. Vac. Sci. Technol. A5 (1987) 3325.

16. M. Bou, J.M. Martin. Th. Le Mogne, Appl. Surf. Sci., 47 (1991) 149-161.

17. F.S. Ohuchi and S.C. Freilich, J. Vac. Sci. Technol. A4, 1039 (1986).

18. J. Riga, J.J. Pireaux, J. Verbist, Molecular Physics, 1, 34 (1977) 131-143.

XPS study of the first stages of copper growth on PPQ (Polyphenylquinoxaline)

J.J. Pireaux[1], Ch. Grégoire[1], R. Giustiniani[2], A. Cros[2]

[1]FUNDP, Laboratoire LISE, 61, rue de Bruxelles, B-5000 NAMUR (Belgium)
[2]Université d'Aix-Marseille II, Faculté des Sciences de Luminy, Département de Physique, Case 901, F-13288 MARSEILLE Cédex 9 (France)

ABSTRACT : Monochromatized X-ray Photoelectron Spectroscopy has been used to characterize the copper deposition on four types of polyphenyl quinoxaline, depending on their different oxygen content. In a coverage range from 0 to 20 Å, no particular (strong) chemical reaction of copper is evidenced, suggesting that bonding with PPQ and oxidized PPQ is weak : this confirms that copper has a low reactivity towards PPQ, as was also noticed for other polymers.

INTRODUCTION

In the microelectronic industry, the copper-polymer interface is a key question, as a device contains many applications of polymer dielectric films between copper tracks. Especially today the realization of multilevel interconnections in ULSI technology supposes the existence of stable metal-polymer junctions, notwithstanding the extreme physical and/or chemical conditions imposed by a long and complex process, or by a long term use. Polyphenyl quinoxaline (PPQ) can be classified into the polyimide series : this is a thermostable polymer (decomposition temperature above 550°C in air), with good dielectric properties, as its ε ranges from 2.6 to 2.8 (0-300°C, 1-10 KHZ), with a dielectric loss below 10^{-3}. On the contrary to other polyimides (and PMDA-ODA particularly), virgin PPQ does not contain any single oxygen atom; its chemical formula is presented in figure 1.

Figure 1 : Chemical formula of pristine polyphenylquinoxaline. Note that the polymer contains only carbon, nitrogen and (not shown) H atoms

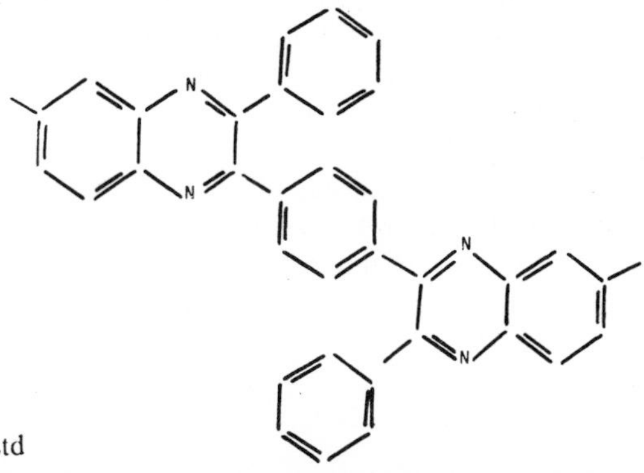

The adhesion and XPS (X-ray Photoelectron Spectroscopy) literature agrees now that metal atoms (eg, Al, Co, Cr, Ni, Ti,....) interact strongly with - and form a well adherent layer on - a polymer surface, <u>especially if</u> oxygen atoms are present in the polymer : the metal interaction with this (these) oxygenated functionality(ies) - of alcohol, carbonyl, acid ... types - leads to the formation of a metal-oxygen-polymer complex at the interface (Burkstrand, 1979). This matter is particularly well documented, e.g. for Cr and Al metallization of polyimide/PMDA-ODA (see for example Atanasoska et al, 1987; Ho et al, 1985).

As the PPQ monomeric unit contains only phenyl rings and the quinoxaline units (figure 1), and as copper is known to have a low reactivity and adheres weakly to a polymer material, different PPQ samples were studied for their different surface oxygen content [0 - 5 % - 10 % - 20 % (atomic percent)] that could indeed be modified by subsequent chemical or physical process. The purpose of this work is to study with X-ray photoelectron spectroscopy the early stage of copper growth on virgin and modified PPQ. It is hoped that the characterization of this interface will shed some light on the adhesion properties of the device, which are to be determined later on.

EXPERIMENTAL SECTION

The monochromatized AlKα (1486.6 eV) XPS measurements were performed on an X-probe 206 ESCA spectrometer (Surface Science Instruments, Ca), with a 600 μm lateral resolution. The analyzer pass energy was either 150 eV (survey scan) or 50 eV (core levels study). Core levels binding energies were referred to the neutral carbon position at 285.0 eV. Charging effect on the PPQ films (2.5 μm thick) was acute; the electron flood-gun used for charge compensation had constantly to be retuned, when testing for sample lateral homogeneity, or especially during the XPS measurements alternating with the successive copper evaporations. XPS spectra did not evidence any radiation damage on the samples; however, for some of them, visual inspection revealed a color change in the X-ray exposed area.

PPQ films were produced by spin coating a solution in metacresol/xylene mixture and annealed (400°C, in a nitrogen flown furnace, during about one hour). Four materials were studied : *sample A* : PPQ film on Si_3N_4, containing no oxygen after anneal; *sample B* = PPQ film on chromium nitride with about 10 % surface oxygen content; *sample C* = same as B with 5 % oxygen (different preparation); *sample D* = deliberately oxidized (20 % oxygen surface content) PPQ film by a 2 h exposure in air to an unfiltered UV lamp.

Cu (99.999 % purity) was evaporated *in situ* (vacuum ~ 1 x 10^{-10} Torr) in the submonolayer regime from a carefully outgassed Knudsen effusion cell. A very low evaporation rate (between 1 and 6 Å/min) as measured with a quartz microbalance was used. The metallic coverages are expressed in Å (microbalance reading), that can be converted in Cu atoms per cm^2, with the following assumption : the geometry of the copper (100) cfc unit cell with a = 3.61 Å predicts 1.535 10^{15} atoms per cm^2 = 1 monolayer ; given the Cu density (8,96 gr/cm^3) and atomic weight (63,546 a.m.), one angström reading on the microbalance is equivalent to 8.5 10^{14} atoms/cm^2. Supposing that the PPQ monomer unit is lying flat in a plane, one can crudely estimate the covered area (~ 100 $Å^2$, neglecting hydrogen), and the number of monomers per cm^2 to be of the order of 1 x 10^{14} : roughly speaking, a coverage of one Cu atom per PPQ unit cell would correspond to a monitored Cu thickness of about 0,12 Å.

CHARACTERIZATION OF THE POLYMERS

The as received PPQ films were analyzed by XPS : selected C_{1s}, N_{1s} and O_{1s} spectra are presented on figure 2 (left panel); relevant data are compiled in Tables 1 and 2.

<u>Figure 2</u> : **Carbon 1s, nitrogen 1s and oxygen 1s spectra of the four studied samples (see text), as received (left panel) and after 300°C anneal (righ panel). Relative intensities of different core level lines are not relevant on this display; see Table 2**

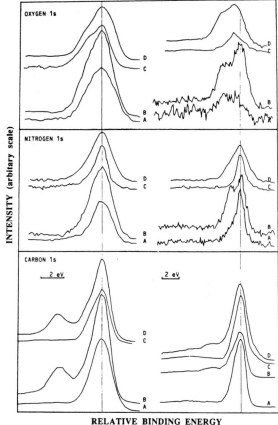

RELATIVE BINDING ENERGY

Stoichiometry : according to its chemical formula (figure 1), the pure PPQ atomic content should be C/N/O = 0.895/0.105/0.0; obviously the material should not contain any oxygen, which, if detected, depends on the polymer process or post-treatment. The attribution of the core level peaks is straightforward for the pure material : the <u>carbon 1s</u> peak at 285.0 eV (set for calibration) fingerprints all the aromatic species, except the neighbours to nitrogen atom, located at 286.0 eV; the wide but low intense band at ~291.5 eV is the shake-up satellite characteristic of the delocalized electrons in the unsaturated polymer; other peaks in the 290-286 eV interval are associated with carbon atoms bond to oxygen. Particularly, the structure at 289.0 eV may correspond to an acid function (O=C-O), or to a triple ether bond. The <u>nitrogen 1s</u> region contains a single peak at 399.2 eV, characteristic of the quinoxaline entity; for sample D especially a distinct peak at 1.5 eV higher binding energy indicates that some of the nitrogen sites have been oxidized. As for the <u>oxygen 1s</u> peak that is exclusively of "contaminating" type, an attribution of all the structures is less evident : from literature data, peaks at 531.5 eV and 533.2 eV are expected to correspond to carbonyl (O=C) and ether (O-C) oxygen respectively; hydroxyl-like oxygen should appear at still higher binding energy; the oxygen induced chemistry will be discussed later on. Let us conclude this section by noting that the polymer stoichiometry, as usual calculated from peak areas, cross sections and instrumental sensitivity factors, (Table 1) indicates that - besides the added oxygen content - the polymer surface composition is as expected. If some part of the deliberate polymer process did include oxygen functionalities, this is not introducing other carbon contamination, nor reducing the nitrogen content.

As already mentioned, charging effect on the polymer surface was sometimes difficult to handle, depending on the studied sample, and the analyzed area. Therefore, we suspected a non uniform thickness or composition of the thick polymer layer, and tested for its homogeneity by recording spectra from different 600 μm-wide spots.

<u>Table 1</u> :   PPQ stoichiometry (atomic percent) measured from core level peak areas for the four samples (see text), before and after 300°C anneal.

	C/N/O	
	BEFORE	AFTER ANNEAL
Sample A	85.7/6.7/7.6	90.6/7.8/1.6
Sample B	66.8/10.9/22.2	79.0/10.6/10.4
Sample C	84.9/6.7/5.8	87.9/9.6/2.5
Sample D	70.3/10.2/19.5	79.5/10.5/10.0

	C/N*	
	BEFORE	AFTER ANNEAL
Sample A	92.7/7.3	92.1/7.9
Sample B	86.0/14.0	88.2/11.8
Sample C	92.7/7.3	90.1/9.9
Sample D	87.3/12.7	88.3/11.7
Average A-->D	89.7/10.3	89.7/10.3
Theoretical	89.5/10.5	89.5/10.5

* Oxygen signal substracted

<u>Table 2</u> :   Core level binding energies measured for the four PPQ samples (see text), before and after 300°C anneal

		C	N	O
Sample A	before	285.0* + ...[a]	399.2	532.1
	after	285.0*	399.1-401.0	531.3-533.2
Sample B	before	285.0* + 288.7	400.4	533.0
	after	285.0 + 288.8	399.25-400.5	531.9-534.0
Sample C	before	285.0-286.1	399.2	532.3-533.6-534.8
	after	285.0*	399.3	532.2
Sample D	before	285.0*-286.3-287.7-288.9	399.9-401.3	531.1-532.1-533.5
	after	285.0* + ...[a]	399.4	533.6-532.0

* :   calibration reference
...[a] :   numerous unresolved structures of low intensity

Homogeneity : On the studied spatial scale, the carbon concentration is measured constant to ± 2,5 % for (at least) four different probed area on samples A-->C; this leads to reproducibility on nitrogen and oxygen signals of about ± 5 to 10 %. On sample D on the contrary, measurements were sensibly more reproducible (± 1.0 % on carbon) suggesting that the UV treatment in air, in addition to the grafting of oxygen functionalities on the polymer (see below), concurred also to homogenize its surface composition.

Oxygen functionalities on the polymer surface are here studied for their possible capacity to enhance the adhesion of a copper metallic film (to be deposited afterwards). Oxygen content : The polymer surface composition shows before a 300°C anneal (table 1) an oxygen content ranging from 5 % to 22 % : the oxygen species are specifically bond to carbon atoms for samples A-->C, as extra peaks appear only on the C_{1s} core level; for sample D (UV treatment), oxidized nitrogen species (20-25%) are also detected. However, a more detailed study is necessary. A closer look to the XPS results for the C_{1s} peaks (Figure and Table 2) indeed reveals that :

Sample A   contains less than 10 % of oxidized species before anneal, with an ill-defined composition, as no clear peak can be resolved in the 290-287 eV region;

Sample B   contains about 20 % of oxidized carbon, at a binding energy of 288.7 eV typical of multiple C-O, or of C=O species; whereas

Sample D   with a similar oxygen content, presents a much richer C_{1s} structure.

The nitrogen 1s signals are single peaked (though broad) for samples A and B, whereas sample D clearly shows a 20 % contribution of highly oxidized nitrogen with a peak at 401.3 eV. The binding energy table (Table II) suggests that both samples B and D (the most oxidized ones) are dominated by another type of nitrogen : the position of these broad peaks higher than 400 eV clearly indicates oxidized nitrogen species, with the "quinoxaline" like species at ~399.2 eV.

THERMAL TREATMENT

As it is part of the process to anneal the polymer before and after the metallization, it is necessary to consider the question of the stability of the oxidized species, added on the polymer surface in order to promote adhesion. Therefore, a gentle heating in vacuo up to 300°C for one hour was performed in order to free the polymer from any residual solvent or adsorbed water, and to test the oxidation stability.

From Table 1 and figure 2, it appears that in general more than 50 % of the oxygen is desorbed; the carbon 1s and nitrogen 1s spectra confirm that most of the oxygen functionalities disappeared. It is very interesting to note on N1s data the presence of two components, unambiguously identifying a "normal" nitrogen species at ~ 399.2 eV (quinoxaline type), and oxidized sites at 401 eV. The C, O and N1s features are sensibly much better resolved than for the non-annealed polymer, suggesting that the original oxygen chemistry was not very selective, ending with a multitude of different functionalities; only the most stable ones resist the anneal treatment. If oxygen is rare on the surface, it disappears quickly and almost completely during the annealing.

PPQ METALLIZATION

During the copper deposition on the polymers (as received or after annealing), up to a total thickness of 20 Å (equivalent thickness measured with the quartz microbalance), the XPS core level peaks, C1s-N1s-O1s and Cu2p - did not reveal any strong and particular chemical information. No large peak shift, no wide intensity variation (except the normal overlayer screening effect) is detected ... except for samples B and D. But these results need first to be presented and studied separately, before to draw general comments.

Sample A   (PPQ-Si_3N_4) : in the 0-20Å coverage (thereafter denoted θ) range, no particular event could be detected. Although a very precise calibration is difficult to obtain on low intensity signals with charging effect, we measured for the C1s-$Cu2p_{3/2}$ binding energy difference a higher value (647.9 eV and more) for Cu coverage below 1 Å. For θ = 5 to 10 Å, the overal C1s, O1s and N1s signal intensities decreased significantly.

Sample B   (PPQ-CrN) : during the copper layer growth, a slight but significant intensity decrease of the oxidized carbon species above 286 eV suggests that copper may bound to these sites. At low coverage ($\leq$ 1 Å) the Cu2p3/2 peak is pointed at slightly higher binding energy (> 932.9 eV) than for high coverages (932.5 eV), whereas, the C1s $\pi \rightarrow \pi^*$ shake up transition is seen to disappear rapidly (figure 3). Overal intensity decrease of the polymer XPS peaks is observed for larger ($\geq$ 5 Å) Cu coverage. Above 10 Å, the Cu $2p_{3/2}$ signal FWHM sharpens (from 2.2 eV at low θ, to 1.6 eV).

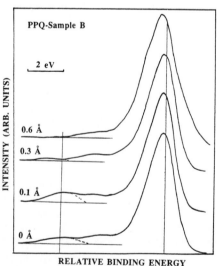

Figure 3 : Carbon 1s spectra of sample B, during Cu deposition. Note the high binding energy feature (π-π^* shake satellite) that seems to disappear

Sample C   (PPQ CrN, other recipee) : over the 0-10 Å coverage range, no particular event could be noted, except that for $\theta \leq 1$ Å, the $Cu2p_{3/2}$ peak is broad (2.1 eV) and at 0.5 eV higher energy than for larger θ.

Sample D   (PPQ UV) : when the metallization is performed on the "as received" sample, no particular feature is detected for a coverage up to 10 Å. However, if the polymer film is annealed (300°C, in vacuum) before the copper deposition, one notes (figure 4) that some oxidized carbon species decrease in intensity. As pointed out previously, the $Cu2p_{3/2}$ is recorded at ~ 0.5 eV higher binding energy for $\theta \leq 1$ Å, it is broad, and remains broad up to $\theta > 10$ Å. This strongly suggests that the copper atoms are bound to oxygen ones.

Figure 4 :   Carbon 1s, nitrogen 1s, oxygen 1s and copper 2p spectra of sample D (after 300° anneal) during metallization

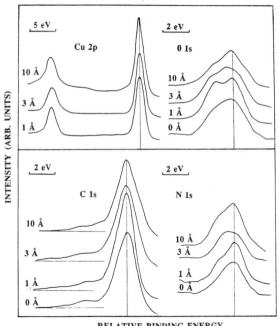

RELATIVE BINDING ENERGY

INTERPRETATION

A weak interaction of copper with the polymer was expected : this is indeed what is observed as, only for the PPQ with highest oxygen content, some interaction with the carbonyl- and ether-like carbon are evidenced; but essentially no further chemical shift was appearent on the C_{1S} spectra. At very low coverage ($\theta \leq 1$ Å) some perturbation of the weak shake-up structure has been noted, but for high coverage this structure can be seen again. This behaviour has already been noted for poly dimethyl phenyloxide and should fingerprint the still high mobility of the copper atoms or clusters (Grégoire et al, 1991). As for the Cu signal, a slighly higher binding energy is recorded for low θ (≤ 1 Å), whereas its width decreased from 2 to 1.5 eV for large θ values (≥ 5-10 Å), except for sample D.

For all the studied Cu-PPQ interfaces, the intensity of the Cu3p$_{3/2}$ signal versus coverage follows the same linear curve : on a microscopic point of view, all the materials behave similarly, and the sticking coefficient of Cu on the polymer is thus supposed constant. The fact that the polymer XPS core level signals become attenuated only for Cu coverage in the 10 Å range indicates heterogeneous surface coverage. Therefore, it is suggested that the Cu condenses of the PPQ polymer as small clusters, followed by a growth of these islands. On a weakly interacting substrate, the cluster core level binding energy is expected to *decrease* to the bulk value as the cluster size increases (Mason, 1983); in this experiment, the very small clusters are identified by the higher binding energy of the Cu 2p$_{3/2}$ peak ($\theta \leq 1$ Å), and it is possible that these copper atoms cluster around/on the aromatic rings (affecting the C1s shake-up satellite); for sample D (and perhaps B), the metal atoms are anchored around oxygen sites of the polymer substrate. Around $\theta \pm 10$ Å, the islands should coalesce to the metallic state, as the Cu2p$_{3/2}$ peak returns to a nominal binding energy (932. 5 eV); as for sample D, no similar line width evolution is recorded, suggesting that the metal-polymer interaction develops through some oxygen-copper charge transfer : copper is partially oxidized, but not as CuO, as no typical Cu shift and shake-up satellite is recorded; as well, no oxygen species characteristic of an oxide is detected below 531 eV. It is expected a lengthier report (to be published) will provide more detailed interpretation of these data.

CONCLUSION

In this study, we have shown that oxygen, although not present in the polymer repeating unit, plays a crucial role in the properties of the PPQ surfaces. It probably comes from the water intake by the polymer and its concentration at the surface is usually in the range of several atomic per cents. In some cases it may be larger than 20 % and there high concentrations are associated to carbon atoms oxidized in the form of C=O and O-C=O species. Upon UHV annealing, the oxygen on these highly oxidized surfaces is not desorbed as easily as on the standard surfaces.

Copper deposited on PPQ does not form strong chemical bonding, as was also noticed for other polymers (Kim et al 1987). As deduced from the rapid decrease of the C(1s) shake up satellite corresponding to π-π* transitions, copper atoms preferentially bond to the aromatic rings at the very first stages of interface formation. It is also suggested that Cu condenses on the polymer surface as small clusters.

REFERENCES

ATANASOSKA L.J., ANDERSON S.G., MEYER H.M. III, LIN Z., WEAVER J.H., J. Vac. Sci. Technol. (1987), A5, 6; A5, 3325

BURKSTRAND J.M. (1979), Phys. Rev. B20, 4853

GREGOIRE Ch., Noël Ph, Caudano R, and Pireaux J.J., these Conference Proceedings.

HO P.S., HAN P.O., BARTA J.W., RUBLOFF G.W., LE GOVES F.K., SILVERMAN B.D., J. Vac. Sci. Technol. (1985), A3, 3
KIM Y.-H., WALKER G.F., KIM J. and PARK J., J. Adh. Sci. Technol. (1987), I, 331

MASON M.G., Phys. Rev. B (1983), 27, 748

Angle resolved x-ray photoelectron spectroscopy (XPS) of aluminium/polyethylene terephthalate interfacial chemistry

Quoc Toan Le*, M Chtaïb**, J J Pireaux* and R Caudano*

*   Facultés Universitaires Notre Dame de la Paix,
     Laboratoire Interdisciplinaire de Spectroscopie Electronique (LISE)
     rue de Bruxelles, 61, B-5000 Namur, Belgium
** Laborlux S. A., B. P. 349, L - 4004 Esch Sur Alzette, Luxembourg

Abstract

Angle resolved X-ray Photoelectron Spectroscopy (XPS) has been used to study the interfacial chemistry between thermal deposited aluminium (30 - 50 Å) and Mylar® (polyethylene terephthalate, Du Pont de Nemours). Two types of Mylar® of different processing condition were used for measurement. In both cases, the quantity of aluminium oxide and hydroxides decreased as a function of analysis depth.

At the PET/aluminium interface, besides the presence of a weak concentration of metallic aluminium (10 - 18 %), and a mixture of aluminium oxide and hydroxides, the existence of Al - O - C bonds was also detected. The XPS results suggest the preferential reaction sites are the oxygen of carbonyl groups (PET). Aluminium is bonded to oxygen resulting in the appearance of low binding energy components in C 1s and Al 2p spectra (compared to C-C in C 1s and Al oxide & hydroxides in Al 2p spectra). The proportion of these bonds varies from 3 to 9 % of aluminium atoms presented at the interface, depending on the analysis depth (take-off angle) and the type of Mylar®.

I. INTRODUCTION

The application of metallized polymers can be found in different fields, such as food packaging, capacitor fabrication, video and data storage, etc...(Ross 1978, Burell et al 1989). The chemical nature of groups at the polymer / metal interface plays an important role in the adhesion aspect. Several reports involving polymer / metal interfaces have appeared in the literature (DeKoven et al 1986, Anderson et al 1988, Rancourt et al 1990). X - Ray Photoelectron Spectroscopy (XPS) was used to investigate the interaction between metal / polymer (Pireaux et al 1990) and indicated the formation of a metal - oxygen - polymer bond, for example in the case of Cu, Ni, Cr with polyvinyl alcohol, polyvinyl methyl ether, polyvinyl acetate and polymethyl methacrylate (Burkstrand 1979 and 1981), or Al with polyimide (Atanasoska et al 1987). More recently, it was shown, using synchrotron radiation (SR) photoemission and XPS, that the Cu atoms interact with PET to form Cu - O - C complexes at the interface (Chtaïb et al 1991); Gerenser (1990) has observed the formation of Ag - O - C complexes at the PET / Ag interface. XPS was also used to study the adhesion mechanism between Al and PET (Bou et al 1990, De puydt et al 1988, Jugnet et al 1989).

With XPS angle dependent (a non - destructive technique for depth analysis), this work studied the interfacial chemistry between thermally evaporated aluminium and PET.

II. EXPERIMENTAL

The XPS core level spectra were obtained on a Surface Science Instruments SSX - 100 X - Ray Photoelectron Spectrometer. Photoelectrons were excited by monochromatized Al K_α X radiation (1486.6 eV) in an ultra-high vacuum system whose operating pressure was about 1.10^{-9} Torr. For this study, all samples used were analyzed at ambient temperature and no x-ray damage to the samples was detected during measurements. The measurements were performed at various electron take-off angle (TOA, sample - analyzer angle), i.e 10, 20, 35, 55, 70 and 80°. Note that a TOA of 35° corresponds to an analysis depth of ~35 Å.

Reference binding energy was taken at 284.6 eV for the C 1s peak. The resolution on the aromatic C 1s peak of PET was 1.1 eV. XPS survey scan spectra (0 - 1100 eV) and narrow scan spectra (20 eV) were acquired using the x-ray spot size of 1000 μm and 600 μm diameter, respectively.

Two types of Mylar® (12 μm thick PET, Du Pont de Nemours) were used for aluminium deposition. During the process, the first sample (A) was thermally fixed at a temperature higher than that of the second (sample B) (235 °C for sample A and 210 °C for B). Metallization was carried out on a Balzers BAE 370 metallizer under high vacuum conditions. The aluminium (Al thickness ~30 Å) was deposited onto the PET according to the following conditions :

- Pressure : 2.10^{-5} hPa
- Evaporation rate : 10 Å/sec
- Vertical wire _ sample distance : 22.5 cm

III. RESULTS AND DISCUSSION

The following table summerizes the correlation between TOA and the analysis depth
(Cadmann et al 1978) :

T.O.A (°)	App. Depth Analysis (Å)
10	11
20	21
35	34
55	49
70	56
80	59

III. 1. Virgin PET

On the virgin PET surface, the C 1s core lineshape consists of four distinct peaks : the first one corresponds to the carbon atom in the phenyl ring (284.6 eV), the second is attributed to the carbon singly bonded to oxygen (286.2 eV), the third component is the signal of the ester carbon atom (288.6 eV) and the last one of weak intensity at higher binding energy is due to the π - π* shake-up transition (291.1 eV) (Figure 1).

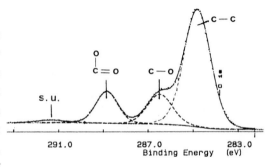

Fig. 1 : C 1s spectrum of virgin PET

The O 1s spectrum of PET exhibits three components : the peak due to the carbonyl (531.3 eV), the ether (532.9 eV) oxygen atoms and the presence of the π - π* transition at higher binding energy (Figure 2).

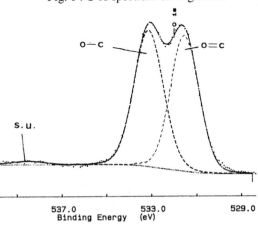

Fig. 2 : O 1s spectrum of virgin PET

III. 2. PET type A

After aluminium deposition , the C 1s and O 1s PET signals were considerably modified, due essentially to the screening effect of the Al layer. Figure 3 and Figure 4 show the evolution of C 1s and Al 2p spectra, respectively, when TOA increases from 10° to 70°. For the TOA higher than 20°, the C 1s photoelectron signal shows the appearance of a new component at low binding energy (~283.6 eV) which is associated with the C - O - Al bond (Bou et al 1990). A 35° TOA should correspond to the beginning of the PET/Al interface. According to the relative concentration of the ester (O = C - O) and ether (C - O) species in Figure 3, we suggest that the potential site for the formation of C - O - Al complexes is the carbonyl group. Furthermore, it is interesting to notice that in Figure 3, we observed a net chemical shift for the C - O group to lower binding energy of about 0.4 eV, this indicates the existence of the interaction between the ether oxygen and aluminium (Gerenser 1990). As can be seen in Figure 4b, the Al 2p spectrum can be decomposed, according to our Al_2O_3 and $Al(OH)_3$ reference, into three distinct components : Al metal (72.2 eV), Al hydroxides ($Al(OH)_3$, $AlO(OH)$, ..., 74.5 eV) and Al oxide (75.1 eV) (Ocal et al 1985, Strohmeier 1989). The formation of the C - O - Al bond was also detected in Al 2p signal by the appearance of the new component at a binding energy around 73.0 eV; again, it is observable for the TOA higher than 20° (Figures 4c, 4d & 4e). Depending on the TOA, the Al metal concentration varies from 7 to 13 %.

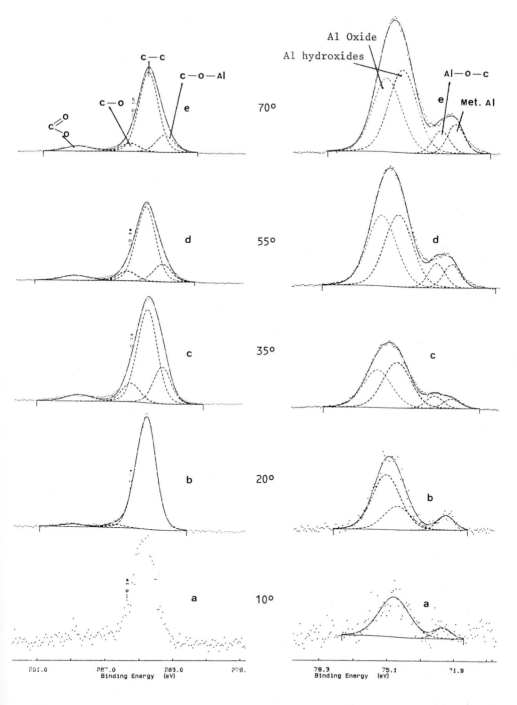

Fig. 3 : C 1s spectra measured at
different TOA (from 10° to 70°)

Fig. 4 : Al 2p spectra measured at
different TOA (from 10° to 70°)

The O 1s photoelectron signal is the signature of many components : O = C and O - C of PET, O - Al from Al oxide and hydroxides (chemical shift ~ 1.5 eV) and Al - O - C; hence the similation of the O 1s spectrum is complex and less accuracy. In Figure 5, we show an example of O 1s spectrum which can be decomposed into three distinct components : O - Al (oxide) at low binding energy, the second one is a mixture of O = C, O - Al (hydroxides) and Al - O - C, the third component is attributed to O - C (PET).

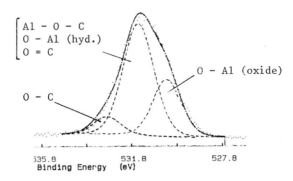

Fig. 5 : O 1s line-shape analysis after Al deposition
on Mylar®

In Figure 6, the evolution of the C = O group proportion taking part in the formation of C - O - Al bond (C 1s spectra simulation) versus TOA is presented (curve a). For 10° and 20° (low analysis depth), no PET/Al interaction was observed; at higher TOA, the C - O - Al complex was detectable and remained constant up to 80° of TOA. Similarly the Al proportion participating in the PET/ Al interaction (Al 2p spectra simulation) as a function of TOA is shown in Figure 6, curve b. Note the similarity of these two curves; the concentration obtained in the case of C 1s (curve a) is roughly double of that found for the Al 2p spectra, suggesting that one Al atom is bonded to two carbonyl groups. Figure 7 indicates the presence of more Al oxide and / or Al hydroxides at the uppermost surface.

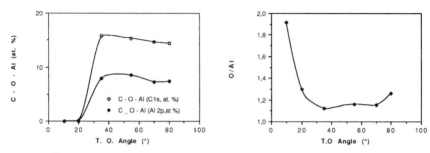

Fig. 6 : Evolution of the C - O - Al bond
proportion as a function of TOA (sample A).

Fig. 7 : variation of O/Al atomic ratio
versus to TOA (sample A).

III. 3. PET type B

For TOA values lower than 35°, no presence of Al - O - C bond is observed; the intensity of ester and ether carbon atoms is weak because of the screening effect of the Al layer. But for this sample, we detect the appearance of a component (chemical shift of 3.0 eV compared with the C - C peak) probably due to the presence of some carbonyl groups of a rupture of polymer chain during Al deposition. For different TOA, the existence of the Al - O - C (~ 73.0 eV) bond in the Al 2p spectra is in agreement with the C 1s spectra.

The variation of Al - O - C bond proportion (deconvolution of C 1s signal, Figure 8, curve a) and simulation of Al 2p spectra (Figure 8, curve b) reaches a maximum around 35° which corresponds to the beginning of the interface. We come to the same conclusion as the previous case. At the outermost suface, the concentration of Al oxide and hydroxides is more important (Figure 9) forming a thicker layer compared to sample A. The Al metal concentration varies from 6 to 18 % in the TOA range studied.

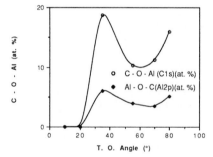

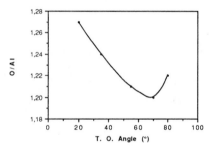

Fig. 8 : Evolution of the C - O - Al bond proportion as a function of TOA (sample B).

Fig. 9 : variation of O/Al atomic ratio versus to TOA (sample B).

IV. CONCLUSION

From our XPS results, we have tried to show the chemical nature of groups and the adhesion mechanism involved at the PET / Al interface by another approach. Al atoms react readily with the exited oxygen of PET to form the Al - O - C species. One Al atom could be bonded to two carbonyl groups of two different PET chains. No significant difference concerning the C - O - Al bond concentration was found between these two samples. In both cases, Al oxide and hydroxides in high concentration were detected on the outermost surface and at the PET / Al interface.

ACKNOWLEGMENTS

We would like to thank Dr. M. Koenig (Du Pont de Nemours, Luxembourg) for providing these samples. This work was supported by the EEC grant N° RI 1B - 0178 (BRITE Program).

REFERENCES

Anderson S G, Meyer H M and Weaver J H 1988 J. Vac. Sci. Technol. A6 (4) pp 2205 - 2212

Atanasoska Lj, Anderson S G, Meyer H M, Zhangda Lin and Weaver J H 1987 J. Vac. Sci. Technol. A5 (6) 3325

Bou M, Martin J M, Le Mogne Th and Vovelle L 1990 App. Surf. Sci. 47 149

Burell M C, Codella P J, Fontana J A, Chera J J and McConnell M D 1989 J. Vac. Sci. Technol. A7 55

Burkstrand J M 1979 J. Vac. Sci. Technol. 16 (2) 362

Burkstrand J M 1981 J. Appl. Phys. 52 (7) 4795

Cadmann P, Gossedge G and Scott J D 1978 J. Elect. Spectr. Rel. Phen. 13 1

Chtaïb M, Ghijsen J, Pireaux J J, Caudano R, Johnson R L, Orti E and Brédas J L 1991 Phys. Rev. B submitted

De Koven B M and Hagans P L 1986 Appl. Surf. Sci. 27 pp 199 - 213

De Puydt Y, Bertrand P and Lutgen P 1988 Surf. Interf. Anal. 12 486

Gerenser L J 1990 J. Vac. Sci. Technol. A8 (5) 3682

Jugnet Y, Droulas J L, Tran Minh Duc and Pouchelon A 1989 post-deadline poster, European Conference on Applications of Surface and Interface Analysis, Antibes, France

Ocal C, Basurco B and Ferrer S 1985 Surf. Sci. 157 233

Rancourt J D, Edie S L and Taylor L T 1990 Polymer Preprints 31 (2) pp 458 - 459

Ross D I 1978 RCA review 39 137

Strohmeier B R 1989 J. Vac. Sci. Technol. A7 (6) 3238

Pireaux J J, Riga J, Boulanger P, Snauwaert P, Novis Y, Chtaïb M, Grégoire C, Fally F, Beelen E, Caudano R and Verbist J 1990 J. Elec. Spect. and Related Phenomena 52 423

Interface effects in metal–polyparaphenylene–metal structures

T P Nguyen[a], H Ettaik[a], S Lefrant[a], G Leising[b]

[a]Laboratoire de Physique Cristalline, Institut des Matériaux, Université de Nantes, 2, rue de la Houssinière, 44072 Nantes Cedex 03 (France)
[b]Institüt für Festkörphysik, Technische Universität Graz, Petergasse 16, A-8010 Graz (Austria)

ABSTRACT : The effect of vapor deposited metal on polyparaphenylene (PPP) surfaces is investigated using X-ray photoemission spectroscopy. Depending on the nature of the metal, changes in carbon and metal spectra are sometimes observed and in these cases, there is oxygen formation on the surface of the polymer film indicating that oxygen-metal-carbon complexes are presumably be formed at the interface. The presence of these complexes, correlated with electrical characteristics of the metal-PPP-metal structures, shows that Schottky barriers are conditioned by the oxidation ability of the metal used as electrodes.

1. INTRODUCTION

It is well known that rectifying or ohmic contact of semiconducting devices can be obtained by choosing metals with appropriate work functions with regard to that of semiconducting materials (Sze 1985). In terms of energy band structure, the barrier height depends on the difference between the work function of the metal ψ_m and that of the semiconductor ψ_s. In actual fact experiments show, in many instances, that real surfaces behave quite differently and in particular, the potential barrier is often found to be independent of the work function of the electrode metal. This can be caused either by surface states or by chemical reactions between the semiconductor and the metal (Henisch 1984).

The electrical properties of metal-PPP-metal structures have been studied previously (Nguyen *et al* 1991). The characteristics current density versus applied field are found to be omhic or rectifying depending on the metal used as electrode. In this work, X-ray photoemission spectroscopy is used to study the electronic structures of metallic overlayers on PPP thin films and from the experimental results, some correlations are drawn between the contacts and the electrical characteristics of metal-PPP-metal structures.

2. EXPERIMENTAL

Thin films of PPP are prepared from a precursor polymer route by casting solutions on copper or stainless steel substrates. After drying under a pure argon gas flow, the precursor films are annealed in a high vacuum furnace at 400°C. Metallic electrodes (Al, Au, Cu, Ni, Cr) are deposited on the polymer films by evaporation under high vacuum conditions ($< 10^{-6}$ Torr), the substrate being kept at ambient temperature. Metal-PPP-metal structures are obtained in the same conditions on glass substrates and the I/V characteristics are measured by a conventional method. The XPS experiments are performed in a

Leybold LH12 ESCA analyser (Université de Nantes-CNRS). The core levels of carbon, oxygen and electrode metals are recorded by using Mg Kα radiation (hν = 1253.6 eV), the pressure of the chamber being kept in the 10^{-9} Torr range during the experiment. Binding energy data are referenced to the Au $4f_{7/2}$ line (84 eV) of a gold plate fixed on the sample holder. The data are processed using computer programs allowing background and satellites substraction, smoothing and integration.

3. RESULTS AND DISCUSSION

To study the metal-polymer interface, we first record the XPS spectra of the surface of the as-deposited sample (polymer covered with metallic electrode). Then, subsequent erosion of the metallic electrode by Ar$^+$ bombardment with controlled flux and accelerating voltage is performed for a given time. Core level spectra of carbon, oxygen and electrode metal are recorded and the process is repeated until reaching the pure PPP film (by observing the intensity of both C 1s and metal lines and by comparing the C 1s position with that of uncovered PPP film).

On uncovered PPP films, the C 1s line is located at 285 eV (curve e, fig.1) with a full width at half maximum (FWHM) of 1.5 eV. Under Ar$^+$ bombardment conditions described above, no change of this line is observed. Therefore, the erosion process does not modify the surface of the films. We find that the films are slightly contaminated by oxygen despite the fact that they are stored *in vacuo* before analyses ; the contaminated O 1s line is located at 533.5 eV but its intensity is very weak.

Figure 1 shows the O 1s, C 1s and Al 2p observed throughout the Al-PPP interface. We can notice an important proportion of oxygen in this region; the O1s peak is shifted from 532.6 eV (O 1s in Al$_x$O$_y$) to 531.8 eV and is very different from the contamination component. The Al 2p spectra show also a continous evolution from 75.6 eV (oxidized aluminium Al$_x$O$_y$) to 74.2 eV whereas the C 1s line located at 284.5 eV throughout the interfacial layer. The shift in binding energy of the carbon line with respect to uncovered PPP films is compatible with findings in metal carbide. On the other hand, aluminium appears in an oxidized state suggesting the formation of an Al-O-C complex.

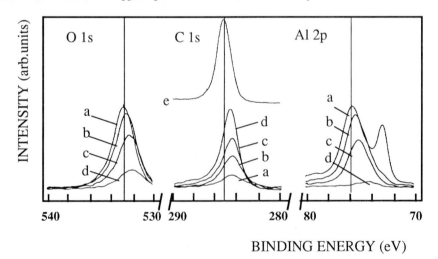

Fig. 1 : XPS spectra of PPP films : (1), pristine PPP film (curve e) ; (2), in the interfacial Al/PPP layer (curves a,b,c,d) with successive removal of the Al layer.

Figure 2 shows the C 1s and Au $4f_{7/2}$ lines observed in the Au-PPP interface. No trace of oxygen is found in this case. We observe that neither the carbon nor the gold spectra have changed in position.

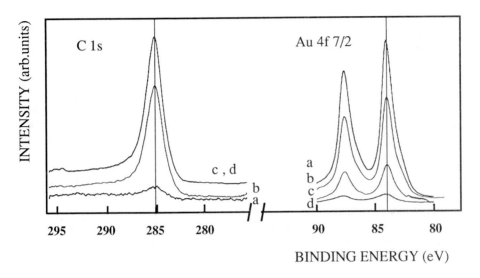

Fig. 2 : XPS spectra of C 1s and Au $4f_{7/2}$ measured in the interfacial Au/PPP layer with successive removal of the Au layer. Notice that curves c and d of the C 1s region are merged together.

This suggests that no reaction between the polymer and the metal has occured : the deposited gold would form clusters embeded in the polymer surface. We can also notice that gold has diffused inside the PPP film. In fact, curves c and d show that whereas the pure polymer is reached (no change in intensity of the C 1s line), the Au $4f_{7/2}$ line is not completely vanished yet. The XPS results for the Cu/PPP and Cr/PPP systems are similar to that of Au/PPP but diffusion of metallic atoms inside the

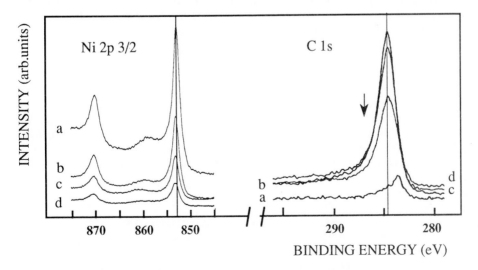

Fig.3 : XPS spectra of C 1s and Ni $2p_{3/2}$ measured in the interfacial Ni/PPP layer with successive removal of the Ni layer.

polymer film is not observed in these cases. Clustering of Cu and Cr seems to occur during metallization of the PPP films.

The spectra of the figure 3 show that the Ni-PPP interface is divided into two different parts : the outer part (corresponding to curves a) with the C 1s peak located at 283.5 eV and the inner one (corresponding to curves b, c and d) with the C 1s peak located at 284.5 eV. In the outer part, the Ni $2p_{3/2}$ peak, located at 852.6 eV, is slightly shifted to a higher binding energy with regard to the pure Ni. This indicates the formation of Ni-C complex on the surface of the metal layer. In the inner part, the Ni spectrum shows a slow evolution towards a final peak position at 853.1 eV. On the other hand, a small shoulder located at 286.6 eV (marked by arrow) is observed in the C 1s spectrum corresponding to this part. Notice that O 1s features appear but their intensity is rather weak. These observations suggest that there is intermixing formation of oxygen-carbon and nickel in the interface near the polymer side.

As one can see, the PPP-metal interface structure depends strongly on the nature of the metal layer. Gold, copper and chromium seem to have no chemical effects on PPP contrarily to nickel and aluminium which form complexes with the carbon in the polymer. The presence of oxygen is observed in the latter cases and its origin is not clear yet. It seems to be due to the oxidation of the electrode on handling and transferring the sample in the spectrometer; the oxidation kinetics of these metals being more important than that of Au, Cu or Cr.

Electrical characteristics of metal-PPP-metal structures show that Al and Ni formed rectifying contacts whereas Au, Cu and Cr formed ohmic ones. Comparing the known work functions of the used metals (Handbook of Chemistry and Physics 1988), we see that the nature of the contact is not in close relation with the value of the work function as found in the Schottky barrier formation in polyacetylene (Waldrop *et al* 1981). Alternatively, transfer of electronic charge between a semiconductor and a metal can be represented by the difference of electronegativity (Kanicki 1986). This approach does not either give satisfactory explanation on the observed experimental contact. On the contrary, XPS data suggest that the electrical behaviour of the structures studied depends directly on the nature of the interfacial layer. Rectifying contacts are apparently due to the formation of complexes at the interface which tends to form potential barriers to the polymer films. The chemical reactions seem to be due to the fact that metals such as Al and Ni are easily oxidized in air and the corresponding kinetics are much faster than that of the other metals.

4. CONCLUSION

We have studied the interfaces of PPP thin films covered with different metals. Using XPS technique, we find that (i) Al and Ni layers have chemical reactions with PPP films and form oxide-metal-carbon complexes in the interfacial layers, (ii) Au, Cu and Cr form clusters in the polymer film, diffusion of gold inside the PPP was also detected. The nature of the interface seems to be the origin of the rectifying contacts observed in metal-PPP-metal structures when using Al and Ni as metal electrodes.

REFERENCES

Sze S M 1985 *Semiconductor Devices* (New York : John Wiley & Sons) 160
Henische H K 1984 *Semiconductor Contacts* (Oxford : Clarendon Press) 25
Nguyen T P, Ettaik H, Lefrant S and Leising G 1991 *Synthetic Metals* **44** pp 45-53
Handbook of Chemistry and Physics 1988 (Cleveland : CRC Press Inc.) E91
Waldrop J R, Cohen M J, Heeger A J and Mac Diarmid A J 1981 *Appl. Phys. Lett.* **38** pp 53-5
Kanicki J 1986 *Handbook of Conducting Polymer* ed T Skotheim (New York : Dekker) ch 17

Polymer–metal interface formation after in-situ plasma and ion treatment

S. Nowak, M. Collaud, G. Dietler, O.M. Küttel and L. Schlapbach

Physics Department, University of Fribourg, Pérolles, 1700 Fribourg, Switzerland

ABSTRACT: This study presents the surface and interface analysis of in-situ plasma and ion-treated polypropylene surfaces with an evaporated magnesium overlayer. The applied surface treatments are found to enhance the sticking probability of the metal vapour on the polymer surface considerably. Depending on the kind of surface treatment, different types of bonding of the Mg with the polymer are observed. The experimental results suggest that the Mg grows in form of clusters.

1. INTRODUCTION

In a number of technical applications of polymers the low intrinsic adhesion properties of these materials require an adequate surface pretreatment. One of the purposes of these treatments is to enhance the generally low surface energy in order to enable wetting of the polymer surface by some kind of coating. Wetting of the polymer is however only a necessary but not a sufficient condition for good adhesion. This goal is only reached if a stable and strong interaction can occur between the polymer and the coating. Plasma and ion treatment of polymers are some of the more recent methods to achieve improved adhesion in systems such as polymer-adhesive (Liston 1989, Occhiello et al. 1991) or polymer-metal interfaces (Gerenser 1990). Although the merits of these kinds of surface treatments are generally accepted there remains a lack of understanding of the detailed microscopic mechanisms responsible for the improved adhesion. The effects of the surface treatment include cleaning by ablation of low molecular weight material, activation of the surface, dehydrogenation, change in surface polarity and wetting characteristics, crosslinking in a surface layer, graphitization and structural modification.

Surface analysis of **plasma** treated polymer surfaces as well as interface formation have been investigated in numerous studies (e.g. Gerenser 1988, Grant et al. 1988, Klemberg-Sapieha et al. 1991). In many cases, however, it is difficult to distinguish between the effects of the surface treatment itself and a reaction of an activated surface with the ambient. Whereas applications such as adhesive bonding mostly involve such an atmospheric contact others such as coating in a vacuum chamber (e.g. metallization) do not require this step. It is therefore important to characterize the surface under well-defined conditions without the exposure to the atmosphere. The ability of **ion** bombardment to improve the adhesion properties in polymer-metal interfaces is also well documented (Bodö et al. 1986; Vasile et al. 1989).

In a plasma, the surface is exposed to a broad spectrum of ions, electrons, neutrals and electromagnetic radiation each of which may have an influence on the interface formation with a subsequent metal deposit. In the case of an ion bombardment this aspect can be reduced to the nature, the energy and the flux of ions reaching the surface. One important difference between a plasma and an ion treatment is the energy of the ions reaching the surface as well as the associated ion current. Whereas the energy is typically much higher in the case of an ion treatment, the current respectively the ion flux is in general higher for a plasma treatment.

In this contribution, we report on the surface modification of polypropylene following an in-situ plasma or low energy ion treatment as well as the interface formed between a treated surface and a subsequent deposition of a thin evaporated Mg film. X-ray photoelectron spectroscopy (XPS) was used to analyse the surface composition before and after plasma treatment and metal deposition.

2. EXPERIMENTAL

The substrate for the present investigation was isotactic polypropylene (PP, monomer: $CH_2=CH-CH_3$) on the basis of Propaten HF 26 (ICI). The base material was mixed and extruded with 0.2% ®Irganox 1010 (Ciba-Geigy) and 0.1% Ca-stearate at 260°C. After cooling, the PP pellets were pressed to plates of 0.8 mm thickness in a heat mold at 230°C.

The polypropylene surfaces were treated in a low-pressure plasma obtained by electron cyclotron resonance (ECR) with permanent rare-earth magnets at 2.45 GHz microwave frequency at pressures ranging from $4 \cdot 10^{-4}$ up to $2 \cdot 10^{-2}$ mbar. The samples were treated in different gases, namely argon, oxygen and nitrogen. The plasma treatment was performed in a high vacuum plasma chamber (base pressure $4 \cdot 10^{-8}$ mbar) connected to the surface spectrometer which enabled in-situ surface analysis, i.e. without atmospheric contact, to be performed. The ion treatment was carried out in the preparation chamber of the surface spectrometer using an argon sputter gun. The ion dose was determined by measuring the sputtering current.

Thin metal films were deposited by thermal evaporation in the preparation chamber of the spectrometer. The presented experiments were performed with magnesium. This element, although not of practical relevance for polymer metallization, is well documented with respect

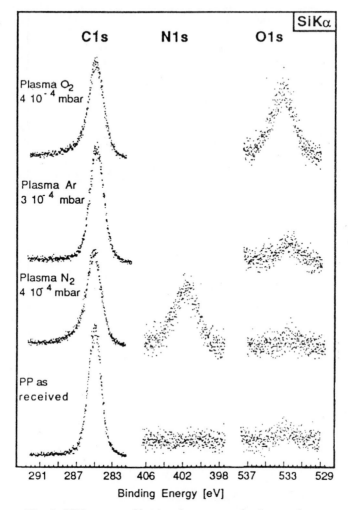

Fig. 1 XPS spectra of in-situ plasma treated polypropylene

to photoelectron spectroscopy (Peng et al. 1988) and allows detailed interaction with the polymer surface to be observed. Typical film thicknesses measured by a quartz crystal microbalance (QCM) were between 1 and 64 monolayers (1 ML = 2.25 Å for Mg), the film thicknesses representing the equivalent thickness to which the polymer surface was exposed. X-ray photoelectron spectroscopy (XPS) was performed in a VG ESCALAB 5 spectrometer at a base pressure lower than 10^{-10} mbar. Experiments were carried out with non-monochromatized Mg Kα (1253.6 eV) and Si Kα (1740 eV) radiation. Charging of the polymer samples was corrected by setting the energy of the main hydrocarbon component of the C 1s peak at 285.0 eV.

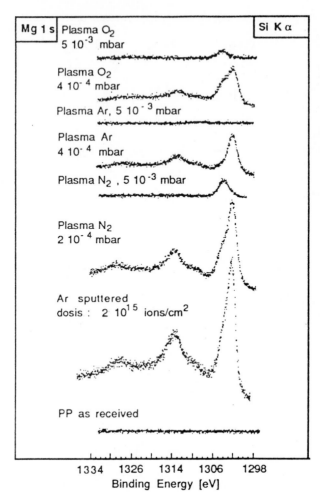

Fig. 2 Mg 1s XPS spectra of polypropylene after different surface treatments and exposure to 13 monolayers of evaporated Mg

3. RESULTS

We shall first consider the effects of the two kinds of surface treatments on the elemental surface composition without any metal deposit. For the case of the plasma treatment, well-defined surface functionalities could be obtained. Depending on the nature of the gas different surface compositions were observed as can readily be seen on the C 1s peak. Whereas a treatment in argon leads to clean hydrocarbon surfaces, plasmas of oxygen and nitrogen result in the according C-O and C-N functionalities (Gerenser 1987) which can be identified by the shoulder of the C 1s peak towards higher binding energies (figure 1). A maximum of surface functionalization was observed for a treatment time of 60 seconds. All of the following results were obtained at this treatment time. The observed C-O and C-N functionalities showed also some variation depending on which gas pressure was used for the plasma treatment. In general, a higher amount of these functionalities was observed at higher pressures. Up to 15 % of oxygen in an oxygen plasma and up to 6 % of nitrogen in a nitrogen plasma were built into the surface. The Ar ion bombardment on the other hand leads to clean but reactive hydrocarbon surfaces. Upon atmospheric contact, all of the treated surfaces oxidize to a certain degree. The capability of in-situ plasma and ion treatment is therefore very useful in producing clean and well defined surface compositions.

Metal evaporation experiments were performed following plasma treatment for two pressure values for each of the gases used. The polymer surfaces were then exposed to a constant amount of Mg (13 monolayers), evaporated at constant rate and analyzed by XPS. A very large variation of the sticking probability of the metal vapour could be observed. As it can be seen from the Mg 1s spectra displayed in figure 2 (bottom trace), a polypropylene surface without any surface treatment, does not acquire any measurable amount of Mg. As we have shown in an earlier contribution (Nowak et al. 1991), the Mg was found to be reflected from the surface. This means that the sticking probability of Mg on as received polypropylene surfaces is low. The plasma treatment leads to increased sticking probabilities depending on the nature of the gas and the pressure used. Oxygen and argon plasmas lead to similar increases of the amount of Mg, whereas a nitrogen plasma leads to the highest amount of Mg of the three gases used. With respect to pressure, the lower value favours the sticking of the Mg for all of the three gases. For comparison, one trace in figure 2 shows the result of a polypropylene sample bombarded with Ar^+ ions at a dose of $2 \cdot 10^{15}$ ions/cm^2. This ion bombardment leads to similar amounts of Mg, in other words to a similar increase in sticking probability as the plasma treatment in nitrogen.

Beside the large variation in the observed sticking probability, the photoelectron spectra also indicate different types of bonding of the Mg at the interface. In the cases where there are no O- or N-functionalities at the polypropylene surface, the Mg is found in a purely metallic state. This is true for the ion bombardment as well as the plasma treatment in argon at the low pressure value. The corresponding Mg 1s photoeletron spectra show the excitation of plasmon losses which is typical for a metallic state of the Mg overlayer. As we will show later, the Mg is directly bonded to a free carbon site. For the cases of a plasma treatment in oxygen or nitrogen, at the higher pressure value, the Mg is found to be in a non-metallic state. The binding energy of the Mg 1s peak is found up to 1.7 eV higher than for the metallic state. The Mg appears to be bonded via the existing C-O or C-N functionalities to form a metal-oxygen-carbon or a metal-nitrogen-carbon complex as observed by Gerenser (1988). In these cases the Mg 1s trace (figure 2) does not show any plasmon excitation which is typical for a non-metallic nature (Mg-O or Mg-N bonding) of the Mg film (Peng 1988). At the lower pressure, the sticking probabilitiy increases, as discussed above, and more Mg is found at the surface. This Mg is found to be partly in a metallic state and partly in a non-metallic state, the metallic state being responsible for the excitation of plasmon losses. Since the amount of Mg is larger in this case, the observed photoelectron spectra can be interpreted in terms of an interface layer in which the Mg interacts with the oxygen respectively with the nitrogen and a metallic top layer.

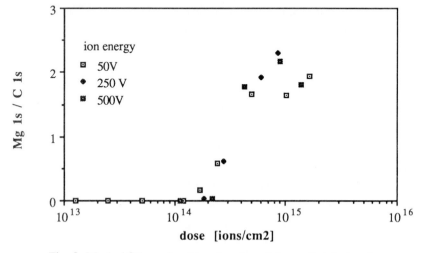

Fig. 3   Mg 1s / C 1s peak ratio as function of the applied Ar ion dose

As pointed out in an earlier contribution (Nowak et al. 1991), the observed increase in sticking probability following an ion treatment requires a minimum ion dose. This behaviour, expressed in terms of the relative increase of Mg to C ratio, i.e. the integrated peak ratio Mg 1s / C 1s corrected for the cross sections, is shown in figure 3 as a function of the ion dose for different ion energies. In the range of the ion energy used for these experiments, the minimum ion dose $2 \cdot 10^{14}$ ions/cm^2 for increased sticking does not depend on the ion energy. On the other hand, the increase in sticking of Mg on the treated polypropylene surface reaches a saturation above an ion dose of $5 \cdot 10^{14}$ ions/cm^2.

In order to follow the evolution of the interface formation of an ion treated surface with evaporated Mg, in figure 4, we present the Mg 1s spectra for different thicknesses of the Mg film, ranging from 1 to 64 ML. For the lowest coverage, the Mg appears to be in a non-metallic state. The energy of the elastic photoelectron peak is situated at 1305 eV which is typical for Mg oxide. This oxide is however not responsible for the bonding with the polymer surface. As we will show later, the oxidation of the Mg appears to be a consequence of the x-ray induced effects in the polymer. For all of the following deposits, the Mg appears to be in a

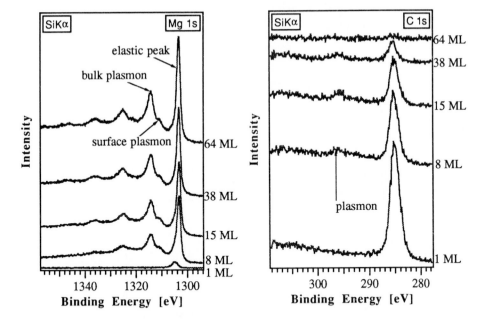

Fig. 4 Mg 1s spectra of polypropylene after Ar ion bombardment (energy = 100 V, dose = 8 · 10¹⁴ ions/cm²)

Fig. 5 C 1s spectra of polypropylene after Ar ion bombardment (energy = 100 V, dose = 8 · 10¹⁴ ions/cm²)

metallic state, as identified by the energy of the elastic peak at 1303.3 eV and the excitation of plasmon losses. As the film thickness increases, up to 4 bulk plasmon peaks can be distinguished in addition to the surface plasmon peak. Figure 5 shows the C 1s spectra going along with the evaporation experiment in figure 4. As expected, with increasing Mg film thickness, the C 1s signal decreases. However, the very slow rate of decrease points to the fact that the surface is not homogeneously covered with Mg. Even at the highest deposition of 64 ML, there remains a small C 1s signal. Two more features may be observed on the C 1s traces. The shape of the photoelectron peak appears to be asymmetric with a shoulder to the low binding energy side, which is characteristic for direct metal-carbon bonds. At coverages between 8 and 38 ML, the C 1s signal also shows a small plasmon feature. The energy difference to the elastic peak is 11 eV which is the value of the plasmon excitation of Mg. This plasmon appearance is a clear evidence for a carbon signal which originates from a region covered with Mg.

Figure 6 shows the C 1s peak before and after ion treatment as well as with a thin Mg film. The ion treatment is found to increase the line width. With Mg, as mentioned above, the C 1s

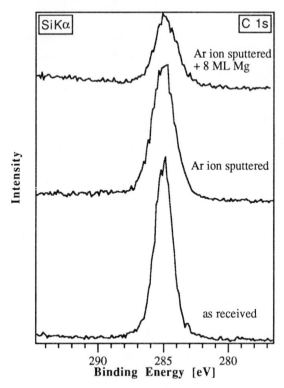

Fig. 6 C 1s spectra for different polypropylene surfaces

becomes asymmetric to the low binding energy side. This comparison clearly demonstrates the carbide-like bonding of the Mg with the polymer.

In figure 7, we plot the photoelectron peak ratios Mg 1s / Mg 2s as well as C 1s / Mg 1s as a function of exposure to Mg vapour. The behaviour of the Mg 1s / Mg 2s data follows what one would expect from simple mean free path considerations. A constant value of this ratio is reached above about 40 ML. This constant value is in good agreement with that of a bulk Mg sample, cleaned by argon sputtering.The C 1s / Mg 1s data prove that the polymer is only partially covered with Mg up to high equivalent film thicknesses.

The characteristics of the plasmon features may also be used to obtain more information on the interface evolution. As can be seen from figure 4, the first plasmon loss has a distinctive signature of a volume plasmon ($h\omega_p = 10.9$ eV) and a surface plasmon ($h\omega_p / \sqrt{2} = 7.7$ eV). It appears that, with increasing film thickness, the volume plasmon increases more rapidly than the surface plasmon. A fit procedure based on the work of Steiner et al. (1978) has been used to fit the measured plasmon losses. In figure 8, the surface plasmon creation rate is plotted as

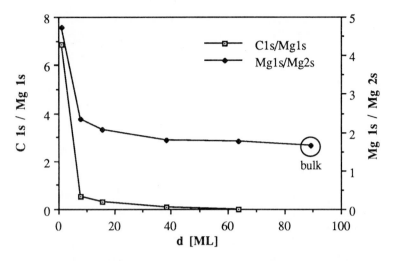

Fig. 7 Mg 1s / Mg 2s and C 1s / Mg 1s peak ratios as function of Mg film
thickness after Ar ion bombardment (energy = 100 V, dose = $8 \cdot 10^{14}$
ions/cm^2)

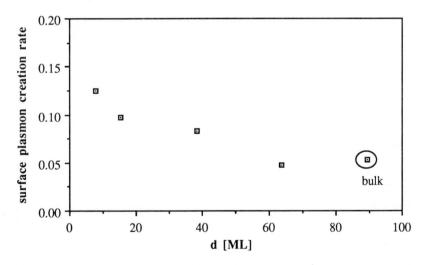

Fig. 8   Surface plasmon creation rate as function of Mg film thickness

a function of film thickness. It can be seen, that with increasing film thickness, this surface
plasmon creation rate decreases to reach the value of a bulk sample.

For a Mg film in a metallic state, the first plasmon intensity of the Mg 1s peak is typically 60
% of the elastic peak, independent of film thickness. In case of plasmon excitation for the C 1s

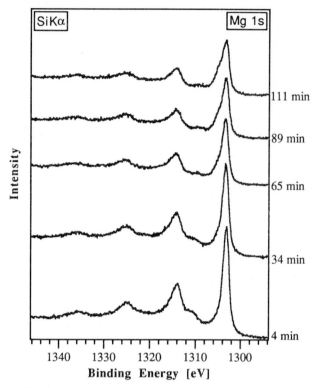

Fig. 9 Mg 1s spectra for prolonged x-ray exposure

peak, the intensity of the first plasmon is only about 20 % of the elastic peak. This means that a large part of the elastic C 1s peak originates from a region not covered by Mg.

In figure 9, the evolution of the Mg film under prolonged x-ray exposure is presented. It can be observed that the Mg gradually oxidizes. Although polypropylene does not contain any oxygen groups this oxidation occurs as a consequence of x-ray induced effects in the polymer. Evidence for these effects were obtained by mass spectrometry which showed strong water desorption of the polymer under x-ray exposure. It is however interesting to note that the oxidation seems to occur at the Mg-vacuum interface and not at the polymer-Mg interface. This can be deduced from the fact that the surface plasmon loss decreases and eventually vanishes with time which can only be explained if the surface becomes of non-metallic nature, e.g. oxidized.

4. DISCUSSION

In our earlier contribution (Nowak et al. 1991) it was found that an increase in sticking probability of Mg on PP occurs for a sufficiently high ion dose down to ion energies of 20 eV which was the minimum energy applied in our experiment. In an ECR-plasma, particle energies are particularly low, since these particles cannot follow the high frequency of the applied field. Typically, ion energies will be in the range of the plasma potential of the order of a few volts. The observation that the sticking probability increases also for a plasma treatment can be interpreted in several ways. Beside the ion dose which was found to be important for the Ar ion bombardment, the chemical and physical modification of the polymer surface by the plasma treatment has to be considered.

As pointed out above, the x-rays induce water desorption of the polymer which leads to an oxidation of the Mg film. This oxidation only occurs for a reactive overlayer and not for the naked polymer surface. In addition, the oxidation is faster for a thin film than for a thicker one. This is the reason why it was not possible to detect the metallic nature of the Mg film for the smallest coverage (1 ML) although this is anticipated to be the case before analysis. On the other hand, for a thick film, the oxidation was observed to be very slow. It appears that the Mg overlayer acts as a diffusion barrier.

The presented results suggest that the growth of the Mg is far from layer-by-layer. A more probable explanation for our observations is a cluster formation of Mg, i.e. insufficient wetting of the polymer by the Mg. Support for this kind of interpretation is obtained by the large amount of Mg necessary to reduce the carbon signal, the large fraction of the elastic C 1s peak compared to the associated plasmon loss as well as by the observed decrease of the surface plasmon creation rate with increasing film thickness.

It is interesting to note different types of bonding of the Mg with the polymer depending on the applied surface treatment. Whereas oxygen and nitrogen functionalized surfaces form the according metal-oxygen- respectively metal-nitrogen-carbon complexes, ion treated surfaces lead to metal-carbon type bonds. The adhesion of the Mg film to the polymer will also depend on the type of bonding at the interface.

The observed increase in sticking probability for the applied surface treatments has to be related to the adhesion in the observed system which has not yet been determined. The sticking probability is a measure of the chance that a metal atom is adsorbed to the surface. However, no direct information is obtained on the strength of the interaction with the substrate material. Nevertheless, a strong correlation of the increase in sticking probability as a function of the gas nature with the results of the adhesion of silver on plasma treated polyethylene by Gerenser (1988) can be observed.

5. CONCLUSION

The possibility of in-situ surface treatment of polymer surfaces allows well-defined and contamination-free surface compositions to be obtained. For evaporation experiments with

Mg, plasma and ion treated polypropylene surfaces show a strong increase of the sticking probability of the metal vapour. Depending on the nature of the applied surface treatment and the associated carbon-functionalities, the interface is characterized by the formation of a metal-oxygen-, a metal-nitrogen-complex or a direct metal-carbon bond. Plasma treatment in nitrogen and Ar ion bombardment lead to the strongest increase in sticking probability of the metal vapour. The growth of the Mg may be characterized by cluster formation due to inadequate wetting.

REFERENCES

Bodö P and Sundgren J E 1986 J. Appl. Phys. 60 1161

Gerenser L J 1987 J. Adhesion Sci. Tech. 1 303

Gerenser L J 1988 J. Vac. Sci Technol. A6 2897

Gerenser L J 1990 J. Vac. Sci Technol. A8 3682

Grant J L, Dunn D S and McClure D J 1988 J. Vac. Sci. Technol. A6 2213

Klemberg-Sapieha J E, Küttel O M, Martinu L and Wertheimer M R 1991 J. Vac. Sci. Technol. A9 in press

Liston E M 1989 J. Adhesion 30 199

Nowak S, Maurom R, Dietler G and Schlapbach L 1991 Metallized Plastics 2 (New York: Plenum) in press

Occhiello E, Morra M, Morini G, Garbassi F. and Johnson D 1991 J. Appl. Pol. Sci. 42 2045

Peng X D, Edwards D S and Barteau M A 1988, Surf. Sci. 195 103

Steiner P, Höchst H and Hüfner S 1978 Z. Phys. B 30 129

Vasile M J and Bachman B J 1989, J. Vac. Sci. Technol. A7 2992

ACKNOWLEDGEMENTS

We acknowledge fruitful discussions with P. Gröning. T. Greber helped us with the plasmon fit procedure. We thank H.P. Haerri (Ciba-Geigy, Fribourg, Switzerland) for providing us with the polymer samples. This work was supported by the Swiss National Science Foundation, NFP 24 and Ciba-Geigy.

AES and SSIMS study of the Al-PP adhesion phenomena: Correlation to peel-test measurements

V. André[*][1], F. Arefi[1], J. Amouroux[1], G. Lorang[2], Y. De Puydt[3] and P. Bertrand[3]

[1] Laboratoire de Chimie des Plasmas-ENSCP (Université Pierre et Marie Curie), 11 rue Pierre et Marie Curie, 75231 Paris Cedex 05, France
[2] Centre d'Etudes de Chimie Métallurgique- CNRS, 15 rue G. Urbain, 94407 Vitry sur Seine, France
[3] Université Catholique de Louvain La Neuve, Unité PCPM, 1 Place Croix du Sud, B1348, Belgium

Adhesion of polypropylene (PP) films with thin aluminium coatings (~20 nm) has been improved by applying a non equilibrium plasma to the polymer, prior to the *in situ* metallization.

Static SIMS has been used to achieve analyses of the non metallized plasma treated substrates. For very short treatment times (23 to 500 ms), a cleaning effect of the N_2 plasma as well as an incorporation of nitrogen and oxygen species has been detected.

AES depth analysis was performed on both treated and untreated metallized plastics to ensure at least a qualitative approach of the chemical composition of the Al-PP interface. It was observed that the charging effects, which usually occur on untreated samples during Auger spectra acquisition, were hardly existant in the case of the N_2 treated Al-PP interface. These results are consistent with an increase of the apparent surface conductivity detected by potential decay measurements.

A quantitative measurement of the adhesion was carried out with a peel-test well adapted to very thin metallic layers deposited on flexible substrates. We have been able to correlate the plasma treatment time to the increase of the peel-strength at the interface. We were then able to determine the optimum treatment time for a noticeable improvement of the adhesion properties.

[*] Current address: BASF AG, Kunststoff Laboratorium, Polymer Physics Division. D6700 Ludwigshafen. Germany

General Introduction

A plasma treatment and *in situ* metallization apparatus was used to improve the adhesion between polypropylene (PP) and a thin aluminium coating (20-25 nm). The process itself was already described [1] and it has been demonstrated that treatment times as short as 23 ms in a N_2 discharge with corona configuration of electrode was enough to improve the adhesion. In order to gain a better understanding of the chemical and physical modifications responsible for the improvement of adhesion, we have studied the surface modifications created by the plasma by an analysis of the non metallized polypropylene. SIMS used in static conditions on polymers, provides a unique "fingerprint" of the probed surface. We have carried the study for such short treatment times (23-400 ms) and have been able to detect the nitrogen containing species not detected by XPS. The aluminium-polypropylene (Al-PP) interface composition and the structure were used with the help of high resolution Auger Electron Spectrometry (AES). In an earlier work, the chemical composition of the interface as well as its physical structure were determined [2,9]. In this paper we will show with the help of AES, up to now not a common technique for insulating substrates, how the electrical conductivity of the polymer-metal interface is modified by the plasma treatment. Those results have been completed by Potential Decay Measurements performed on the treated non metallized polymer films as well as by an ion bombardment of the PP surface simulating a plasma treatment. Finally, we have correlated these results to a quantification of the Al-PP adhesion for different plasma treatment times. This was performed with the help of a U form peel-est adapted to a flexible thin substrate (8 μm) and a very thin coating (25 nm)[16].

I.Experimental

We will only describe here samples treated in a nitrogen low pressure plasma with a corona configuration (hollow electrode-cylinder). The exprimental conditions used for the treatment and the metallization as well as the apparatus are described elsewhere [1,16]. The polypropylene was isotactic (98%) and XPS or ISS analysis showed negligeable amount of oxygen on the virgin surface [2,16].

II. Static SIMS study of non metallized PP films

The static SIMS operating conditions were published elsewhere [2]. All spectra were acquired in the positive mode (ions Xenon, incident energy 4 keV). Direct spectra of treated and non treated PP were shown previously [2]. We will focus our attention on the trends of nitrogen, oxygen containing clusters as well as hydrocarbons. For all peaks the intensity will be divided by the one of the references chosen at 27 a.m.u corresponding to C_2H_3 [3]. The even mass peaks correspond to nitrogen containing clusters as shown by Budzikiewicz et al.[4].

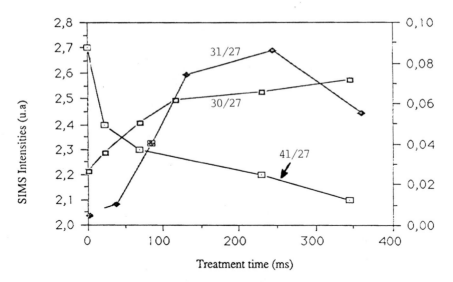

<u>Figure 1</u>. Static SIMS analysis of a N_2 plasma treated polypropylene film. Trend for pure hydrocarbon clusters (e.g 41 amu) oxygen clusters (e.g 31 amu) and nitrogen clusters (e.g 30 amu)in relative intensities (reference peak at 27 amu).

We will look at the trend of pure and oxidized hydrocarbon clusters (Figure 1). When the hydrocarbons are oxygen free clusters (e.g. 41 a.m.u, C_3H_5), we observe a decrease of their relative intensities (with respect to that of the peak at 27 amu) as a function of the treatment time. This could confirm the hypothesis that low molecular weight fragments are cleaned by the plasma as it was proposed previously [1]. When the clusters correspond to both oxidized and non oxidized fragments (43 a.m.u, C_3H_7 or C_2OH_3), the trend is similar which lead us to the conclusion that the cleaning effect is more important than the plasma induced oxidation (Figure 2). The latter is shown to occur by the intensity variations of the oxygen containing clusters with masses which do not interfer with pure hydrocarbon clusters. One hypothesis is that, in the case of amu which could correspond to both oxygen and hydrocarbon clusters (Figure 2), the oxygen containing clusters are masked by the hydrocarbons which always have a much higher intensity. The second suggestion is that the plasma creates relatively few oxidized new species (for treatment times of 23 and 115 ms) and a small part of the ones cleaned were present before the plasma treatment and not only created during the exposure to the discharge. This would explain the difference observed in the case of the nitrogen containing clusters (30 a.m.u, CNH_4), the intensity of which increases with the treatment time whereas no nitrogen species are present on the film prior to the plasma treatment (Figure 1 and 3).

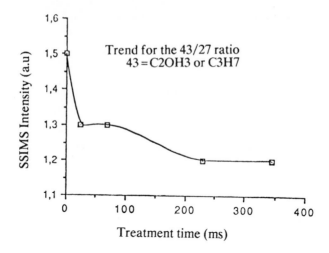

<u>Figure 2</u>.Static **SIMS** analysis of a N_2 plasma treated polypropylene film. Trend for pure and oxidized hydrocarbon clusters (e.g 43 amu) in relative intensities (reference peak at 27 amu).

The cleaning effect of the plasma is very clearly seen for hydrocarbon clusters. However, it is not possible to say if the low molecular weight fragments containing oxygen are a result of the plasma treatment induced oxidation of the light molecular weight surface contamination layer present before the treatment or of the PP film itself. The cleaning effect is more remarkable for treatment times below 0.5 s where the polymer degradation is very limited. The incorporation of nitrogen on the surface of the PP (Figure 1) increases as a function of the treatment time and a plateau is reached for longer treatment times [2]. It is particularly observed for low molecular weight nitrogen containing fragments. The cleaning of low molecular fragments and the nitrogen incorporation are simultaneous phenomena. Quantifications of these SIMS results is not possible nowadays and the amount of nitrogen and oxygen present on the surface of the plasma treated samples can not be deduced, either in an absolute or in a relative manner. A part of the detected oxygen may come from the air contamination [5] although some authors have reported this to be negligeable after an oxygen plasma treatment [6].

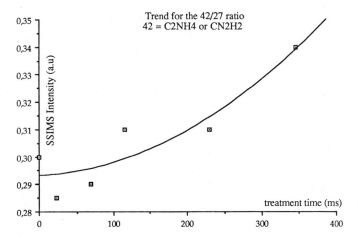

Figure 3. Static SIMS analysis of a N_2 plasma treated polypropylene film. Trend for nitrogen clusters (e.g 42 amu) in relative intensities (reference peak at 27 amu).

III. Potential Decay Measurements

The surface potential of non metallized treated and non treated samples was measured by exposing the samples to the injection of charges produced by a negative corona discharge. The sample was then moved under a Monroe sensor which measured the surface without any contact with the polymer. The method is widely described by Cohelo at al [7,8] and has enabled us to show that in the case of the N_2 treated polymer the surface potential decays while as in the case of the non treated film it remains constant (Figure 4). In the case of the latter, the charges can be localised in the defects of the polymer leading to a constant surface potential. As shown by SSIMS, the N_2 plasma treatment gives rise to a more homogeneous surface [1], leaving a surface less apt to localise charges. In this way the apparent surface conductivity slightly increases.

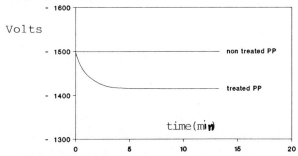

Figure 4. Potential Decay Measurement of a non treated (top) and treated (bottom) PP surface. Plasma conditions: t=0.23s, f=280 sccm, p=200 PA.

IV. Auger Depth Profile Analysis of Aluminium Films on Polypropylene

Depth profiling was performed on metallized polypropylene (PP) films, using high resolution Auger Electron Spectrometry in combination with ion sputtering. An external electron gun (55° incidence with the sample normal) operates with a 9 keV primary electron beam, rastering a sample area of about 250 μm^2 (the focused electron beam having a 1 μm diameter). A semi-dispersive type analyser (Riber Mac II) was employed, its axis making an incident angle of 40° to the surface normal. The resolution ΔE is constant over the whole energy range (1.1 eV FWHM for a 1 keV elastic peak is achieved with 8V energy pass).

Digitized spectra are recorded in the direct mode n (E) between sputtering sequences (in order to avoid low energy spectra distorsions and energy shifts induced by ion bombardment). Auger intensities in sputter profiles are presented after computing both smoothing and differenciation (1.5 eV energy width) on the direct N(E) spectra. If charging effect occur, the transformation of the n(E) in the E. n(E) representation, as in Figure 7 should be corrected from their respective energy shifts. The other experimental conditions have been described elsewhere [9].

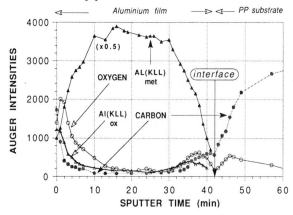

Figure 5. AES sputter profile of a vaporized Al film onto a pretreated polypropylene substrate (0.7 s in a N_2 plasma). Sputter rate:0.5 nm.min^{-1}

Nitrogen plasma treated and untreated polymers were subjected to an *in situ* aluminium deposition in industrial like conditions (10^{-5} mbar); the samples were then transfered to the Auger chamber to be analysed. Figure 5 shows typical AES depth profile carried out on a thin Al film deposited on a pretreated PP substrate (0.7 s in a N_2 plasma). Starting from the external surface, a contaminating carbon layer covers a thin alumina film [Al(ox)+O]. Below this natural layer built during the air transfer, a rather clean aluminium coating aroused showing only some C and O traces, which are obviously related to the residual pressure level existing in the chamber during the metallization.

When approaching the interface, Al(met) transition relative to the deposit quickly falls down, while Al(ox) reappears with a simultaneous oxygen and carbon contamination. A first determination of the interface location can be achieved by extrapolating the Al(met) profile of the deposit or the C profile of the polymer substrate to the zero intensity scale. At the interface (t=42 min), charging effects may occur when the first layers of the polymer substrate are exposed successively to ion and electron bombardments (the surface charging potential, S.P. is shown with a dashed line in Figure 6 a and b, is measured by the energy shift of the whole spectrum). Our treated samples are characterized by very weak or non-existent charging effects: in the present case (Fig.6a), they did not exceed 50 eV (it should be presently remarked that this surface polarization is stable enough to allow several AES acquisitions)

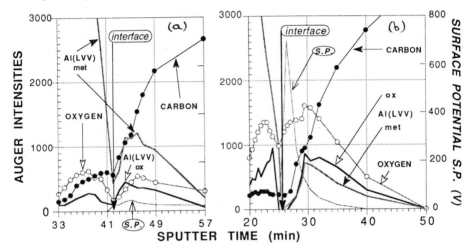

Figure 6.AES depth profile analyses of the Al-PP interface produced after a pretreatment of 0.7s (a), (b) without pretreatment.

Nitrogen was never detected in the polymer by AES; the ion bombardment might be responsible for a desorption of nitrogen since in AES the ion current density is quite high (1000 times higher than in Static SIMS for example). Carbon profile produced by the substrate C atoms sharply increases while a second "oxidized layer" can be observed. LVV transitions of both Al (ox) and Al (met) are depicted in this part of the PP. A real knock-on effect of Al atoms from the film into the substrate caused by the ion sputtering process is not satisfactory to explain the particular shape of Al(ox) and Al(met) sputter profiles in the first polymer layers. This would initiate the formation of small inclusions as proposed by P.S.Ho et al., in, however annealed substrates [10].

The Al(ox) profiles in the polymer side seem to be produced by a further oxidation of the vaporized Al at the contact of remaining oxygen included in the external PP layers. Such a profile should reflect the cleaning effect of the plasma treatment for short treatment times. Larger treatment times (5 s) of the polymer in a N_2 plasma seem unsuitable because they lead to an increased surface roughness of the substrate with the notable broadening of the interface and also to stronger quantities of Al(ox) and oxygen imbedded in the first layers of the polymer.

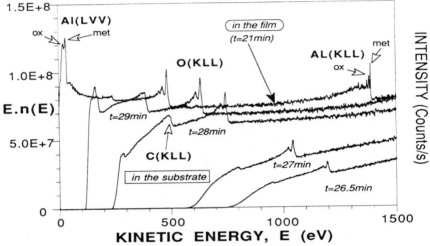

Figure 7. AES spectra carried out at the Al-PP interface of an untreated sample (Auger spectra are classified according to the sputter time scale of Figure 6b).

Reaching the interface of an untreated polymer (t=25.5 min, Fig.6b), considerable charging effects occur which complicates the Auger spectra acquisition. Such effects are displayed in Figure 9 by Auger line distorsions at low energy mainly due to a decrease of the secondary electron yield with the accumulation of negative charges onto the eroded PP surface. For the maximum charging effect (t=26.5 min), the surface potential (Fig.6b) may reach several hundreds of volts: in the present situation Auger and backscattered electrons were impeded to escape from the sample. This is an explanation for the missing points in Figure 6 for sputter profiles of Al (LVV) transitions near the interface (KLL transition was not retained for its lower sensitivity, also the energy shift often prevents us to record spectra in the authorized energy range of 0-2000 eV and creates a significant loss in energy peak resolution). In the interface analyses in Figure 6 Al (LVV) metallic profiles always prevails on the corresponding Al(LVV) ox and O profiles in plasma treated substrates, while a reversed situation arises for untreated specimen which seem to have a higher oxygen content after metallization.

Burkstrand [11] correlated an increased adhesion of metallic coatings (Cr, Ni, Cu) to the formation of a.metal-oxygen-polymer complex. Our most recent work [12] indicates that, in ultra-high vacuum conditions, the deposition of a very pure Al film and the elimination of residual oxygen in the polymer by an intensive ion bombardment (leading to a graphite-like surface), gave a very sharp interface. In this interface we were able to identify Al-C bonds by high resolution AES, in total agreement with the XPS studies of Bödo et al.[13], which concluded to a better adhesion in the case of a direct metal-carbon bonding. In our experiment, the improvement of adhesion was only tested with a scotch-test because of the reduced size of the sample and should be validated by more appropriate adhesion test measurements.

V. Peel-Test Measurements

In order to evaluate the improvement of adhesion, we have used a peel-test adapted to a thin flexible substrates with very thin metallic coatings. One of the difficulties consisted in the impossibility to detach the metal coating from the PP film in the case of a plasma pretreatment. In the absence of treatment the metal would be peeled-off with a scotch but there was no possible quantification of the improvement.

Many references [14] indicated the use of a second hydrophilic polymer melted and pressed on top of the metallized substrate. The sandwich would be then peeled off. We have not used this method because the alien polymer could diffuse to the metal-polymer in the case of which the nature of its interaction with the interface cannot be evaluated. Another possibility was to use a two component resin to fix the metallized substrate to another holder and peel it off, but in that case, the resin diffused in the metal, forming a very hard phase that could not be peeled at all. We were then lead to another solution which consisted in using a high adherence doubled scotch tape placed onto a stainless steel metal sheet (2 mm thick). The metallized polypropylene would be fixed to the metal support with the help of that scotch (the aluminium directly on the tape), as shown on Figure 8. The experimental conditions were then chosen in order to peel the metal without tearing the polymer (table 1). The strength which we can measure includes the elastic deformation of the polymer. However we were not able to separate this factor and we will consider that it is constant for all of the experiments. It is understood that the measurements do not give an absolute value for the strength. For treatment times exceeding 115 ms, we obtain an adhesive failure of the scotch tape-metal interface and not of the polymer-metal interface; we can therefore conclude that the strength is higher than the one measured values.

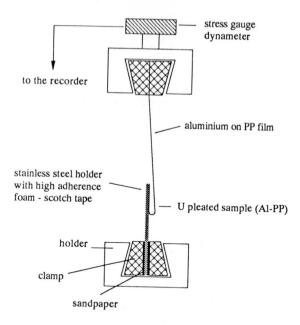

Figure 8. Apparatus used for the peel-test measurements

Table 1. Experimental conditions for the peel-test measurements

peel - test type:	$180°$ U
sample:	metallized PP (25nm)
sample size:	2.54 cm x 22 cm
scotch - tape:	3 M 4945
peeling speed:	0.1 cm/s
sample holder:	stainless steel 2 mm thick
peel - strength range:	0 - 5 dN
sample conditioning:	none
test - temperature:	$20°$ C

The samples were always tested immediately after being taken out of the reactor. For experiments in which the metal was peeled off, we were able to evaluate the amount of aluminium missing and to correlate it to the measured peel strength [16]. The missing metal was measured with the help of a video camera and an image processor using the black and white contrast between the covered and non covered areas. If we look at the trend of the peel strength as a function of the treatment time, we can see that the strength increases up to a treatment time of about 1.38 s (Figure 9). If we compare the peel-strength with the SSIMS results (30 amu on Figure 1) they could mean that the increase of the strength at the Al-PP interface is closely related to the amount of nitrogen incorporated in the polymer. The amount of nitrogen remains very small but this is in good agreement to former results concerning the role of nitrogen in the improvement of metal-polymer structures.

Elsewhere, Y. Novis et al, [15] have also described the role of nitrogen added to the carrier gas which gives rise to an Al-N-PET bonding. other experiments carried out in NH_3 plasmas point out peel strengths two times higher than in the case of a N_2 treatment. Furthermore, the N/C ratio dectected in XPS is also higher than in the case of N_2 treatment [17].

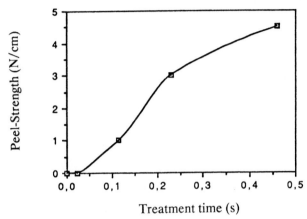

Figure 9 . Peel-strength at the Al-PP interface (f=280 sccm, p=200 Pa, d=7 mm, V=1 kV, I=0.06 A).

General Conclusion

This work has shown that the adhesion of PP films can be improved by a low pressure N_2 plasma for very short treatment times (23 to 500 ms). With the help of Static SIMS, we have shown how the non metallized film is slightly oxidized by the plasma treatment and how the incorporation of nitrogen species increases with the treatment time. The peel-test measurements also show an increaseof the measured peel strength with the treatment time in a similar manner to the nitrogen containing species. The cleaning role of the plasma was observed in terms of the decrease of the low molecular weight hydrocarbons which form a contamination layer on the non treated PP. The evidence of a nitrogen-metal-complex has not been observed by AES in this work, but because of the peel-test results where we can see the print of the nitrogen effect, we assume that it could be present at the Al-PP interface.

The electrical behaviour of the surface is modified by the plasma treatment. Since the surface has less defects, the charges localise less and the apparent surface conductivity is increased. This increase was confirmed by the Auger analysis which showed that the Al-PP interface was more conductive after a pretreatment. The Al-PP interface was founded to be more oxidized in the absence of a plasma treatment: as the non treated polymer is almost free of oxygen, the metallization seems to have a different effect when carried out on a treated or a non treated surface.

Finally, the complete mechanisms of the adhesion improvement cannot be identified but it seems that the nitrogen species play an important role. The role of the increase of the apparent surface conductivity in the adhesion improvement has not been yet elucidated. As for oxygen, negative SSIMS performed on N_2 pretreated films should help us to obtain a better understanding of its role in the adhesion phenomena.

Acknowlegements

This work was supported by Electricité de France and the Université Pierre et Marie Curie is also acknowledged.

V.André wishes to thank W.Wacker for his help with the graphic presentation of this paper.

Literature

[1] V. André, F. Arefi, J. Amouroux, Y. De Puydt, P. Bertrand, G. Lorang and M. Delamar, Thin Solid Films, 181, 451 (1989)

[2] V. André, J. Amouroux, Y. De Puydt, P. Bertrand, J.F. Silvain, ACS Books "Metallization of Polymers" Chap.31 (1990) 423-432

[3] Van Ooij, Surface Interface Analysis, 11 (1988), 430-440

[4] M. Budzikiewicz, C. Djerassi, D.H. Williams in "Mass Spectrometry of Organic Compounds", Ed. Holden-Day, San Fransisco (1967).

[5] Y. De Puydt, paper presented in Namur "Polymer -Solid Interfaces",2-6 Sep.1991

[6] E. Occhiello, M. Morra, G. Morini, F. Garbassi and P. Humphrey, J. Appl. Polym. Sci., 42 (1991) 551-559

[7] R. Cohelo, J. Lévy, D. Serrail, Phys. Stat. Solid, 94 (1986) 595

[8] R.Cohelo, P. Jestin, J. Lévy, D. Serrail, IEEE Transactions on Electrical Insulation E1-22 n°6, Dec.1987

[9] V. André, F. Arefi, J. Amouroux and G. Lorang, 16 (1990) 241-245

[10] P.S. Ho, P.O. Hahn, J.W. Bartha, G.W. Rubloff, F.K. Le Goues and B.D. Silvermann, J. Vac. Sci. Technol., A3(3) 739 (1985)

[11] J.M. Burkstrand, J. Appl. Phys., 52 (1981) 4735

[12] G.Lorang, V. André, F.Arefi, J. Amouroux and J.P. Langeron, ECASIA 91 Conference 14-18 Oct. 1991. Budapest (H)

[13] P.Bodö and J.E. Sundgren, Surf. Interface Anal, 9 (1986) 437

[14] Y. De Puydt, P. Bertrand, P.Lutgen, Surf. Interface Anal, 12 (1988)

[15] Y. Novis, M.Chtaib, R.Caudano, P. Lutgen, G. Feyder, Intern Communication Du Pont

[16] V. André, Doct. Thesis, Pierre et Marie Curie University, Paris le 8 Novembre 1990

[17] M. Tatoulian, F. Arefi, V. André, J. Amouroux and G. Lorang, to be published

Morphology and adhesion of magnesium thin films evaporated on polyethylene terephthalate

J.F. Silvain, A. Veyrat, J. J. Ehrhardt

CNRS Laboratoire Maurice Letort, rue de Vandœuvre, F-54600 Villers-les-Nancy. France

Magnesium films evaporated on PET at 77 and 300K have been studied by transmission electron microscopy (TEM). The epitaxial growth of Mg particles on a MgO layer bonded to the polymer is observed. The high adhesion strength observed on these laminates is attributed to the initial formation of the oxide at the metal-polymer interface. Adhesion strength of Mg evaporated on PET is compare with the adhesion of other metals (Al, Ag, Cu) evaporated on the same polymer.

1. INTRODUCTION

Metallized polymers are commonly used, for audio and video recording tapes, electronic devices, and packaging industry. Knowledge of the properties of the metal-polymer interface is important, particularly to enhance control of the interface chemistry and to promote adhesion. In order to improve the adhesion, several techniques are commonly used before and during metal deposition : corona discharge treatment [1,2], rf ion plating treatment [3], chemical treatments (bromosulfic acid [4], fluorine gaz), and so on.... Another way to modify the adhesion is to change the metal which is evaporated on the polymer. Chromium, for example, is known to have a higher adhesion on polyimide [5] then copper [6].

XPS study on Al evaporated onto PET has shown that the very first layer of the metal in contact with the polymer is fully oxidized [7]. This result seems to be general and leads us to suggest that the nature of the oxide formed at the interface is very important for the adhesion point of view. In that sense, we have chosen to study the properties of thin Mg films evaporated onto PET. Inded, Mg form an oxide (MgO) with a high heat of formation (ΔH_{MgO} = -144 kcal/mole).It has to be notice that, in the case of the formation of MgO, Ag_2O, and CuO at the metal/PET interface, each oxide molecule is formed with one PET oxygen atom. In opposite, if we suppose the formation of Al_2O_3 during Al deposition on PET, three PET oxygen atoms have to be used to formed one aluminium oxide molecule. The microstructure of the metal/PET interface is correlated with the adhesion of the metallic film.

2. EXPERIMENTAL DETAILS

Metallizations have been performed in an evaporation chamber at a pressure in the low 10^{-6} mbar range. A heat shielded Knudsen cell has been used for the metal considered here. The metallic vapor flux was monitored with a quartz balance and kept constant at 3 nm/mn. The distance between the Knudsen cell and the surface of the polymer was about 50 cm allowing an uniform deposition rate on the entire surface of the substrate. The thickness of the evaporated metal film was around 50 nm. Thin metallic films have been evaporated on commercial PET (Du Pont de Nemours (Luxembourg) SA (Mylar™)). The PET temperature was kept constant during deposion at either room temperature (293K) or liquid nitrogen temperature (77K).

These metal/PET layers were examined either by conventional or cross-section TEM in a JEOL 200 Cx microscope. For the cross sections, the samples were prepared by bonding two Mg/PET samples with Mg surfaces in contact with eachother with an epoxy resine and curing the resultant couple under moderate conditions temperature (40°C) and pressure. Thin slides were then cut perpendicular to the metal surfaces and thinned by mechanical grinding to approximately 50 µm followed by further thinning with an ion miller, utilizing a collimated beam of argon ions accelerated at 6 kV. The plane section samples were prepared by dissolving the PET in trifluoroacetic acid for 5 to 10 min.

For the adhesion measurements, test pieces consisted of an Al support (1 mm thick) / double sided tape (Permacel P-94) / PET (12µm) / evaporated Mg / ethylene acrylic acid (EAA) copolymer film. These laminates were prepared for the peel test by compression under $1.3 \; 10^5 \; N/m^2$ at 120°C for 10 seconds. The peel test was performed by peeling the EAA copolymer sheet from the laminate in an Instron tensile tester at 180° peel angle and 5 cm / min peel rate.

3. RESULTS AND DISCUSSION

The Mg/PET adhesion, as measured by the peel test method, is always very high whatever the deposition temperature. It is of about 1000 g/inch to 1200 g/inch i.e. higher than the value measured for Al deposited on as produced PET, corona treated PET, and fluorinated PET (see table 1). We can notice in this table that, for Al films evaporated onto PET, a good correlation can be established between the decrease of the grain size and the increase of adhesion.

Properties	Control PET	Corona-treated PET	Fluorinated-treated PET
Grain size (nm^2)	1230 ± 70	870 ± 50	470 ± 30
Adhesion (g/inch)	130	200	700

Table 1 : Average grain size and adhesion measured on 30 nm Al thin film evaporated on untreated PET, corona treated PET, and fluorinated PET [8].

3.1. Mg films evaporated on PET : TEM study

The Mg films evaporated onto PET at 77K and 293K have been observed by TEM. The most significant pictures are presented in figure1. A plane section bright field image for a Mg film deposited at 77K (figure 1a) shows large grains whose size range from 0.01 μm^2 to 1 μm^2. Similar pictures have been obtained for Mg deposited at room temperature.

Diffraction diagrams (figure 1b)on a particle (A area figure 1a) and at the border (figure 1c) of the same particle (B area figure 1a) are presented. Figure 1b shows the spots arranged in a hexagonal array which corresponds to (0001) plane of Mg giving evidence of a Mg single crystal formation with its c axis normal to the surface. In the diffraction pattern figure 1c, the 200*, 220*, and 400* spots are consistent with the MgO fcc {111} plane and the 10.0, 20.0, and 30.0 with the (0001) plane of Mg. The dark geometric grains and the uniform bright background of figure 1a are attributed respectively to the Mg metal and the MgO oxide.

Cross section bright field images obtained from Mg layers deposited at 293K and 77K are presented in figure 1d and 1e respectively. The bottom and the upper parts are related respectively to the polymer and the epoxy used for the cross section samples preparation. Several features can be observed in these two figures:

(1) the thickness of the film is around 50 nm as expected from flux measurements,

(2) the Mg films seems to be continuous and no grain boundaries can be observed throughout the metallic layer. The absence of grain boundaries is these two figures (d, and e) is related to the huge size of the Mg particules whit respect to the small area analysed in these cross section pictures.

(3) for the Mg film deposited at room temperature, an interface layer (about 6 nm) is visible between the PET and the Mg film; this is less clear for the film deposited at liquid nitrogen temperature.

Our results can be interpreted in terms of the epitaxial growth of Mg particules onto a MgO layer covering the polymer. The shape of the particles characteristic of the growth morphology, as observed in figure 1a, is consistent with the diffraction diagram (three fold symetry) (figure1b). This result is reminiscent of previously published micrographs of fine Mg particles prepared by smoke formation in argon [10,11]. Furthermore, the epitaxial relationship between the MgO oxide and the Mg particles is equivalent ($(0001)_{Mg}$ // $(111)_{MgO}$).

The presence of an oxide layer covering the polymer is certainly responsible for the high adhesion of the magnesium film. The strength of the organometallic binding may be related to the high heat of formation of the oxide.

3.2. Mg, Al, Ag, and Cu films evaporated on PET : correlation between the adhesion and the heat of formation of their oxide

Table 2 shows the adhesion strength values measured at 77K and 293K and the heat of formation of the oxide for four different metals evaporated onto PET.

Two classes of oxide can be observed:

1) On the first one, the oxide presents onlt one oxygen in his structure (MgO, Ag_2O, and CuO). One oxygen atom of the PET is so necessary to form the metallic oxide and the equilibrium distance between the oxygen and the metallic atom should be respected. For this class of oxide, the adhesion strength follow pretty well the heat of formation of the oxide. In that sense the failure during the peel-test could be adhesive, i.e., rupture of the bonds between the oxygen PET atoms and the metal.

	Adhesion (g/inch) 77K	Adhesion (g/inch) 273K	ΔH oxide (Kcal/mole)	Oxide
Mg	1000	1300	-144	MgO
Al	390	400	-399	Al_2O_3
Ag	200	140	-39	Ag_2O
Cu	45	40	-7	CuO

Table 2 : Correlation between the adhesion and the heat of the oxide formation for Mg, Al, Cu, and Ag evaporated onto PET.

2) For the second class of oxide, more than one oxygen atoms of the PET has to be used to form the steochiometric oxide. The typical oxide of this class is Al_2O_3. Because of the fixed position of the oxygen in the monomer of the PET, the distance between the oxygen and the aluminium atoms has to be much larger than in the bulk Al_2O_3. As a consequence of this geometrical consideration, polymer chain accomodations may be induced leading to chain breaking and fragmentation of the outer skin of the PET. This may explain a smaller adhesion strength in comparison with the magnesium one in spite of a larger heat of oxide formation. The cohesive failure observed after peeling the aluminum film reinforced the idea of the degradation of the polymer during Al evaporation.

ACKNOWLEGMENTS

We would like to thank Dr. M. Koenig (Du Pont de Nemours, Luxembourg) for providing the polymer samples and for many helpful discussions.

REFERENCES

1. B. Leclercq, M. Sotton, A. Bazkin, and L. Minnasian-Saraga, Polymer, *18*, 675 (1977)
2. J. Amouroux, M. Goldman, and M.F. Revoil, J. of Polymer Science, *19*, 1373 (1982)

3. K. Suzuki, A.B. Christie, and R.P. Howson, Vacuum, *34*, 181 (1984)

4. L. Placzed, Coating, *10,* 328 (1978)

5. M.P. Andrews, ACS symposium series 440, *18*, 243 (1990)

6. J.F. Silvain, J.J. Ehrhardt, and P. Lutgen, Thin solid films, 195, L5 (1991)

7. J.F. Silvain, A. Arzur, M. Alnot, J.J. Ehrhardt, and P. Lutgen, Surface Science, In press

8. J.F. Silvain, J.J. Ehrhardt, A. Picco, and P. Lutgen, ACS symposium series 440, *33*, 453 (1990)

9. J.F. Silvain, C.L. Bauer, A.M. Guzmna, and M.H. Kryder, IEEE Trans. Mag., *22(5)*, 1296 (1986)

10. K. Kimoto, Y. Kamiya, and M. Nonoyama, Japan J. Appl. Phys., *Vol. 2, No 11*, 702 (1963)

11. K. Kimoto, and I. Nishida, Japan J. Appl. Phys., *Vol. 6, No 9*, 1047 (1967)

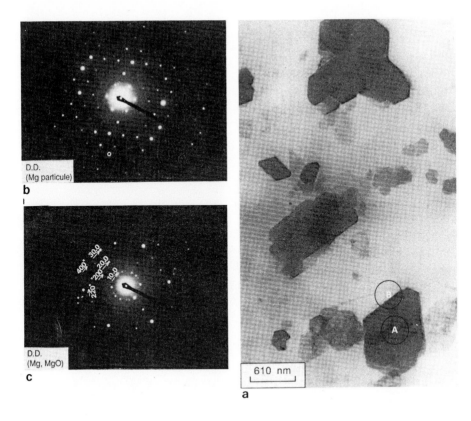

D.D.
(Mg particule)

b

D.D.
(Mg, MgO)

c

610 nm

a

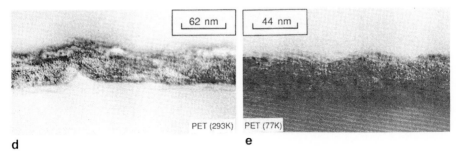

62 nm

44 nm

PET (293K)

PET (77K)

d

e

SECTION III

POLYMERS :
BULK/SURFACE/INTERFACE
PROPERTIES

Electronic states of polyethylene as a realizable model of one-dimensional solids and band-structure effects on hot-electron injection across the interface

N. Ueno

Department of Materials Science, Faculty of Engineering, Chiba University, Chiba 260, Japan

ABSTRACT: This paper summarizes the electronic structure of crystalline and amorphous parts of polyethylene measured by angle–resolved photoemission, secondary–electron emission, and low–energy–electron transmission experiments. Further it is shown that the efficiency of hot–electron injection into the film across the surface reflects the density–of–states of the conduction bands into which the electrons are injected. An anomalous band–gap current was observed for the electron injection into the band gap. Two possibilities which might explain the band–gap current are proposed. Finally, it is shown that there is a large temperature dependence of the electronic structure in polyethylene, which is a characteristic of polymers and is not observed in model compounds.

1. INTRODUCTION

As the thickness of a dielectric film deposited on a metal substrate is reduced, the carrier transport in the film becomes to be dominated by the motion of hot carriers. The motion is in principle expressed by the electronic states where they propagate. Further, the injection efficiency of hot carriers such as photoexcited electrons into the film across the interface can also be understood by obtaining information about the electronic states of the materials which build up the interface.

Polyethylene and its model compounds (n–alkanes, fatty acid etc) have been actually used in various field. An interest of this class of materials is that they can be considered as a physically realizable model of one–dimensional solids in crystalline state (Bloor 1980), since the alkyl chains show extended zigzag structure in the crystalline state and the interchain interaction is considerably smaller than the intrachain interaction arising from strong chemical bonds.

We have carried out following three types of experiments for crystalline and molten thin films of long alkyl molecules, which are idealized models of the crystalline and amorphous parts of polyethylene respectively, and real polyethylene samples with different crystallinities; (a) angle–resolved ultraviolet photoemission (ARUPS) measurements with synchrotron radiation for the study of the valence bands, (b) secondary electron emission (SEE) experiments for the conduction bands, and (c) low–energy electron transmission (LEET) measurements for the conduction bands and the hot–electron injection into the thin films of these materials. The experiments on both crystalline and molten films give useful data from which we can obtain important information on the electronic states and the hot–carrier injection across the interface. This article gives brief summary of our recent work on the electronic structure, both valence and conduction bands, of polyethylene, and on the hot electron injection into this class of materials. Detailed results are described in literatures (Ueno et al 1986, 1990a, 1990b, 1991a).

2. EXPERIMENTAL

The ARUPS measurements were carried out using ARUPS apparatus at the beam line BL8B2 of the UVSOR storage ring at Institute for Molecular Science (Seki et al 1986a, 1986b). In order to minimize the radiation damage of the specimen, the incident light intensity was reduced by introducing a nickel mesh of 50% nominal transmission between the monochromator and the measurement chamber when the storage ring current was greater than 50mA. Further, we changed sampling position of the specimen after every 2 or 3 spectra. By such careful measurements, we could obtain the spectra amenable for the subsequent analysis.

A SEE spectrum is given by the energy distribution curve (EDC) of secondary electrons emitted and inelastically scattered from a thin sample film owing to a monoenergetic electron impact at the normal incidence. The SEE spectrometer consisted of a spherical electron collector and an electron gun (Ueno et al 1980). The spectra were measured as the first derivative of the collector current against the retarding potential V_R applied between the collector and the sample. The spectra were recorded under the conditions of a total collector current of $5x10^{-11}$ or $1x10^{-10}$ A. At these conditions of the incidence, we could measure the spectra of organic thin films without radiation damage. The pressure of the spectrometer chamber was 10^{-8} Torr during the measurements.

The LEET spectrum is basically similar to the energy dependence of secondary electron yield at low−energy region (Hiraoka and Hamill 1972, Sanche 1991). In this measurements, a monoenergetic electron beam impinges perpendicularly on a thin film deposited on a metal substrate, and the electron current transmitted through the film (I_t) is measured as a function of the incident−electron energy (E_i) by keeping the incident current (I_i) to be constant. A typical I_i was 1×10^{-11} ($\pm 3 \times 10^{-13}$) A, and the number density of the incident electron at the sample surface was about $(3-5) \times 10^{-8}/\text{Å}^2\text{s}$. The ultralow density of the incident electron at the film surface enable us to measure the spectra without radiation damage.

The thin films of long alkyl molecules and polyethylene were prepared on metal substrates by spin−casting of 2000−2800 rpm and vacuum evaporation under ultrahigh vacuum.

3. ONE−DIMENSIONAL VALENCE BAND DISPERSION ALONG ALKYL CHAIN

For perpendicularly oriented thin films of long alkyl molecules, we observed that the binding energy of valence band structure depended on photon energy ($h\nu$), and could determine the dispersion relation of valence bands along the chain from $h\nu$ dependence of normal emission ARUPS spectra. An example of $h\nu$ dependence of the spectra is shown in Figure 1 for perpendicularly oriented thin films of pentatriacontan−18−one $[CH_3(CH_2)_{16}CO(CH_2)_{16}CH_3]$.

For the determination of the energy band dispersion $E=E(k)$ along the alkyl chain from the normal−emission spectra, we follow the procedure described by Seki et al(1986c). The results are shown in Figure 2 together with the previous results, where the results for hexatriacontane (Seki et al 1984, 1986c, Fujimoto et al 1987) and Cd arachidate (Ueno et al 1985, Seki et al 1986c) are shifted by about 2 eV in the binding energy for a fit with those of pentatriacontan−18−one. In the figure, the result of theoretical band calculation for an ideal infinite alkyl chain by Karpfen (1981) is also shown for comparison. The energy scale of the theoretical result is contracted by 0.8 times for the best fit with the experimental results.

It is notable that the theoretical result for the ideal one−dimensional chain agrees excellently with the experimental results. The good agreement shows that quasi−one−dimensional energy band dispersion characterized by the momentum exists in the alkyl chain even for the molecule with 16 CH_2 units. In passing, the calculated band structure with extended Hückel method (McCubbin and Manne 1968) also agrees fairly with the experimental results (for comparison with other theoretical results, see for example Seki et al 1986c).

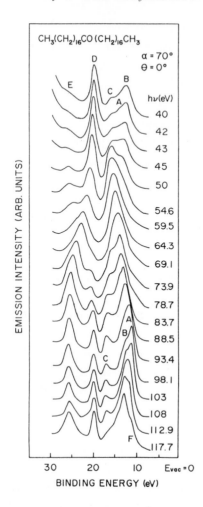

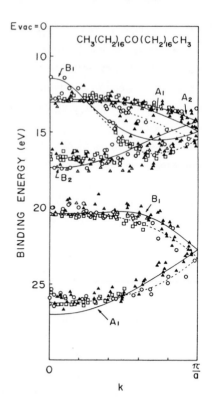

Fig. 1. Photon energy dependence of the normal–emission photoelectron spectra of vertically–oriented thin films of pentatriacontan–18–one, $CH_3(CH_2)_{16}CO(CH_2)_{16}CH_3$. Incidence angle of photon $\alpha = 70°$ and take–off angle of photoelectron $\theta = 0°$.

Fig. 2. Experimental energy band dispersion. o :pentatriacontan–18–one, $CH_3(CH_2)_{16}CO(CH_2)_{16}CH_3$. △ :hexatriacontane, $CH_3(CH_2)_{34}CH_3$ (Seki et al 1984, 1986c). ▲ :hexatriacontane (Fujimoto et al 1987). □ :Langmuir–Blodgett film of Cd–arachidate (Ueno et al 1985, Seki et al 1986c). The theoretical results (Karpfen 1981) are shown by solid curves. The dotted curves indicate the experimentally deduced dispersion curves for the A and B bands.

4. STRUCTURE OF CONDUCTION BANDS

As an example of crystalline part of polyethylene, the EDC's of backscattered and emitted electrons from crystalline tetratetracontane (n–$C_{44}H_{90}$) in retarding potential (V_R) scale are shown in Figure 3 as a function of temperature. Here, the origin of the kinetic energy of electrons (E_k) is taken at the sharp cutoff near V_R=0. These spectra were obtained using the electron beams of E_i=19.1eV. They are mainly excited by electron–electron inelastic scattering, since the threshold energy of the inelastic scattering E_{th} is 7.4 eV. Five stationary features (SEE features) A–E which are independent of the incident energy E_i were observed even at a temperature just below the melting point (T_m=85.6°C). These features originate from the structure of the conduction bands, since they were also observed in the spectra excited by electron–phonon inelastic scattering (Ueno et al 1986). Here we note that the density of scattered electron in the film increases rapidly near the bottom of the conduction bands because of the lack of decay channel to the lower–energy states (Ueno et al 1986). As a result, the peak A corresponds to the bottom of the conduction bands as shown later.

On melting, a drastic smearing out of the SEE features was observed. Similar results were obtained for the spectra excited with electron–phonon inelastic scattering. The molten–state spectra show no peak near V_R=0 which reflects the bottom of the conduction bands. The results indicate that the the DOS oscillation smears out and the bottom of the conduction bands shifts below the vacuum level due to melting.

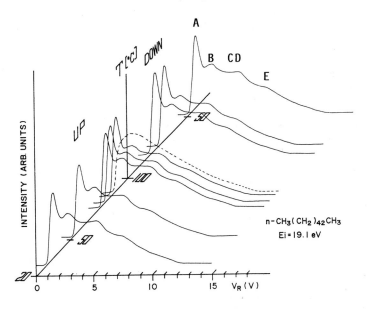

Fig. 3. Temperature dependence of SEE spectra for n–$C_{44}H_{90}$. The incident electron energy E_i=19.1 eV. Molten–state spectrum is shown by dotted curve.

5. HOT ELECTRON INJECTION INTO THE FILM

At first we show some examples of the LEET spectra in Figure 4, where the temperature dependences of the LEET spectra of $n-C_{44}H_{90}$ and low–density polyethylene (LDPE–1, T_m=128°C) are displayed. The LEET spectra of crystalline $n-C_{44}H_{90}$ show 8 features A–F. Similar LEET features were also observed for other long alkyl molecules independent of the film thickness, indicating the fine features are produced at the electron injection through the surface. For LDPE–1, overall structure of the LEET spectra is similar with $n-C_{44}H_{90}$, but the fine features B–F are considerably diffused and the feature A is intense in comparison with the spectra of $n-C_{44}H_{90}$. On melting, the fine features disappear and the transmitted current below E_i=1 eV increases drastically. Such a sharp spectral change due to crystal–melt phase transition can be ascribed to the disappearance of the band structure characteris-

tic of the crystalline state. Further, as shown in the next section the LEET peak A in crystalline $n-C_{44}H_{90}$ almost disappears at a temperature just below T_m. This means that the electrons can not be injected into the film for E_i< 0.5 eV. Therefore, we conclude that the bottom of the conduction bands of crystalline long–chain alkanes is located 0.5 eV above the vacuum level. The result agrees excellently with that obtained by Vilar et al(1988). Thus the increased transmitted current at E_i<1 eV for the molten films can be understood by a large tailing of the electronic states from the conduction bands into the band gap. These results correspond well with those obtained by the SEE experiments.

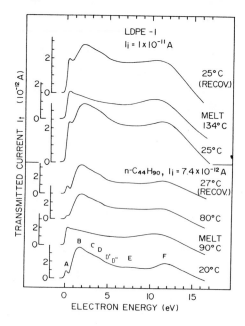

Fig. 4. Temperature dependence of LEET spectra for $n-C_{44}H_{90}$ and LDPE–1. Recovered spectra by recrystallization after melting are shown in the upper parts.

In Figure 5, the LEET spectra of crystalline $n-C_{44}H_{90}$ are compared with some other experimental results, the C1s energy loss spectra (Ritsko 1979), the UPS spectra (Ueno et al 1980) and the SEE spectra, which reflect the DOS of the conduction bands. We also compare the curves of the intensity ratio between the SEE spectra for crystalline and molten films (I^{RT}_{SEE} /I^{MELT}_{SEE}) for E_i =6.6 and 19.3 eV. These curves approximate the oscillating part of the DOS of the conduction bands. As can be seen from the comparison, there is a direct correlation in energy positions between LEET maxima (minima) and DOS maxima (minima). The intense peak A in the SEE and UPS spectra shows an excellent correspondence with the bottom of the conduction bands. These results indicate that the efficiency of the hot–electron injection into the film correlates strongly with the DOS of the conduction bands into which the hot electrons are injected. Such a result was also observed for oriented thin films of copper phthalocyanine (Ueno et al 1991b).

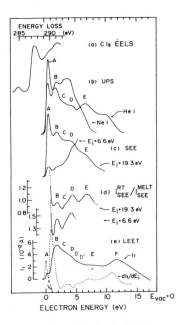

6. ORIGIN OF THE BAND–GAP CURRENT

The bottom of the conduction bands of crystalline long chain alkane and crystalline part of polyethylene is located above the vacuum level. Therefore, we can directly inject electrons into the band gap from vacuum as observed in the LEET spectra (LEET feature A). The observed gap current is more than 10 % of

Fig. 5. Comparison of LEET spectra, SEE spectra, and other spectroscopic results. (a) Carbon 1s energy–loss spectrum (EELS) of polyethylene (Ritsko 1979). (b) He I and Ne I photoemission spectra (UPS) of n–$C_{44}H_{90}$ (Ueno et al 1980). (c) SEE spectra of n–$C_{44}H_{90}$. (d) DOS oscillation approximated by I^{RT}_{SEE}/I^{MELT}_{SEE}. (e) LEET spectrum of n–$C_{44}H_{90}$ and its derivative.

the incident current for a film of about 100 A thick. Such a large current cannot be explained by a simple tunneling mechanism due to the large potential barrier.

From an intuitive point of view, one expects that defect states in the band gap play an important role and the gap current is dominated by electron diffusion through these gap states combined with electron inelastic scattering (Caron et al 1986). In order to clarify this point, LEET experiments were performed for two types of amorphous films (solid amorphous and molten amorphous films) of n–$C_{44}H_{90}$, and the spectral changes due to annealing and crystallization of the films were carefully investigated. In Figure 6, the temperature dependences of the intensities of the LEET peak A (gap current) and the LEET peak B (conduction–band current) are shown for the amorphous film deposited on the cooled copper sub-

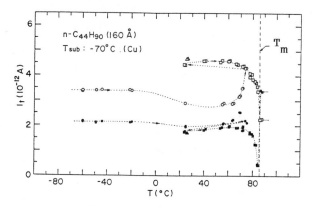

Fig. 6. Temperature dependence of the transmitted current through the band gap and the conduction band. ●,■,▲ :intensity of feature A (band–gap current). ○,□,△ :intensity of feature B (conduction–band current). ○,● were obtained during annealing (first heating) of the solid amorphous film to 74° C. □,■ were obtained during second heating to 88°C. △,▲ are for the melt–grown polycrystalline films. The bulk melting point T_m is indicated by the vertical dashed line.

strate (–70°C). It should be notable that the gap current does not decrease by the annealing which increases crystallinity of the film, while the conduction–band current increases remarkably. These facts cannot be explained by mechanisms based on the electron diffusion through the gap states.

For the origin of the gap current, we note that a potential dip appears at the outside of the film surface due to the image potential and the negative electron affinity of the film (Cole and Cohen 1969). Such a potential is schematically shown in Figure 7. In this case, the incident electrons can be scattered inelastically into the potential dip by exciting molecular vibrations or phonons at the surface. At lower temperature,

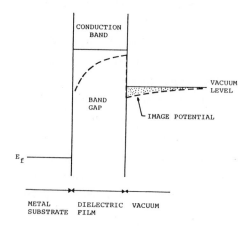

Fig. 7. Potential dip produced by the image potential and the negative electron affinity. The energy level after the modification by the image potential is shown by the dashed line. The potential dip is indicated by the dotted region.

the trapped electrons in the potential dip cannot escape from it and finally contribute to the gap current. At higher temperature, where the molecular vibrations at the surface are considerably excited (soft phonon), the electron–phonon scattering may excite the trapped electrons into vacuum and the excited electrons can go back not to contribute to the gap current. There is another possibility that the injected electron may deform the lattice to create electronic states below the bottom of the conduction bands through which the electron can transport through the band gap. However, we need further investigations, both experimental and theoretical, for unified conclusion. The anomalous temperature dependence of the gap current will be a key phenomenon to clarify the origin of the anomalous gap current.

7. DIFFERENCE BETWEEN THE MODEL COMPOUNDS AND POLYETHYLENE

The use of the model compounds of the crystalline and amorphous parts of real polyethylene can basically give the electronic structure of each part. However, there is an important characteristic in real polyethylene which reflects the coexistence of both crystalline and amorphous parts. In Figure 8, we show the temperature dependence of the SEE spectra of LDPE–1. It is seen that all SEE features of LDPE–1 become diffuse significantly with temperature even below the T_m (128°C), while those of crystalline n–$C_{44}H_{90}$ can be observed clearly even at a temperature just below T_m. Similar temperature dependence of SEE spectra was observed for various polyethylene samples (Ueno et al 1991). In figure 9   the intensity

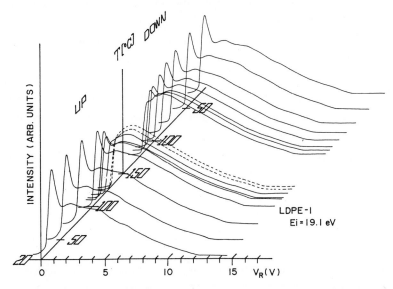

Fig. 8. Temperature dependence of SEE spectra of LDPE–1. The molten-state spectra are shown by dashed curves.

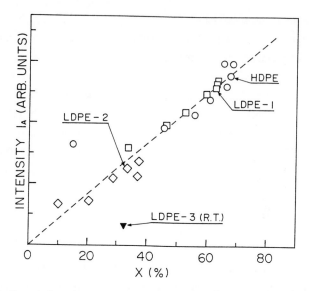

Fig. 9. Correlation between the intensity of SEE peak A (I_A) and the crystallinity X for high–density polyethylene (HDPE) and three low–density polyethylene (LDPE-1, LDPE-2 and LDPE-3). All results for LDPE-3 located below the dashed line, and an example is shown by a result (▼) at room temperature (RT).

of the SEE peak A (I_A) is plotted for various polyethylene as a function of the crystallinity measured by x–ray diffraction. Here, I_A and X vary depending on both temperature and polymer sample. It is found that the SEE intensity I_A shows a strong correlation with the crystallinity X, indicating that the large temperature dependence of the conduction bands of polyethylene below T_m originates in that of the specific volume of the crystalline parts. That is, the increase of specific volume of the amorphous parts by heating increases the density of gap states considerably. This type of temperature dependence of the electronic states is an important characteristic of polymers of two–phase structure, and it must be taken into account in analyzing the temperature dependence of the carrier–transport properties in polymer solids.

7. SUMMARY

The electronic structure of polyethylene and the hot–electron injection into the films were investigated by means of ARUPS, SEE and LEET spectroscopy using its model compounds and real polyethylene.

For valence bands, the one–dimensional energy band dispersion [E=E(k)] was observed along the alkyl chain of successive 16 CH_2 units. For conduction bands, the bottom of the conduction bands was found to be located at 0.5 eV above the vacuum level for the crystal-line parts, and below for the amorphous parts.

It was found that the efficiency of hot–electron injection into the films reflects the DOS of the conduction bands to which the electrons are injected. Further, an anomalous temperature dependence of the band–gap current was observed when the electrons were injected directly into the band gap. Two possibilities which might explain the band–gap current are proposed in conjunction with the surface state which is produced by the image potential and the nega-tive electron affinity, and with the lattice deformation due to the electron injection into the band gap.

Finally, we found a large temperature dependence of the electronic structure in polyethylene. It was not observed in model compounds and is a characteristic of polymer solids. Such a large temperature dependence of the electronic structure of polymers should be taken into account in analyzing the temperature dependence of their carrier–transport properties.

ACKNOWLEDGMENTS

The author is very grateful to Prof. K. Seki of Nagoya University and Prof. H. Inokuchi of Institute for Molecular Science for their useful comments, discussion and constant encour-agements. He also thank to Prof. K. Sugita of Chiba University for his encouragement.

This work was partly supported by a Grant–in–Aid for Scientific Research from the Minis-try of Education, Science and Culture, Japan.

REFERENCES

Bloor D 1980 Chem. Phys. Lett. 40 323
Caron L G, Perlusso G, Bader G and Sanche L 1986 Phys. Rev. B33 3027
Cole M W and Cohen M H 1969 Phys. Rev. Lett. 23 1238
Fujimoto H, Mori T, Inokuchi H, Ueno N, Sugita K and Seki K 1987 Chem. Phys. Lett. 141 485
Hiraoka K and Hamill W H 1972 J. Chem. Phys. 57 3870
Karpfen A 1981 J. Chem. Phys. 75 238
McCubbin W L and Manne R 1968 Chem. Phys. Lett. 2 230
Ritsko J J 1979 J. Chem. Phys. 70 5343
Sanche L 1991 in Excess Electrons in Dielectric Media, edited by Ferradini C and Jay-

Gerin J –P(CRC, Boca Raton, FL), and references therein

Seki K, Karlsson U, Engelhardt R and Koch E E 1984 Chem. Phys. Lett. 103 343

Seki K, Nakagawa H, Fukui K, Ishiguro E, Kato R, Mori T, Sakai K and Watanabe M .lm6 1986a Nucl. Instr. Method A246 264

Seki K, Fujimoto H, Mori T and Inokuchi H 1986b UVSOR Activity Reports 11

Seki K, Ueno N, Karlsson U O, Engelhardt R and Koch E E 1986c Chem. Phys. 105 247

Ueno N, Fukushima T, Sugita K, Kiyono S, Seki K, and Inokuchi H 1980 J. Phys. Soc. Jpn. 48 1254

Ueno N, Gaedeke W, Koch E E, Engelhardt R, Dudde R, Laxhuber L and Moehwald H 1985 J. Mol. Electron. 1 19

Ueno N, Sugita K, Seki K and Inokuchi H 1986 Phys. Rev. B34 6386

Ueno N, Seki K, Sato N, Fujimoto H, Kuramochi K, Sugita K and Inokuchi H 1990a Phys. Rev. B41 1176

Ueno N and Sugita K 1990b Phys. Rev. B42 1659

Ueno N, Seki K, Sugita K and Inokuchi H 1991a Phys. Rev. B43 2384

Ueno N, Sugita K and Shinmura T 1991b Phys. Rev. B44 6472

Vilar M Rei, Blatter G, Pfluger P, Heyman M and Schott M 1988 Europhys. Lett. 5 375

Gas phase molecular models for the study of the electronic structure of polymers

A. Naves de Brito, N. Correia and S. Svensson

Department of Physics, Uppsala University, Box 530, S-751 21 Uppsala, Sweden

ABSTRACT: The use of gas phase model molecules in the spectroscopical study of the polymer electronic structure is reviewed. In particular core hole photoelectron spectroscopy, core hole shake-up/shake-off satellites and NEXAFS spectroscopy are discussed. Theoretical methods used in the analysis of photoelectron of core states and the associated shake-up states are also presented. The n-alkenes series, C_nH_{n+2}, and the oligomer series of PMMA, poly (3-alkylthiophene) and polyacetylene are given as an examples of gas phase model molecules.

1. INTRODUCTION

A number of different spectroscopical methods have been used on gas phase model molecules to provide information about the polymer electronic structure and other closely related physical properties such as geometrical conformation. A model molecule is a system where the infinite polymer chain is broken and replaced by a suitable substituent group. In many cases the first members of the series of such molecules representing monomer, dimer, trimer, etc, sub-units of the polymer constitute a good basis for extrapolation of the properties to the "infinite" quasi one dimensional system, i. e. to the polymer chain.

Furthermore a number of undesirable side effects take place together with a specific phenomena of interest in the polymer which are not present in the gas phase models. In the special case of x-ray and UV- photoelectron spectroscopy (XPS and UPS, respectively) of polymer surfaces, the experimental problems are largely connected to charge broadening of the spectral lines, difficulties with the energy calibration, effects due to surface orientation as well as possible effects due to radiation damage. Inelastic scattering of the outgoing photoelectron in core hole XPS also makes an assignment of the weak shake-up structures rather difficult.

An interpretation of the polymer electronic properties calls for quantum mechanical calculations in order to establish links between measured quantities. The use of gas phase model molecules facilitates this task since their relative small size makes it possible to reduce greatly the number of approximations required for the solid polymer.

If on one side the gas phase models represent a good basis for the understanding of the polymer electronic structure it is important to identify the properties of the polymer which can not be represented by the gas phase modeling such as inter-chain interaction and long range intra-chain correlations. However, if one makes a restriction to phenomena mainly localized to a certain polymer atom the presence of such collective effects are minimized thereby increasing the reliability of the gas phase model predictions.

Several processes can be described as localized phenomena: (1)Core hole state photoionization; (2)Shake-up satellites; (3)Auger decay processes; (4)Photoexcitation processes observed by Near Edge X-ray Absorption Fine Structure (NEXAFS).

However, it is also useful to study the valence energy band structure of the solid polymer by observing the successive development of the valence orbital region in a series of oligomers. The valence electrons are delocalized, but the successive steps in the development of the energy band in the solid are clearly distinguished in the separate model molecules.

In the present paper we shall discuss the above topics in the context of gas phase models applied to the study of the polymer electronic structure. *In section 2* we present a number of core hole photoelectron spectra where the use of gas phase model molecules has shown to provide information about the core photoelectron spectra of the polymer chain. *Section 3* deals with shake-up/shake-off satellites in model molecules; valence spectra are also discussed. *In section 4* we briefly summarize the theoretical methods used in the analysis of core and shake-up photoelectron lines. *In section 5* NEXAFS spectroscopy applied to model molecules are briefly reviewed.

2.CORE PHOTOELECTRON SPECTROSCOPY

One of the early examples of this type of work is the study of core-electron relaxation energies and valence-band formation of linear alkanes studied by X-ray photoelectron spectroscopy by Pireaux et al. (1976). In this study the C1s binding energy was found to decrease with increasing size of the molecule as is shown in Fig. 1.

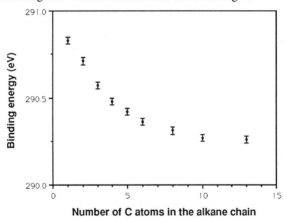

Fig. 1. C1s binding energy (Pireaux et al 1976) (centroids of the peaks) for linear alkanes.

The asymptotic trend of the binding energies in the alkane series provided a good estimation of the C 1s binding energy for the infinite linear chain referred to the vacuum level. Recently it was observed (Nordfors and Ågren 1991) that an empirical rule (Gasteiger and Hutchings 1984), could be used to predict the aliphatic C1s binding energy with a very good accuracy. An almost linear relation was found when the connectivity number was plotted versus the binding energy . In the definition of the connectivity number for a given atom A, one simply adds a contribution of $(0.5)^{n-1}$ from each atom in the molecule. n is the number of bonds connecting this atom to atom A. This simple rule has shown to provide a good basis for extrapolation of the binding energy of a particular atom in a homologous oligomer series to that of an infinite polymer chain.

The electronic structure of the polymer polymethylmethacrylate (PMMA) has recently been studied in great detail using model molecules (Brito et al 1991a). A detailed knowledge of the electronic structure of the pure material is required for a number of applications, e.g. in the product control of intraocular lenses where a clean PMMA surface is required in order to avoid undesirable foreign-body reactions in the eye. In this case XPS has become a standard technique for routine quality control (Lindberg et al 1988). In this technique one usually deconvolutes the O 1s and C 1s polymer spectrum assuming stoichiometric ratios as a routine process for detecting adventitious species containing those elements. However, from a study involving gas phase model molecules a non-stoichiometric ratio as well as asymmetric line profiles have been observed. The series of model molecules used is shown in Figure 2 together with their name, acronym and atom position. Mi-B is the closest model to the polymer chain.

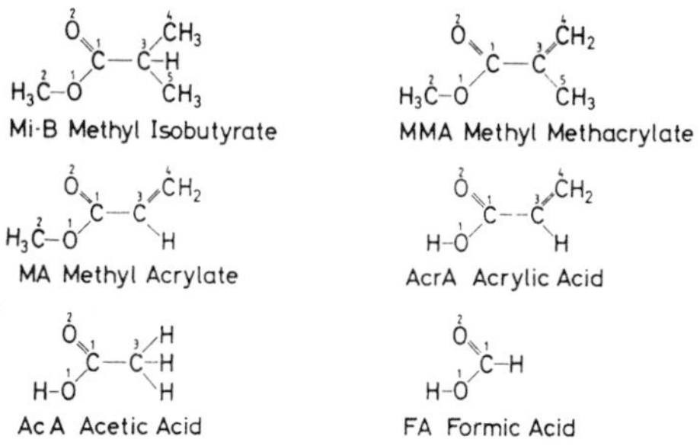

Fig. 2. Structure formulae for model molecules of PMMA (Brito et al 1991a). The atom position and the acronyms are given.

In this series the group O1,O2,C1 is common to all molecules. By comparing the core electron binding energies of these atoms in different molecules the dependence of energies with the size of the molecules can monitored.

A similar asymptotic decreasing of the binding energies with the increasing size of the molecules as in the alkane series was observed. A clear illustration of the presence of asymmetric peak shapes can be followed in the Fig. 3. This figure shows the C1s photoelectron spectra of the model molecules. The peak at the highest binding energy corresponds to carbon C1. Carbon C2 in Mi-B, MMA and MA can also be identified as the peak lying around 292.5 eV binding energy. C3 in AcA is associated with the peak at around 291.5 eV. The C1s line of the other carbons was only assigned after a curve fitting procedure and with the use of binding energies obtained from a large scale *ab initio* calculation. Following the variation in the binding energy of the peak assigned to carbon C1, the above mentioned asymptotic trend can also be observed. A fairly large asymmetry of the line profile associated to carbon C2 in Mi-B and C3 in AcA is observed. This asymmetry is associated to vibrational excitation within the methyl groups and is, to a large extent, independent of molecular size.

The intensity ratio between O1 and O2 in the model molecules was also found to be significantly larger than 1 for all molecules studied. As we will discuss in more detail in Section 2 the shake-up and shake-off probabilities are larger for oxygen O2 which is doubly bonded. Accordingly, for the well resolved peak associated to to carbon C1, the ratio C1 to C_{tot} is larger than the stoichiometric ratio (C_{tot} being the sum of the C1s main line intensity of all the carbons) . Also in this case the larger shake-up/shake-off intensity for the doubly bonded carbon C1 explains the deviation from the stoichiometric ratio.

Using the above mentioned empirical relation involving the connectivity number (Nordfors and Ågren 1991), the chemical shifts for the polymer could be estimated. Based on the results from these gas models a physically meaningful fit of the polymer PMMA spectra could be obtained.

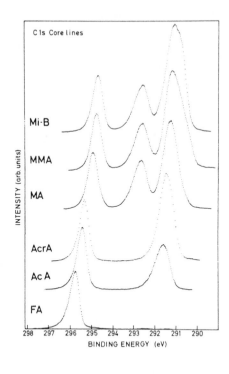

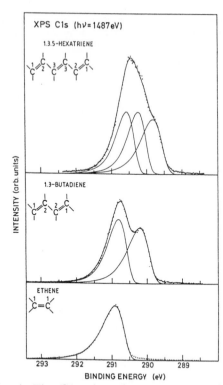

Fig. 3.The C1s photoelectron spectra of the model molecules for PMMA after Brito et al (1991a).

Fig. 4. The C1s photoelectron spectra of ethene, 1,3-Butadiene and 1,3,5-hexatriene taken from Brito et al (1991b).

In a parallel study a high resolution monochromatized X-ray study on spin cast thin films of PMMA has been performed (Beamson et al 1991a). The results of this study confirm the non-stoichiometric ratio observed in the gas phase model molecules.The results also show an extra structure in the peak corresponding to C3,C4,C5 carbon atom which, according to the gas phase results, should correspond to chemically shifted carbons C3. In a further study Beamson et al (1991b) report high resolution C1s spectra for polyethylene, polypropylene, hexatriacontane (n-$C_{36}H_{74}$) and polystyrene. In the first 3 of these compounds asymmetric C1s core line shapes were observed. This asymmetry was attributed to the excitation of the C-H stretching.

The PMMA polymer backbone is formed by saturated carbons. However, in most conducting polymers, the underlying structure is an alternating conjugated singly/doubly bonded chain. Polyacetylene is the simplest and one of the most studied of these materials. As the gas phase models for polyacetylene; ethene, 1,3-butadiene and 1,3,5-hexatriene have been used to represent the monomer, dimer and trimer sub-units Brito et al (1991b).

In figure 4 we present the gas phase C1s photoelectron spectra of these three molecules.The spectra show partially resolved lines for 1,3-butadiene and 1,3,5-hexatriene systems. A large scale ΔSCF with core relaxation (Brito et al 1991b) has been used for the assignment. The terminal carbon C1 in 1,3-butadiene was assigned to the peak at the lowest binding energy. According to the calculation the C1 in 1,3,5-hexatriene is assigned to the peak at the lowest binding energy, C2 to the peak at the highest binding energy and C3 to the peak in between.

The perturbation caused by the broken symmetry at the terminal carbon of the chain gives rise to an oscillatory behaviour of the C1s shifts.

Using a fitting procedure the chemical shifts have been determined. Based on this shifts and on the connectivity number described above, the binding energy of the terminal carbon named C1, the second carbon, C2 and the third carbon C3 in the "infinite" quasi one dimensional system was extrapolated. Furthermore using the quasi one dimensional property of the system, the binding energy of the series benzene(C_6H_6) and cyclooctatetraene (C_8H_{10}) could be taken as model molecules for a system where the broken translational symmetry at the edge was eliminated. The C1s binding energy of a carbon in the middle of the very long chain could thus be extrapolated.

The carbon C1 in a semi-infinite polymer chain is the zero dimensional boundary (the "surface") to the quasi one dimensional chain. The broken symmetry leads to the shift between this carbon and the central carbon (the "bulk"). The relationship to surface shifts observed on solids using UPS and XPS is evident. The result of the above described extrapolations are displayed in figure 5.

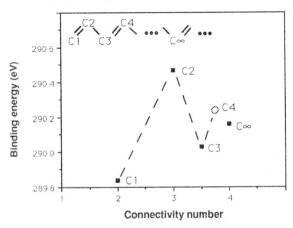

Fig. 5. The extrapolated binding energy of C1, C2, C3 and C∞ in a semi-infinite polymer chain.

As was found for 1,3,5-hexatriene the oscillatory behaviour of the chemical shifts according to the extrapolation remain in the semi-infinite polymer chain.

3. SHAKE-UP/SHAKE OF SATELLITES OF MODEL MOLECULES

Another field where model molecules have been applied is the study of the shake-up satellites associated with core hole photoionization. An example of this concerns the study of the thermochromism (a reversible colour change as a function of temperature) observed in the polyalkylthiophenes. In an earlier XPS,UPS and optical absorption study of poly (3-alkylthiophene) (Salaneck et al 1988, Salaneck et al 1989 and Thémans et al 1989) and this colour change could be linked in a first stage to a temperature-dependent electronic change. From the XPS and UPS studies a π localization with increasing temperature was found. With the help of valence effective Hamiltonian calculations the electronic changes could be associated to geometry changes in the polymer backbone on increasing temperature. The chemical structure of poly (3-alkylthiophene) is shown in figure 6.

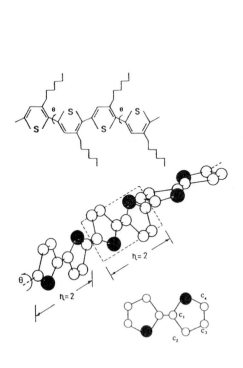

Fig. 6. In the top the structure formula of poly (3-alkylthiophene) (Keane et al 1990), middle : The dimer model where pair of coplanar thiophene rings are twisted through a torsion angle of 45 degrees.Bottom the bithiophene chemical formula and carbon numbering.

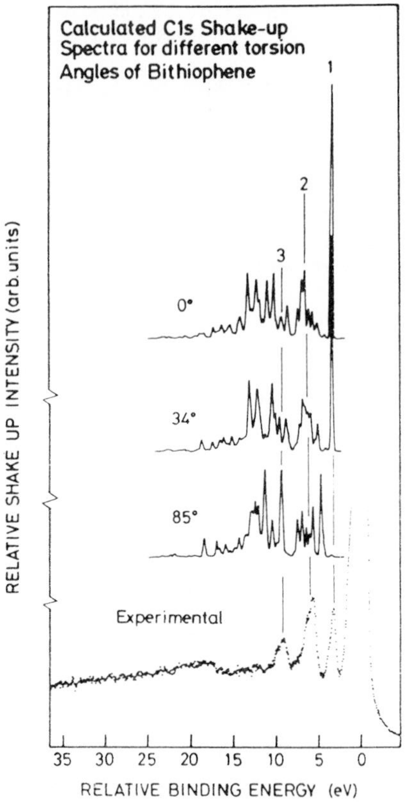

Fig. 7. Experimental and calculated shake-up C1s spectra of Bithiophene dimer at torsional angles 0⁰,34⁰,85⁰ (Keane et al 1990)

The shake-up intensities as well as peak positions in these unsaturated systems are very sensitive to the polymer geometrical configuration, since the overlap between the occupied and unoccupied π orbital depends strongly on the geometry. A small twist through a finite torsion angle between planar segments of the polymer backbone is observable in the spectrum. The use of model molecules is motivated by two reasons: *i)* the assignment of the shake-up spectrum requires calculations which are difficult to make for the solid materials *ii)* gas phase studies can be performed at a higher resolution and the inelastic peaks may be eliminated by using a low gas pressure.

Figure 7 shows the shake-up spectrum of bithiophene together with calculated spectra for different torsion angles namely 0⁰,34⁰,85⁰. While the calculated spectra shows no major change in going from 0⁰ to 34⁰, the first shake-up peak a at 85⁰ practically disappears.

By comparing the C1s shake-up spectra of the gas phase model molecule with the corresponding spectra of the solid polymer, shown in Fig. 8 below, a great similarity can be observed. A major contribution to the C1s shake-up spectra of the polymer seems to originate from shorter conjugated segments of the polymer chain in agreement with the concept of a soft conformation mechanism for the thermochromism in polymer material (Inganäs et al 1988).

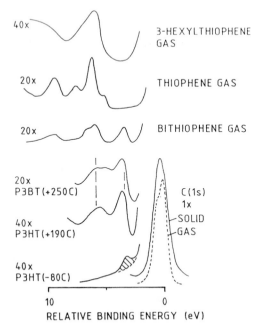

Fig. 8. The C1s shake-up spectra of the model molecules together with the corresponding spectra of poly(3-alkylthiophene) . The cross-hatched peak is interpreted as an energy loss peak (Keane 1990).

Bearing in mind the trends observed in the core electron binding energies when going from the monomer model molecules to the infinite system one could also ask if the same behaviour occurs when considering the first shake-up peak in the oligomer series. Fig. 9 shows the shake-up binding energies plotted versus the oligomer size. The same saturation trend is found for the first shake-up peak relative to the main line as was found for the C1s photoelectron peak in the alkane series (Pireaux et al 1976).

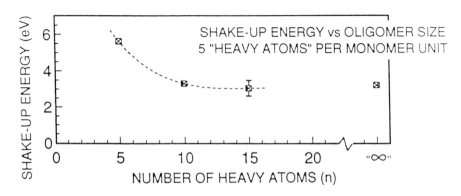

Fig. 9. Shake-up energy versus oligomer size (Keane et al 1991a)

High resolution XPS spectra of the core hole, valence and shake-up from polyacetylene as well as from the series of model molecules ethene, 1,3-butadiene and 1,3,5-hexatriene have been reported (Keane et al 1991a and Brito et al 1991b). In Table 1 we show calculated energies and intensities for the C1s shake-up lines together with experimental results (Keane et al 1991b).

	After CI		Experimental		Assignment
Atom position	Energy[a] (eV)	Intensity (%)	Energy (eV)	Intensity (%)	
Trans					
C1	3.46	18.5	3.22	6.4	$3\pi \rightarrow 4\pi^*$
	5.82	1.0	9.4	4.1	$3\pi \rightarrow 5\pi^*$
C2	4.18	0.05			$3\pi \rightarrow 4\pi^*$
	6.16	2.4			$3\pi \rightarrow 5\pi^*$
C3	4.00	5.9	4.50	3.3	$3\pi \rightarrow 4\pi^*$
	7.34	2.5			$3\pi \rightarrow 5\pi^*$

[a]A chemical shift of -0.56 eV for C1, 0.16 eV for C2 and -0.18 eV for C3 has been added to the shake-up energies.

Table 1 The calculated shake-up intensities for 1,3,5-hexatriene. Note the oscillating behaviour.

The same asymptotic trend as for the C1s shifts occur also in the case of the first shake-up energy of the model molecules for polyacetylene. A plot of the first shake-up energy versus the connectivity number is shown in figure 10.

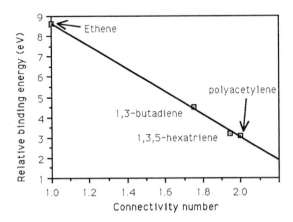

Fig. 10. Plot of the first shake-up peak of the oligomer series of polyacetylene versus the connectivity number.

As a final example of shake-up study in model molecules we show the O1s shake-up spectrum of formic acid. For the models of PMMA the doubly bonded oxygen was found to have less intensity than the single bonded oxygen. Difference in shake-up probability probably causes this difference. In order to verify this a calculation has been made of the O1s shake-up spectrum of formic acid shown in Figure 11 (Sjögren et al 1991). The contribution from each oxygen is given in separated spectra so that a comparison can be made. Adding up the shake-up intensity due to each oxygen and subtracting from the total intensity (O1s plus shake-up) the calculated ratio O1/O2 is 1.09 as compared to the experimental intensity ratio 1.05. The difference in the intensity of the π - π^* transitions are found to be the main reason for the deviation from the stoichiometric ratio.

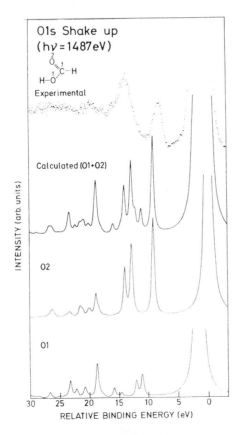

Fig. 11. The experimental and calculated O1s shake-up spectra of formic acid. Bottom the O1 and O2 shake-up spectra.Top the experimental spectra together with the sum of the calculated spectra.

4 CALCULATION OF CORE PHOTOELECTRON SPECTRA AND SHAKE-UP SATELLITES IN MODEL MOLECULES.

The assignment of partially resolved structures in the core hole photoelectron spectra calls for quantum mechanical calculations, as pointed out above. In terms of ab-initio calculations the Koopmans energies correspond to the simplest approximation. This model, however, is not suitable for calculating small chemical shifts when the relaxation in the core hole state is large. For the alkane series mentioned above (Pireaux et al 1976) a simple Koopmans model gave a reversed ordering of the chemical shifts to that observed in the spectra. A wrong ordering was also obtained for 1,3,5-hexatriene. A ΔSCF calculation gives a better result, but due to computational difficulties the absolute energies can still be out by as much as 2 eV for standard ΔSCF calculations, as compared with the very small shifts of tenths of eV that can be determined from high resolution XPS spectra. The problems occur due to the fact that in order to avoid the variational collapse of the energy, the core hole orbital has normally to be kept " frozen " while the rest of the orbitals are allowed to relax in the ΔSCF procedure. To solve this problem optimization algorithms have recently been developed that obtain local convergence of the core hole orbital while avoiding the variational collapse of the total energy (Jensen et al 1986) (when this core hole relaxation is included the ionization energies can be calculated within a few tenths of an eV). Both static and dynamical correlation as well as relativistic corrections can be neglected due to a cancellation of errors.

In figure 12 we show the calculated versus experimental binding energies of the model molecules of PMMA taken from (Brito 1991c). The calculated energies are in good agreement with the experimental values.

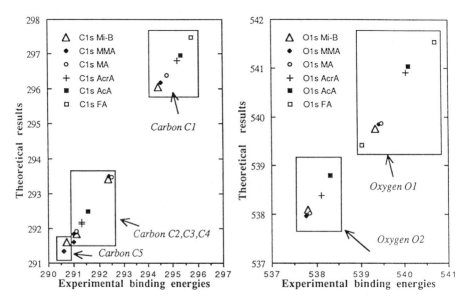

Fig. 12. Correlation between ΔSCF and experimental binding energies of carbon and oxygen.

In most of the shake-up calculations presented here the INDO (intermediate Neglect of the Differential Overlap)-CI (Ridley et al 1973, Bacon et al 1979, Zerner et al 1980 and Lunnel et al 1988) method has been used. The intensities are calculated in the sudden approximation (Åberg 1967). This method has a number of limitations. The equivalent core hole approximation treats the core electrons as fixed charges. Therefore, no intensity is assigned to the parent triple coupled states. Also the limited basis set used in this calculation gives rise to a quite good description of the lower shake-up structures. Less reliable results are obtained above this region.

5. NEXAFS SPECTRA OF MODEL MOLECULES.

While the core hole photoelectron spectra and the shake-up satellites give information about the electronic structure of singly ionized core electron states the NEXAFS spectra of solid 1,3-butadiene and 1,3,5-hexatriene, presented in figs 13 and 14 (Brito 1991d), give information about the neutral core excited systems. According to a ΔMCSCF calculation the structure around 284 eV corresponds to the core excited C1 to the lowest unoccupied orbital (LUMO) and the partially resolved peak around 285 eV corresponds to the core excited C2 to

LUMO excitation. Also according to the ΔMCSCF calculation the peaks at around 288 can be identified to transition from the chemical shifted carbons to the other unoccupied π orbitals. This result is in agreement with the C1s XPS for 1,3-butadiene.

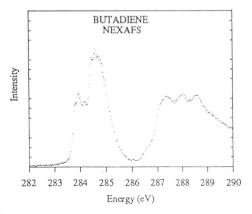

Fig. 13 NEXAFS spectra of 1,3-butadiene.

The assignment of the 1,3,5-hexatriene NEXAFS spectrum, shown in Fig. 14 can be made immediately from the results obtained for 1,3-butadiene. The dominant structure around 284 eV originates from excitation of core electron on the different carbon atoms to the lowest π orbital. Thus the observed fine structure is a consequence of a chemical shift. It should be noted that additional even smaller structures can be seen in Figs 13 and 14 which are due to vibrational effects.

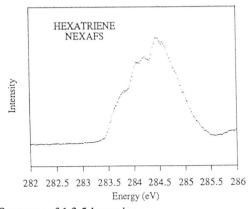

Fig.14   NEXAFS spectra of 1,3,5-hexatriene

References:
Åberg T 1967 Ann. Acad. Sci. Fenn. A 6 308
Bacon A D and M.C. Zerner M C 1979 Theor. Chim. Acta 53 21
Beamson G, Bunn A and Briggs D 1991a Surf. Interface Anal. 17 105
Beamson G, Kendrick J, Clark D T and Briggs D 1991b to be published
Brito A N de, Keane M P, Correia N, Svensson S, Gelius U and Lindberg B J 1991a
    Surf. Interface. Anal. 17 94
Brito A N de , Svensson S, Keane M P, Karlsson L, Ågren H, Correia N 1991b To be
    published
Brito A N de, Correia N, Svensson S, Ågren H 1991c J. Chem. Phys. 00

Brito A N de et al 1991d To be published.

Gasteiger J and Hutchings M G 1984 J. Am. Chem. Soc. vol 106 no. 22 pp 6489

Inganäs O, Salaneck W R, Österholm J-E and Laakso J 1988 Synth. Met. 22 395

Jensen H J Aa and Ågren H 1986 Chem. Phys. Letters 104 229

Keane M P, Svensson S, Brito A N de, Correia N, Lunell S, Sjögren B, Inganäs O and
    Salaneck W R 1990 J. Chem. Phys. 93 6357

Keane M P, Salaneck W R, Svensson S, Brito A N de , Correia N and Lunell S 1991a
    Proceedings ICSM-90, Tubingen, Synthetic Metals, in Press.

Keane M P, Brito A N de, Correia N, Svensson S, Karlsson L, Wannberg B, Gelius U,
    Lunell S, Lögdlund M, Swanson D B and MacDiarmid A G 1991b Submited for
    publication in Phys. Rev B

Lindberg B J, P. Måansson P, Hartman E-L and Öhrlund Å 1988 Surf. Interface Anal. 12
    469

Lunnel S and Keane M P 1988, UUIP Report No.1183,Institute of Physics, Uppsala
    University

Nordfors D and Ågren H 1991 J. Electron Spectrosc. 56 1

Österholm J-E, Thémans B, Brédas J-L and Svensson S 1988 J. Chem. Phys. 89 4613

Pireaux J J, Svensson S, Basilier E, Malmqvist P-Å, Gelius U, Caudano R,and
    Siegbahn K 1976 Phys. Rev. A 6 2133Salaneck W R, Inganäs O, Nilsson J-O,

Ridley J and Zerner M C 1973 Theor. Chim. Acta 32 111

Salaneck W R, Inganäs O, Nilsson J-O, Österholm J-E, Thémans B, and Brédas J-L
    1989 Synth. Met . 28 C377

Sjögren B,Brito A N de , Svensson S, Correia N, Keane M P 1991 To be published

Svensson S, Keane M P, Correia N, Karlsson L, Brito A N de, Sairanen O-P,
    Kivimäki A and Aksela S 1991 To be published.

Thémans B, Salaneck W R, and Brédas J-L 1989 Synth. Met. 28 C359

 Zerner M C, Loew G H , Kirchner R F and Mueller-Westerhoff V T 1980 J.Am.Chem.
    Soc. 102 589

Comparison of the XPS valence spectra of low and high molecular weight poly(oxyethylene) films: experimental and theoretical studies

P. Boulanger[*], J. Riga, J.J. Verbist and J. Delhalle[+]

Laboratoire Interdisciplinaire de Spectroscopie Electronique and
Laboratoire de Chimie Théorique Appliquée[+], Facultés Universitaires N.D. de la Paix,
61, rue de Bruxelles B-5000 Namur (Belgium)

ABSTRACT: X-ray photoelectron spectroscopy measurements of low and high molecular weight poly(oxyethylene) samples are reported in order to search for conformational signatures in their valence band spectra. Comparison with ab initio calculations on two conformations of CH_3-CH_2-O-$(CH_2$-CH_2-$O)_4$-CH_2-CH_3 point to a dominantly gauche conformation at the surface both for low and high molecular weight samples. Infrared measurements, on the contrary, indicate conformational differences in the bulk depending on the molecular weight of the samples.

1. INTRODUCTION

Poly(oxyethylene) (POE) is one of the polyethers most used in industry because of particular properties such as low melting point T_m, high solubility in common solvents, low elastic modulus, crystallinity, etc. Depending on the catalyst used, the molecular weight M_w can vary from 10^2 to 10^6 which allows interesting modulations of these physico-chemical properties.

Vibrational spectroscopy studies carried out in solid phase (Matsuura et al 1973) on the CH_3-$(O$-CH_2-$CH_2)_n$-O-CH_3 oligomers, n = 1-7, and on POE (M_w = 7500) (Yoshirara et al 1964 ; Miyazawa et al 1962 ; Matsuura et al 1973) indicate that these molecules adopt a gauche (TGT) conformation around the O-C-C-O bonds. Rao et al (1985) have followed the evolution of the infrared absorbance in the spectral region 1600 to 800 cm^{-1} of a thin POE film (M_w = 20×10^3) annealed at 55°C during 6 h and interpreted in terms of conformational changes the IR spectral modifications observed during the crystallization.

(*) Permanent address : GLAVERBEL S.A., Centre de Recherche de Jumet
Rue de l'Aurore, 2   B-6040 Jumet (Belgium)

Recent infrared measurements, Figure 1, by Boulanger (1991b) on POE films (1-2 μm thick ; $M_W = 5{\times}10^3$, $20{\times}10^3$ and $5{\times}10^6$) point to a noticeable dependence on M_W of the characteristic lines of the conformations. Low molecular weight POE ($M_W = 5{\times}10^3$) mainly exhibits the caracteristic lines (1358, 1280, 1235 and 945 cm^{-1}) of the TGT conformation (Yoshirara et al 1964 ; Matsuura et al 1973). As M_W grows the intensity of these lines decreases with a parallel increase of those (1342, 1241 and 960 cm^{-1}) typical of planar zig-zag (PZZ) segments. The intensity ratio I_{1342}/I_{1358} varies from 0.5 to 3.7 as M_W increases from $5{\times}10^3$ to $5{\times}10^6$. Thus, it is reasonable to conclude that films obtained by evaporation of a solution of high molecular weight POE ($M_W = 5{\times}10^6$) are rich in PZZ segments, while low molecular weight films ($M_W = 6{\times}10^2$) mainly contain TGT segments.

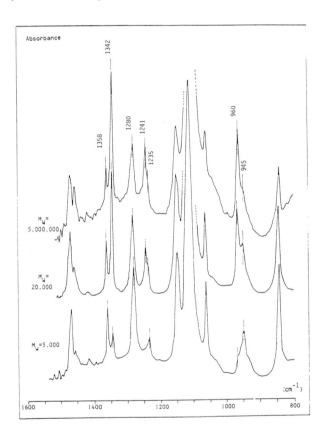

Fig.1. Infrared spectra of POE films (1-2 μm thick) of increasing molecular weight ($M_W = 5{\times}10^3$, $20{\times}10^3$ and $5{\times}10^6$).

Much in the same line as in the study of two crystalline varieties of poly(oxymethylene) (Boulanger et al 1989), the purpose of this contribution is to search for differences in the XPS valence spectra of low and high molecular POE films that could account for conformational changes of the type discussed above. Section 2 contains the details on the theoretical calculations carried out to interpret the XPS valence spectra, while section 3 provides information on the experiments. The results are reported in section 4 and the paper ends with a few remarks on the interest of XPS valence spectra of polymers (section 5).

2. THEORETICAL CALCULATIONS

The use of theoretical calculations in predicting and explaining features of the XPS valence spectra of polymers in connection with their primary and secondary structures has already been discussed, e.g. see (Delhalle et al 1974, 1977, 1979, 1987) and (Boulanger et al 1989, 1991), accordingly we limit the content of this section to a minimum.

2.1. Model Systems

X-ray diffraction studies by Tadokoro et al (1964) on POE suggest that the polymer crystallizes in the monoclinic system where chains tend to adopt a 7/2 helical conformation (Figure 2a). In this model, torsion angles around **C-O-C**-C and O-**C-C**-O bonds are 188.3° and 65°, respectively. More recent measurements by Takahashi et al (1973a) have pointed out slight distorsions in the conformation with respect to the ideal 7/2 helix (Figure 2b). Upon mechanical treatments (Takahashi et al 1973b), POE crystals can change to triclinic symmetry and the chains switch to planar zig-zag conformation (Figure 2c). The main difference between molecular parameters of 7/2 helix (TGT) and the planar zig-zag (PZZ) conformations is in the value of the torsion angle around the O-**C-C**-O bond, 65° and 180°, respectively.

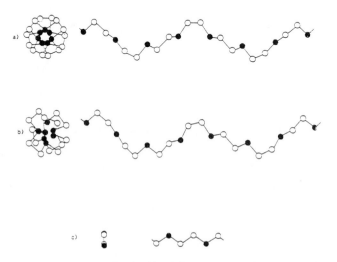

Fig.2. Representation of POE chains in (a) the regular 7/2 helical conformation, (b) the distorted 7/2 helix and (c) in the planar zig-zag conformation.

On the basis of these X-ray data, we choose to compare XPS measurements with results of calculations on the two conformers of CH_3-CH_2-O-(CH_2-CH_2-O)$_4$-CH_2-CH_3 shown in Table I, called TGT and PZZ, and modelling the chains of low (7/2 helix, $M_w = 6 \times 10^2$) and high molecular weight (planar zig-zag, $M_w = 5 \times 10^6$) POE samples, respectively.

TABLE I. Geometrical parameters used in model systems of POE, lengths in Å and angles in degrees.

Conformer	Bond length	Bond angle	Torsion angle
TGT	C-C : 1.54	OCC : 110.0	COCC : 188.3
	C-O : 1.43	COC : 112.0	OCCO : 65.0
	C-H : 1.08	CCO : 110.0	
PZZ	C-C : 1.54	OCC : 110.0	COCC : 180.0
	C-O : 1.43	COC : 112.0	OCCO : 180.0
	C-H : 1.08	CCO : 110.0	

2.2. Theoretical Details

The calculations reported in this work have been carried out at the ab initio level (STO-3G basis) using the GAUSSIAN 82 series of programs (Binkley et al 1982). The use of the minimal STO-3G basis is imposed by computer limitations and the size of the model oligomers needed to simulate polymer chains realistically. The qualitative ordering of the STO-3G levels has been checked against experimental and more refined theoretical results on systems closely related to POE (Boulanger et al 1991a). In order to have a more convenient comparison between theory and experiment, all the ab initio one-electron energies ε_i (in eV) are contracted and shifted on the energy scale according to : $\varepsilon_i' = 0.82\varepsilon_i - 2.90$. The validity of Koopmans' theorem in the interpretation of XPS spectra is assumed and the Gelius intensity model (Gelius 1974) is used in constructing the theoretical XPS valence spectra. They result from the addition of the theoretical peak intensities at the corrected occupied valence one-electron energies ε_i'. Each peak is represented by an equal weight linear combination of a lorentzian and a gaussian curve, each having a full-width at half-maximum (fwhm) of 1.5 eV. The relative atomic photoionization cross sections used for O_{2s}, O_{2p}, C_{2s}, C_{2p}, and H_{1s} are 1.400,

0.159, 1.000, 0.077 and 0.000, respectively. It should be stressed that the calculations are performed on isolated systems, assuming that the interchain interactions affect evenly the position and shape of the inner as well as the outer valence levels.

3. EXPERIMENTAL

The low molecular weight polymer (M_W =600, T_f = 60 °C) is prepared by Aldrich Chemical Industries, while the high molecular weight (M_W =5×10^6, T_f = 65 °C) has been obtained from Union Carbide. The purity of the samples is better than 97%.

The photoelectron spectra are recorded with a Hewlett-Packard 5950A spectrometer using the Al Kα monochromatized radiation (hv = 1486.6 eV). Films of the polymer samples, obtained from solvent casting, are deposited on a gold substrate and introduced in a vacuum chamber (pressure below 10^{-8} Torr). The temperature is maintained at 273 K by cooling with liquid nitrogen. During the analysis, the positive electrostatic surface charge left at the film surface due to emitted photoelectrons is kept constant by the use of an electron flood gun. The final experimental valence bands are obtained by adding up several recordings, each accumulated for 6 h. The binding energy of the valence band lines is referenced to the C_{1s} peak of the CH_2 groups. The C_{1s} peak is itself calibrated by mixing poly(trifluoroethylene) with the polymer, the C_{1s} core energy levels of poly(trifluoroethylene), CF and CF_2, are fixed at 289.4 and 291.6 eV, respectively. The radiation damage at the surface, checked by monitoring the C_{1s} peak after the analysis, was minimized by cooling and operating with short accumulation periods.

4. RESULTS

The core levels of both samples are identical in low and high molecular weight samples and require no special comment apart from indicating that the C_{1s} level involved in an ether linkage is located at 286.1 eV and that of O_{1s} at 523.4 eV. The theoretical and experimental valence binding energies for the PZZ and TGT conformations and for high (M_W =5×10^6) and low (M_W = 600) molecular weight POE samples are collected in Table II, the corresponding spectra are shown in Figure 3.

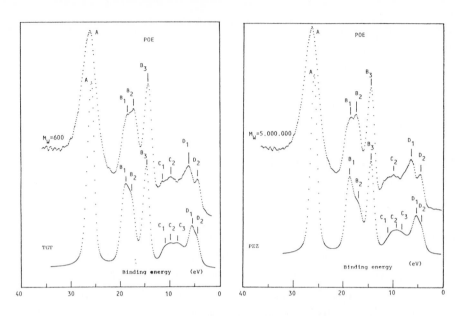

Fig.3. Theoretical XPS valence spectra (TGT) and (PZZ) and the corresponding experimental spectra for POE films, M_w = 600 and 5×10^6, respectively.

TABLE II. Theoretical $(- \varepsilon_i')$ and experimental binding energies (in eV) for TGT and PZZ conformers and POE films, M_w = 600 and 5×10^6, respectively.

Line	TGT	$M_w = 600$	PZZ	$M_w = 5 \times 10^6$
A	25.97	26.4	26.17	26.5
B_1	18.98	18.8	19.23	18.8
B_2	17.54	17.4	16.67	17.4
B_3	14.67	14.5	14.81	14.5
C_1	11.17	11.6	11.83	-
C_2	9.84	-	10.14	9.4
C_3	8.19	9.8	8.6	-
D_1	5.42	6.3	5.80	6.3
D_2	4.49	4.5	5.00	4.6

A rough inspection reveals that there is a satisfactory overall agreement between the theoretical and experimental XPS valence spectra shown in Figure 3. They can be decomposed in four regions, referred to as A, B, C and D, they correspond to molecular orbitals of which dominant contributions are O_{2s}, C_{2s}-O_{2s}, C_{2p}-O_{2s}-H_{1s} and O_{2s}-O_{2p}, respectively. The nature of these peaks has already been described (Boulanger et al 1991a), therefore we comment on possible spectral modifications due to conformational changes.

Peaks of region A correspond from molecular orbitals of the type σ_{C-O} wherein O_{2s} atomic orbitals are dominant, they mainly describe the C-O bond. In the theoretical as well as in experimental spectra there is no detectable dependence on conformational changes in that region.

In the four spectra, region B is composed of two main bands corresponding to bonding and antibonding σ_{C-C} levels, all of which having a small admixture of antibonding σ_{C-O}. The first band contains two peaks, B_1 and B_2, broader and less intense than the antibonding region characterized by the peak B_3. A comparison of the B regions for the theoretical simulations points to differences that can only be attributed to conformation. First, the width of the bonding σ_{C-C} band, $B_1 + B_2$, is larger in the trans (PZZ) conformer (fwhm = 3.08 eV) than in the gauche (TGT) form (fwhm = 2.78 eV). Second, the intensity ratio, I_{B2}/I_{B1} between the two peaks B_2 and B_1 of the bonding σ_{C-C} band is 0.70 for the PZZ form and 0.88 for the TGT form. Finally, peaks B_1 and B_2 are somewhat better resolved in the PZZ form than in the TGT isomer. This can be due a larger overlap between the bonding σ_{C-C} and antibonding σ_{C-O} levels in the gauche form (TGT). These differences are <u>not</u> observed in the experimental spectra of POE films, M_w = 600 and 5×10^6, respectively assumed to correspond to TGT and PZZ conformers.

The resolution of the spectral features in the C region of the theoretical and experimental spectra is so low that it is not possible to reliably relate small changes in the relative intensities of peaks C_1, C_2 and C_3 to conformational effects.

The structure of region D is similar in the four spectra shown in Figure 3 : an intense peak, D_1, centred at 5.6 eV in the theoretical spectra and at 6.3 eV in the experimental ones. This peak corresponds, in the molecular orbital language, to σ_{C-O} levels involving O_{2s}, O_{2p} and C_{2p} atomic character (Boulanger et al 1991a; Boulanger 1991b). D_2, the second peak, is also evident in the four spectra, it describes the oxygen lone pairs n_O. In the theoretical spectra it is located at 4.7 eV, while it is found at 4.5 eV in the experimental ones. The intensity ratio between D_1 and D_2 peaks seem unaffected by the conformational modifications, but a close inspection of the positions of these two peaks in the theoretical XPS valence spectra (TGT and PZZ) reveals a shift towards higher binding energies on going from the gauche (TGT), D_1 = 5.80 and D_2 = 5.00 eV, to the trans (PZZ), D_1 = 5.42 and D_2 = 4.49 eV, forms. Thus, predicted stabilization of these peaks on going from the gauche to the trans forms is of the order 0.38 eV and 0.51 eV, respectively for D_1 and D_2. A more detailed analysis of these energy shifts in terms of the molecular orbital theory can be found in (Boulanger 1991b). Except for a small difference, 0.1 eV, between the peak D_2 of the low, $D_2 = 4.47$ eV, and the high, $D_2 = 4.57$ eV, molecular weight POE films, the regions D in <u>experimental spectra are essentially similar</u>.

5. DISCUSSION

From the above results, it appears that calculations predict small but observable differences in the spectral features of simulated XPS valence spectra of TGT and PZZ, two conformers of $CH_3\text{-}CH_2\text{-}O\text{-}(CH_2\text{-}CH_2\text{-}O)_4\text{-}CH_2\text{-}CH_3$ used to simulate low and high molecular weight POE films, respectively, for which infrared measurements indicate substantial differences of conformational origin. Yet, the measured XPS valence spectra of low and high molecular weight POE samples come out almost identical and thus do not reveal conformational dependence. It is obviously tempting to incriminate the theoretical results for the discrepancy. However, if there are various reasons that could be advanced to criticize these theoretical predictions on a quantitative ground (poor basis set, Koopmans' theorem, molecular models, absence of intermolecular interactions, etc.) it nonetheless remains that the qualitative trends in regions B and D are consistent with the conformational changes imparted to the model molecules. Furthermore, similar predictions (Boulanger et al 1986) in the case of hexagonal and orthorhombic forms poly(oxymethylene), a closely related system to POE by its chemical nature, have been experimentally verified (Boulanger et al 1989). In particular, the changes in the relative intensity of peaks B_2 and B_1 as well as the shift in the position of peak D_1 have been found in excellent agreement with experimental measurements on pressed pellets of hexagonal and orthorhombic crystalline powders.

Since the infrared measurements on the various POE samples, Figure 1, match the results of previous investigators, it is also hard to question the dependence on the molecular weight of the conformational characteristics of POE chains. The sampling depth of infrared measurements is much more important than photoelectron spectroscopy, a typical surface technique. Thus the infrared results on the conformation pertain to the bulk rather than to the surface. Noting the closer similarity of the theoretical XPS valence spectrum of TGT with the two experimental spectra, almost identical, we tentatively conclude that the POE chains adopt preferentially a gauche conformation at the film surfaces of both low and high molecular weight samples. In the bulk, due to intermolecular interactions, the chains will elect a different mode of organization depending on their molecular weight.

As it has already been conjectured (Delhalle et al 1987), we have probably shown in this work that XPS valence band spectra, which hold information on the secondary structure of polymers (Delhalle et al 1979), can be used to discriminate between bulk and surface polymer conformations. It must be stressed that, to evidence surface conformations, it is essential to assist XPS valence band measurements with model calculations and results of various spectroscopies to cross-check the results. Work is now in progress along these lines to ascertain the tentative conclusions of this work.

Acknowledgments. The authors acknowledge with appreciation the support of this work within the Science EEC program n° SC1-0016-C. They also thank the Belgian National Fund for Scientific Research (FNRS), IBM-Belgium, and FUNDP for the use of the Namur Scientific Computing Facility.

REFERENCES

Binkley J S, Frisch M J, De Frees D J, Raghavachari K, Whiteside, Schlegel M B, Fluder E M and Pople J A 1982 GAUSSIAN82 (Carnegie-Mellon University, Pittsburgh)
Boulanger P, Lazzaroni R, Verbist J J and Delhalle J 1986 Chem. Phys. Lett. 129 275
Boulanger P, Riga J, Verbist J J and Delhalle J 1989 Macromolecules 22 173
Boulanger P, Magermans C, Verbist J J, Delhalle J and Urch D S 1991a Macromolecules 24 2757
Boulanger P, 1991b Relations entre la Structure Electronique et la Geométrie des Chaînes Moléculaires dans les Polymères Oxygénés, Ph D Thesis (Facultés Universitaires N.D. de la Paix, Namur, 1991).
Delhalle J, André J M, Delhalle S, Pireaux J J, Caudano R and Verbist J J 1974 J. Chem. Phys. 60 595
Delhalle J, Delhalle S, André J M, Pireaux J J, Riga J, Caudano R and Verbist J J 1977 J. Electron Spectrosc. Relat. Phenom. 12 293
Delhalle J, Montigny R, Demanet C and André J M 1979 Theor. Chim. Acta (Berlin) 50 343
Delhalle J, Delhalle S and Riga J 1987 J. Chem. Soc. Faraday Trans. 2 83 503
Gelius U J 1974 J. Electron Spectrosc. Relat. Phenom. 5 985
Matsuura H, Miyazawa T and Machida K 1973 Spectrochimica Acta 29A 771
Miyazawa T, Fukushima K and Ideguchi Y 1962 J. Chem. Phys. 37 2764
Rao G R, Castiglioni C, Gussoni M and Zerbi G 1985 Polymer 26 811
Tadokoro H, Chatani Y, Yoshihara T, Tahara S and Murahashi S 1964 Makromol. Chem. 73 109
Takahashi Y and Tadokoro H 1973a Macromolecules 6 672
Takahashi Y, Sumita I and Tadokoro H 1973b J. Polym. Sci. Polym. Phys. 11 2113
Yoshirara T, Tadokoro H and Murahashi S 1964 J. Chem. Phys. 41 2902.

Valence electronic structure of PMMA: Theoretical analysis of experimental XPS data

E. Ortí[1], R. Viruela[1], J.L. Brédas[2], and J.J. Pireaux[3].

[1] Univ. Valencia, Dept. Química Física, Dr. Moliner 50, E-46100 Burjassot (Spain).
[2] Univ. Mons-Hainaut, Service de Chimie des Matériaux Nouveaux et Département des Matériaux et Procédés, Place du Parc 20, B-7000 Mons (Belgium).
[3] FUNDP, Laboratoire Interdisciplinaire de Spectroscopie Electronique, Rue de Bruxelles 61, B-5000 Namur (Belgium).

ABSTRACT : We present a theoretical description of the valence band XPS spectrum of poly(methyl methacrylate) using valence effective Hamiltonian (VEH) calculations. Our goal is to illustrate the role that theoretical calculations could play in elucidating the photoemission spectral features of polymers. The VEH electronic band structure calculated for syndiotactic PMMA is analyzed in detail and correlated with the XPS spectrum. The agreement between theory and experiment is very good and allows for a complete interpretation of the photoemission bands. Contributions from the main chemical environments are clearly identified in the light of the VEH results.

1. INTRODUCTION

Metallization of polymers is an extensively used way of developing materials for a wide variety of applications ranging from integrated circuits, data storage or recording media, to packaging [Burrell (1989), Feast (1987), Mittal (1989)]. The design and fabrication of metal-polymer interfaces showing the appropriate properties to satisfy technological demands from these different industries necessitates for a comprehensive and fundamental understanding of the molecular events that influence the nature of the adhesive joint. Among the great variety of surface analysis techniques, photoelectron spectroscopy has been shown to be a powerful tool to investigate the nature of chemical bonding at the interface. The studies using X-ray photoelectron spectroscopy (XPS) or ultraviolet photoelectron spectroscopy (UPS) techniques have been conducted primarily by evaporating metals on polymer substrates in a high vacuum environment and monitoring the changes induced by metal deposition in the polymer spectra [Burkstrand (1979, 1982), Ho (1988), Jordan (1987), Kowalczyk (1988), Ohuchi (1986), White (1987)]. To attribute these changes to well-defined interactions between the deposited species and some clearly identified local group of the polymer, a detailed understanding of the electronic structure of the polymer constitutes a preliminary requirement. In this context, theoretical studies have been shown to be very useful since they allow for a full characterization of the photoelectron band structure of the polymer, and thereby shed more

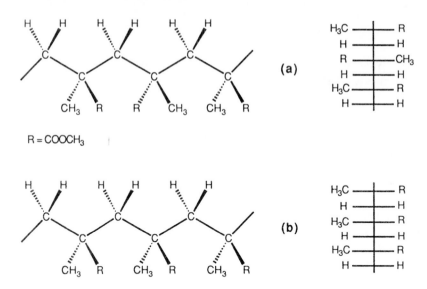

R = COOCH₃

Figure 1. Syndiotactic (a) and isotactic (b) configurations of poly(methyl methacrylate), presented (left) in the fully extended (all-trans) planar zig-zag conformation, (right) as the corresponding Fischer projection.

light on the specific interaction at the interface [Ho (1988), Kowalczyk (1990)]. We have recently performed a joint theoretical and experimental photoemission study on the copper/poly(ethylene terephthalate) interface [Chtaïb (in press)].

In this work we present a detailed interpretation of the valence electronic structure of poly(methyl methacrylate) (PMMA) as an example to illustrate the role that theoretical calculations could play in elucidating the photoemission spectral features of polymers. PMMA is one of the most widely used polymers due to its good mechanical, thermal, and optical properties and is employed in the microelectronics industry as a photoresist and insulator material. From a structural point of view, PMMA can be visualized as a polyethylene chain substituted by a methyl group and an acetate group on every other carbon, its monomeric unit corresponding to $[CH_2-C(CH_3)(COOCH_3)]$. Nowadays, PMMA can be prepared as crystalline samples of syndiotactic (s) or isotactic (i) chains with a high degree of stereoregularity. Figure 1 shows these two main configurations for PMMA chains in an ideal fully extended (all-trans) planar zig-zag conformation. While the syndiotactic configuration (Figure 1a) locates methyl and ester groups alternatively above and below the plane defined by the polyethylene chain, the isotactic configuration (Figure 1b) places ester groups on one side and methyl groups on the other side of the chain.

In this paper, we mainly focus on the valence electronic structure of syndiotactic PMMA. We have performed band structure and density of states calculations to understand the electronic and spectral features in the valence region of the XPS spectra of PMMA. This information will facilitate the interpretation of the electronic structure changes induced by the formation of metal-PMMA interfaces. The theoretical approach employed is the valence effective Hamiltonian (VEH) quantum-mechanical technique, which has been shown to provide an accurate picture of the valence electronic structure for a wide variety of polymers [Brédas (1982, 1983) (1985, 1986), Chtaïb (in press), Kowalczyk (1990), Lazarroni (1990), Orti (1989, 1990), Wu (1987, 1988)]. The interaction of acrylic polymers with aluminium has been recently studied by Chakraborty et al. [Chakraborty (1990)] on the basis of molecular orbital calculations on small PMMA oligomers.

2. METHODOLOGY

2.1 Experimental XPS Technique and measurements

Photoelectron spectra were recorded with a HP 5950 A spectrometer (monochromatized AlK_α radiation), with conventional stabilization of any charging effect with the flood gun technique. Energy calibration of the spectra was obtained by indirect internal referencing: by physically mixing powders, poly(trifluoroethylene) -P3FE- was calibrated against polyethylene (284.6 eV), in order to precisely locate the PMMA carbon 1s peak at 285.0 ± 0.15 eV against P3FE. Polymer samples (Strathclyde University) were prepared either as thin films cast from solution, or as pressed powder pellets. In order to avoid radiation damage, valence band spectra were recorded during a few hours on different fresh samples; the collected data were then added.

2.2 Theoretical Details

The electronic band structure calculations presented here have been performed in the framework of the nonempirical VEH technique [Nicolas (1979, 1980)], as extended for the treatment of stereoregular polymeric systems [André (1979)]. The VEH method takes only into account the valence electrons and is well documented in the literature [André (1979), Brédas (1981), (1982), 1893, Kowalczyk (1990), Nicolas (1979, 1980)]. We just recall here that it is parameterized to provide one-electron energies of Hartree-Fock ab initio double-zeta quality. The VEH atomic potentials used in this work are those previously optimized for hydrogen, carbon and oxygen [Brédas (1981), Themans (1985)].

The density of valence states (DOVS) has been calculated from the electronic band structure following the methodology of Delhalle and Delhalle [Delhalle (1977)]. To compare the resulting DOVS curves to the experimental photoemission spectra, a three-step procedure is generally used [Brédas (1982, 1983), (1985, 1986), Chtaïb (in press), Kowalczyk (1990), Orti (1989, 1990), Lazzaro (1990), Wu (1987, 1988]) and consists in: (i) The convolution of the bare DOVS curves by a Gaussian function of adjustable full width at half maximum (FWHM). This is a way of introducing in our theoretical simulation the static and dynamic disorder effects that exist in the solid and cause the broadening of the photoemission linewidths with respect to the gas phase. (ii) The application of a rigid shift of the DOVS curves toward lower binding energies in order to deal with the solid-state polarization energy due to interchain relaxation effects. (iii) The contraction of the VEH-DOVS curves along the energy scale in order to correct for too wide a valence band that Hartree-Fock ab initio calculations and, therefore, VEH calculations provide.

To obtain a theoretical simulation of the XPS spectra, it is necessary to take into account an additional factor in order to include the effect of relative photoionization intensities of the crystal orbitals in the DOVS curves. As in previous works [Chtaïb (in press), Kowalczyk (1990), Orti (1989, 1990)], we have used the model of Gelius [Gelius (1972)] extended to polymers [Delhalle (1975)], which relates the experimental photoelectron intensity of the nth one-electron level to its Mulliken gross population multiplied by an experimental σ factor. The values used are $\sigma(O,2s) = 1.4$, $\sigma(O,2p) = 1.4/8$, $\sigma(C,2s) = 1$, $\sigma(C,2p) = 1/13$, and $\sigma(H,1s) = 0$.

X-ray diffraction data indicate the presence of a helical conformation with a large number of units per turn (37/4) in crystalline s-PMMA [Kusuyama (1982), (1983)]. Experimental evidences show that in this helical structure the conformation of the backbone chain is very near that of the trans-trans staggered conformation. Indeed, theoretical calculations found that for s-PMMA this conformation is energetically the most favoured [O'Reilly (1981), Sundarajan (1974), (1986), Vacatello (1986)]. VEH band structure calculations on s-PMMA have been therefore performed assuming an all-trans zig-zag conformation for the polyethylene backbone (Figure 1a). It is to be noted that the VEH calculation scheme takes explicitly into account the mirror symmetry presented by s-PMMA along the planar chain. This results in the fact that only the $C_5O_2H_8$ unit cell depicted in Figure 2 is to be considered instead of the $C_{10}O_4H_{18}$ translational unit cell.

The structural parameters (bond lengths and bond angles) used to build up the unit cell of PMMA for VEH calculations are shown in Figure 2. The bond lengths and bond angles of the ester group are taken from the work of Vacatello et al. [Vacatllo (1986)]. This group is assumed to be in a planar conformation with the carbonyl oxygen oriented cis with respect to the C1-methyl group.

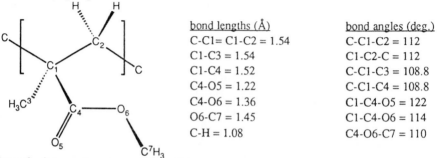

bond lengths (Å)	bond angles (deg.)
C-C1= C1-C2 = 1.54	C-C1-C2 = 112
C1-C3 = 1.54	C1-C2-C = 112
C1-C4 = 1.52	C-C1-C3 = 108.8
C4-O5 = 1.22	C-C1-C4 = 108.8
C4-O6 = 1.36	C1-C4-O5 = 122
O6-C7 = 1.45	C1-C4-O6 = 114
C-H = 1.08	C4-O6-C7 = 110

Figure 2.   Schematic representation of the structural unit cell used for VEH band structure calculations on s-PMMA. The atom numbering and the various geometrical parameters are shown.

The bond angles in both the ester- and C1-methyl groups are adapted to be tetrahedral. All the bond angles along the main polyethylene chain are considered to be identical and equal to 112°. This assumption disregards the fact that interactions between C1 substituents on adjacent cells are considerably larger than those between C2 hydrogens, but it is a necessary requirement to obtain a linear polymer and be able to carry out the band structure calculation. The adopted value of 112° is intermediate between the 106° and 124° values reported by Vacatello et al. [Vacatello (1986)] for C-C1-C2 and C1-C2-C angles, respectively. We have confirmed that band structure calculations using main-chain angles ranging from 109.5 to 124° introduce only slight changes in the electronic valence structure calculated for s-PMMA, since, as we will show in the following section, the electronic structure is mainly determined by the pendant groups.

3. RESULTS AND DISCUSSION

3.1 Theoretical Electronic Structure

Figure 3 displays the valence band structure, DOVS curves, and convoluted DOVS curves calculated for a regular chain of s-PMMA using the VEH method and the geometrical parameters described above. The band structure contains twenty occupied valence bands which have been numbered starting from the bottom of the valence band. All the crossings between these bands are forbidden for symmetry reasons. The convoluted DOVS curves (Figure 3c) have been computed using Gaussian functions with a FWHM value of 0.7 eV and the photoionization cross-section model of Gelius [Delhalle (1975), Gelius (1972)]. As explained above, this model provides theoretical simulations of the experimental XPS spectra. We now turn to a detailed analysis of the characteristics and atomic compositions of the bands in order to assign the atomic parentage of the peak structures in the DOVS curves.

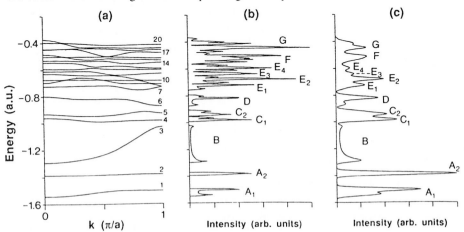

Figure 3. VEH calculated band structure (a), density of states (b), and convoluted density of states (c) for s-PMMA.

Starting from the high binding energy side (thus going from the bottom to the top of the valence band), peak structure A in Figure 3 results from bands 1 and 2 which are mainly built from oxygen 2s atomic orbital contributions. Peak A_1 originates in band 1 which corresponds to the bonding interactions of oxygen atoms with the 2s orbital of the C4 atom. The width of this band (1.18 eV) and, therefore, the structure of peak A_1 is quite sensitive to geometry modifications due to the fact that it involves significant contributions from all the carbon atoms. On the contrary, band 2 is almost totally localized on oxygen atoms and shows a dispersion of only 0.33 eV, giving rise to the highest peak A_2.

The broad peak structure B results from the highly dispersed band 3 (7.29 eV). This band corresponds to the deepest-lying C_{2s} band of polyethylene [Delhalle (1974)] since it mostly involves the bonding interactions of the 2s orbitals of the carbon atoms forming the main backbone of PMMA. It starts at -1.29 a.u., i.e. 0.09 a.u. lower than it is calculated for polyethylene (-1.20 a.u.), due to the stabilizing effect of O5 2s contributions. Compared to polyethylene, the main-chain C-C band 3 is in fact disrupted by bands 4 and 5 in PMMA. Band 4 is mainly due to C7 2s contributions and gives rise to peak C_1, while band 5 originates in both C3 and C2 2s contributions and leads to peak C_2. Both bands involve some C-H bonding interactions.

Peak structure D comes from band 6. This band is mainly localized on the transverse chain formed by the C1 atom and its methyl and ester substituents, the main contributions corresponding to the 2s orbitals of the -C4O5O6- carboxylate group. Band 6 shows a significant dispersion of 1.94 eV; it involves some antibonding interactions between the 2s orbitals of the carbon atoms forming the main chain of PMMA in clear correspondence with the antibonding upper part of the so-called "C-C" band of polyethylene [Delhalle (1974)].

Above band 6, we find a compact group of bands (7 to 20), very close in energy, and with rather small widths (<1.50 eV) due to the avoided crossings. These bands define the outer part of the valence band of PMMA and mainly imply contributions from carbon and oxygen 2p orbitals and hydrogen 1s orbitals. Seven of these bands (7 to 13) are involved in the peak structure labelled E which consists of four different features E_1 to E_4. The lowest of these bands gives rise to peak E_1 and defines the C4-O6-C7 σ−bonding. The intense peak E_2 derives from the very flat regions of bands 8 to 10 and actually corresponds to the lone pairs of carbonyl oxygen atoms. Peak E_3 results from bands 10 and 11 and mainly involves $C7_{2p}$-H_{1s} bonding interactions with some contributions from O6 2p orbitals. The higher energy peak structure E_4 involves both $C3_{2p}$-H_{1s} and $O6_{2p}$-$C7_{2p}$-H_{1s} σ−bonding interactions.

Peak structure F results from bands 13 to 16 and contains three main peaks (Figure 3b). The atomic nature of the first two peaks is difficult to disentangle and they mainly involve contributions from C_{2p}-H_{1s} σ-bonding and carbonyl π-bonding. The most intense peak within structure F corresponds to the σ-contributions to the lone pairs of O6.

Finally, band G presents a double peak structure (Figures 3b and 3c). The most intense peak derives from the flat regions of bands 17 to 19, mostly involving the lone pairs (π-contribution) of O6. This peak presents a shoulder on its high energy side that mainly results from band 20 and originates in the O5 lone pairs. It is to be noted that the top of band 20 at the center of the Brillouin zone (k=0), gives rise to a small density of states in Figures 3b and 3c and corresponds to the uppermost occupied band of polyethylene. In contrast to polyethylene and due to the avoided band crossings present in PMMA, the wave function characterizing this band (C_{2p} interactions along the main chain) is shared by several bands (20,19,17,15).

3.2 Correlation with Experimental XPS Data

The convoluted DOVS curves displayed in Figure 3c are presented in Figure 4 together with the experimental XPS spectrum recorded for highly crystalline samples of s-PMMA. In order to compare with the experimental data, the theoretical simulation has been contracted by a factor of 1.25 (i.e., the whole valence-band width is reduced by a factor of 1.25). Such a value is the same used for other oxygen containing polymers [Chtaïb (in press), Kowalczyk (1990)] and is consistent with that used in most VEH calculations reported previously [Brédas (1982, 1983), (1985, 1986), Lazzaroni (1990), Orti (1989, 1990), Wu (1987, 1988)]. Moreover, a rigid shift to lower binding energies has been applied to the VEH spectrum in order to align peak E_2 with

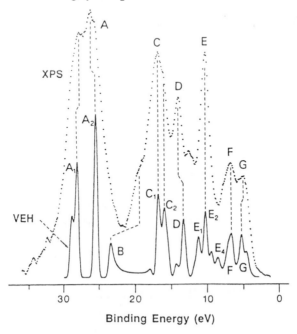

Figure 4. Theoretical VEH and experimental XPS valence band spectra for s-PMMA. Labels correspond to those displayed in Figure 3.

the experimental photoemission peak E. The experimental XPS spectrum corresponds to data averaged on many spectra from many samples after subtraction of the background (Shirley type function) [Shirley (1972)] and subsequent smoothing (9 points, second-order polynomial). Both theoretical and experimental valence band peaks are labelled in the same way as the peak structures discussed in Figure 3. The binding energies and atomic composition of the most salient features of theoretical and experimental XPS spectra are summarized in Table 1 to facilitate a detailed analysis of the correspondence between theory and experiment.

Table 1. Calculated (VEH) and measured (XPS) binding energies (eV) for the most salient features in the valence band of s-PMMA.

Feature[a]	ε_i^b	ε_i^c	XPS
	41.87 (sh)	29.10	29.0 (sh)
A_1 (O5-C4-O6 2s bonding)	40.95	28.37	27.85
A_2 (O 2s)	37.67	25.74	26.3
			25.6 (sh)
B (PE C_{2s}-C_{2s} bonding band)	35.00	23.61	19.6
C_1 ($C7_{2s}$)	26.68	16.95	17.0
C_2 ($C2_{2s}$, $C3_{2s}$)	25.60	16.09	16.2 (sh)
D (O5-C4-O6 2s antibonding +	23.53 (sp)	14.43	
PE C_{2s}-C_{2s} antibonding band)	22.26	13.42	14.05
E_1 (C4-O6-C7 2p σ-bond)	19.67	11.34	12.5 (sp)
E_2 (O5 lone pairs)	18.44	10.36	10.36
E_3 ($C7_{2p}$-H_{1s})	17.41 (sp)	9.54	
E_4 ($C3_{2p}$-H_{1s})	16.25 (sp)	8.61	7.8 (sh)
F (O6 "σ" lone pairs)	13.88	6.71	6.8
G (O6 "π" lone pairs)	12.10	5.29	4.8
(O5 lone pairs)	11.30 (sh)	4.65	

a   Labels correspond to those used in Figures 3 and 4. The main atomic contributions are indicated in parentheses according to Section 3.1 and the atomic numbering used in Figure 2. PE denotes significant contributions from polyethylene chain.

b   VEH binding energies calculated from an isolated chain of s-PMMA.

c   VEH binding energies after a contraction of 1.25 of the energy scale and a rigid shift of the valence band to align the theoretical peak E_2 with the experimental peak at 10.36 eV.

d   The following labels denote: sh: shoulder; sp: small peak.

As can be seen from Figure 4 and Table 1, the relative peak positions and intensities predicted by the VEH simulation are found to be in very good agreement with the experimental XPS data. As mentioned before, the theoretical peak E_2, which is shown to derive from carbonyl oxygen lone pairs in Section 3.1, has been chosen to do the fitting between theoretical and experimental spectra and has been aligned with the sharp experimental peak E located at 10.36 eV. This assignment is performed on the basis of previous results on poly(ethylene terephthalate) (PET) [Chtaïb (in press)]. For this polyester, for which the fitting between theory and experiment was done using the "fingerprints" of the benzene moiety, the deeper-lying lone pairs of carbonyl oxygens calculated at about 10.30 eV (18.40 eV before contraction and shifting) are shown to be in excellent correlation with the sharp photoemission band observed at 10.26 eV. All these binding energies, both theoretical and experimental, are consistent with those obtained here for PMMA (see Table 1).

Going from lower to higher binding energies, the first experimental band measured at about 4.8 eV should be correlated with the theoretical peak structure G. The main peak of this structure comes from the π-lone pairs of the methoxy oxygens and is calculated at 5.29 eV, overestimating the experimental binding energy by ≈ 0.5 eV. This result is consistent with those obtained for small ester compounds like methyl acetate (CH_3COOCH_3), for which VEH calculations also overestimate the second ionization potential resulting from the methoxy oxygen lone pairs by ≈ 0.5 eV (VEH value = 11.71 eV, experimental He II gas-phase value = 11.2 eV [Carrington (1985)]). This concordance in the VEH predictions clearly supports the alignment performed here between the theoretical and experimental spectra of PMMA. The theoretical peak G exhibits a shoulder on its low binding energy side that derives from carbonyl oxygen lone pairs and is not clearly seen in the experimental spectrum.

The second photoemission band observed at 6.8 eV is in excellent correlation with the theoretical peak F calculated at 6.71 eV and mainly coming from the σ-contribution to the lone pairs of methoxy oxygens. The experimental band shows a shoulder on its high binding energy side (7.8 eV) that has no clear correspondence in the theoretical spectrum. In Table 1, this shoulder has been associated to the small theoretical peak E_4 but it could be also assigned to the smooth shoulder that peak G presents about 7.20 eV which implies carbonyl π bonds.

The experimental XPS band E (used in the fitting between theory and experiment) results from carbonyl oxygen lone pairs. The high intensity of this band is due to the important contributions from O5 2s orbitals involved in theoretical peak E_2. As discussed in Section 2.2, these orbitals show larger photoionization cross-sections than O 2p orbitals. The photoemission band D can be associated with the highest peak of the theoretical structure D. This peak mostly derives from the 2s orbitals of carboxylate groups, but it also encloses the antibonding C_{2s} interactions along the polyethylene chain. Peak D is calculated to be too low in binding energy due to an overestimation of the antibonding C_{2s}-O_{2s} interactions. A small feature is observed between bands D and E in the experimental XPS spectrum. This feature has no clear correlation in the theoretical spectrum and, if exists, could be only associated with peak E_1 which defines the C4-O6-C7 2p σ-bonding.

The photoemission band C perfectly correlates with double-peak theoretical structure C and can be envisioned as a carbon 2s band. As discussed above, theoretical peak C_1 is localized on the C7 atom and is calculated at 16.95 eV in excellent agreement with the main experimental peak measured at 17.0 eV. Peak C_2 at 16.09 eV results from C2 and C3 atoms and perfectly matches the shoulder that the experimental band presents at 16.2 eV. Band C also shows a broad shoulder on its high binding energy side about 19.6 eV. This shoulder should be assigned to theoretical peak B which corresponds to bonding C_{2s} interactions along the polyethylene chain and is calculated at 23.61 eV. This is an expected result since the VEH method was mainly parameterized to reproduce the top of the valence band [Brédas (1981), (1982, 1983)]. Indeed, VEH calculations on polyethylene lead to a binding energy of 22.64 eV for the photoemission peak corresponding to the bonding C_{2s} band, overestimating by ≈ 3.8 eV the experimental value of 18.8 eV [Pireaux (1977)]. If we subtract the 3.8 eV factor from the VEH energy of peak B, we obtain a very good agreement between theory (19.77 eV) and experiment (19.6 eV).

Finally, the highest intensity band A clearly results from oxygen 2s electronic levels, giving rise to theoretical peaks A_1 and A_2. As for PET [Chtaïb (in press)], the A_2 peak is calculated to lie too low in binding energy. On the contrary, theoretical peak A_1 is calculated ≈ 0.5 eV too high in binding energy due to an overestimation of the bonding C_{2s}-O_{2s} interactions. This same feature produces an opposite effect on peak D, as mentioned before. Peak A_1 shows a shoulder at 29.10 eV that perfectly correlates with the experimental shoulder observed at 29.0 eV.

4. SUMMARY AND CONCLUSIONS

Experimental XPS measurements and theoretical VEH electronic band structure calculations on syndiotactic poly(methyl methacrylate) have been correlated in the hope of disentangling the electronic valence band of the polymer. The VEH-DOVS curves convoluted to simulate XPS spectra are found to be fully consistent with the experimental spectrum and an excellent quantitative agreement between theory and experiment is obtained when comparing the positions of the main peaks. A complete assignment of all the photoemission bands appearing in the XPS spectrum is performed on the basis of a detailed analysis of the atomic composition of the electronic states giving rise to the different VEH-DOVS structures.

The VEH assignment illustrates that the valence XPS spectrum of s-PMMA is dominated by the contributions of the pendant groups. It has allowed us to identify the electronic "fingerprints" of the main chemical environments such as : oxygen lone pairs (differentiating between carbonyl and methoxy oxygens), carboxylate σ-bonding, polyethylene C_{2s}-C_{2s}

band (locating bonding and antibonding contributions). The carbonyl π bond is not easily identified because it actually lies parallel to the backbone. These results constitute a preliminary step in understanding the chemical bonding at the metal-PMMA interface since they will facilitate the identification of the specific metal-polymer interactions by following the changes observed in the valence band of PMMA upon metallization, in the same spirit as what has been done for the metal/PET interface [Chtaïb (in press)].

Finally, we would like to mention that, when studying interfacial properties, it is of utmost importance to determine the conformation of the polymer chain at the material surface and the orientation of the chemical groups relative to each other. Valence band photoemission spectra have already proved their potentiality in differentiating configurations (isotactic or syndiotactic) and conformations (zig-zag planar or helical) for a variety of polymers [Orti (1989, 1990), Pireaux (1990)]. In the case of PMMA, preliminary investigations on isotactic PMMA using both XPS experimental and VEH theoretical approaches evidence that the valence electronic structure of PMMA experiences noticeable changes depending on tacticity and conformation. The theoretical results point out that certain photoemission bands of i-PMMA are especially sensitive to steric contacts between the pendant groups and that a very detailed knowledge of the geometrical structure of the polymer is necessary to obtain definitive results. Molecular orbital calculations including geometry optimizations for long oligomers of PMMA are now in progress.

ACKNOWLEDGEMENTS

We thank the CIUV (Centro de Informática de la Universitat de València) for the use of their computing facilities. This work has been partly supported by DGICYT Projects PS88-0112 and OP90-0042 and the Belgian "Pôle d'Attration Interuniversitaire en Chimie Supramoléculaire et Catalyse".

REFERENCES

André J M, Burke L A, Delhalle J, Nicolas G and Durand Ph 1979 Int. J. Quantum Chem. Symp. 13 283

Brédas J L, Chance R R, Silbey R, Nicolas G, and Durand Ph 1981 J. Chem. Phys. 25 255

Brédas J L , Chance R R, Silbey R, Nicolas G, and Durand Ph 1982 J. Chem. Phys. 77 371 ; Brédas J L, Chance R R, Baughman R H, and Silbey R, ibid. 1983 78 5656

Brédas J L 1985 Chem. Phys. Lett. 115 119; Brédas J L and Salaneck W R 1986 J. Chem. Phys. 85 2219

Burkstrand J M 1979 Phys. Rev. B 20 4853; 1982 J. Vac. Sci. Technol. 20 440

Burrell M C, Codella P J, Fontana J A, Chera J J and McConnell M D J. Vac. 1989 Sci. Technol. A 7 55

Carrington P H and Ham N S 1985 J. Electron Spectrosc. Rel. Phenom. 36 203

Chakraborty A K, Davis H T, and Tyrrell M 1990 J. Polym. Sci. A 28 3185

Chtaib M, Ghijsen J, Pireaux J J, Caudano R, Johnson R L, Ortí E and Brédas J L in press Phys. Rev. B

Delhalle J, André J M, Delhalle S, Pireaux J J, Caudano R, Verbist J J 1974 J. Chem. Phys. 60 595

Delhalle J, Delhalle S and André J M 1975 Chem. Phys. Lett. 34 430

Delhalle J and Delhalle 1977 S Int. Quantum Chem. 11 349

Feast W J and Munro H S 1987 Polymer Surfaces and Interfaces (Chichester : Wiley)

Gelius U 1972 Electron Spectroscopy (North-Holland, Amsterdam : Shirley) pp 311

Ho P S, Hahn P O, Bartha J W, Rubloff G W, LeGoues F K and Silverman B D 1985 J. Vac. Sci. Technol. A 3 739

Ho P S, Silverman B D, Haight R A, White R C, Sanda P N and Rossi A R 1988 IBM J. Res. Dev. 32 658

Jordan J L , Kovac C A, Morar J F and Pollack R A 1987 Phys. Rev. B 36 1369

Kowalczyk S P, Kim Y-H, Walker G F and Kim J 1988 Appl. Phys. Lett. 52 375

Kowalczyk S P, Staftröm S, Brédas J L, Salaneck W R and Jordan-Sweet J L 1990 Phys. Rev. B 41 1645

Kusuyama H, Takase, Higashihata M Y, Tseng H T, Chatani Y andTadokoro H 1982 Polym. 23 1256

Kusuyama H, Miyamoto N, Chatani Y and Tadokoro H 1983 Polym. Comunn. 14 119

Lazzaroni R, Sato N, Salaneck W R, Dos Santos M C, Brédas J L, Tooze B and Clark D T 1990 Chem. Phys. Lett. 175 175

Mittal K L and Susko J R 1989 Metallized Plastics. Fundamental and Applied Aspects, (New York : Plenun Press); Mittal K L 1989 Proceedings of the Second Symposium on Metallized Plastics : Fundamental and Applied Aspects (Chicago : Electrochem. Soc. Meeting)

Nicolas G and Durand Ph 1979 J. Chem. Phys. 70 2020; 1980 ibid. 72 453

Ohuchi F S and Freilich S C 1986 J. Vac. Sci. Technol. A 4 1039

O'Reilly J M and Mosher R A 1981 Macromolecules 14 602

Ortí E, Stafström S and Brédas J L 1989 Chem. Phys. Lett. 164 240; Ortí E, Brédas J L, Pireaux J J and Ishihara N J 1990 Electron. Spectrosc. Rel. Phenom. 52 551

Pireaux J J, Riga J, Caudano R, Verbist J J, Delhalle J, André J M and Gobillon Y 1977 Phys. Script. 16 329

Pireaux J J, Riga J, Boulanger P, Snauwert P, Novis Y, Chtaib M, Grégoire C, Fally F, Beelen E, Caudano R and Verbist J J 1990 J. Electron Spectrosc. Rel. Phenom. 52 423

Shirley D A 1972 Phys Rev. B5 4709

Sundarajan P R and Flory P J 1974 J. Am. Chem. Soc. 96 5025

Sundarajan P R 1986 Macromolecules 19 415

Themans B, André J M and Brédas J L 1985 Mol. Cryst. Liq. Cryst. 118 121

Vacatello M and Flory P J 1986 Macromolecules 19 405

White R C, Haight R, Silverman B D and Ho P S 1987 Phys. Rev. B 36 1369

Wu C R, Nilsson J O, Inganäs O, Salaneck W R, Osterholm J R and Brédas J L 1987 Synth. Met. 21 197 ; Lazzaroni R, Riga J, Verbist J, Brédas J L and Wudl F 1988 J. Chem. Phys. 88 4257

Surface composition of poly(styrene-b-paramethylstyrene) copolymers determined by static secondary ion mass spectroscopy

S Affrossman[1], F Hindryckx[3] , R A Pethrick[1] and M. Stamm[2]

1. Department of Pure and Applied Chemistry, University of Strathclyde, Glasgow G1 1XL, Scotland,
2. Max-Planck-Institut fur Polymerforschung, 6500 Mainz, Germany
3. Now at University of Liege, Liege, Belgium

ABSTRACT: The surface compositions of a series of random and block copolymers of styrene and paramethylstyrene have been examined by static SIMS. For random copolymers, the SIMS data are shown to bear a quantitative relationship to the bulk compositions. In contrast, the surface of block copolymers exhibited a strong surface segregation of the paramethyl-styrene component. A deuterated styrene block was used to enhance the SIMS sensitivity to the minor surface component. The data show that the surface segregation of the paramethyl-styrene block, observed previously in the near surface region, by neutron and nuclear reaction techniques, extends to the outermost layer, resulting in a surface which approaches 99% paramethylstyrene.

1. INTRODUCTION

The composition of a multicomponent system in the vicinity of an interface is determined ultimately by the surface energies of the components. For a polymer system, the shape and relatively large size of polymer molecules will affect the composition profile, normal to the surface, in three ways. Firstly, the perturbing influence of the surface will be apparent for distances of at least one, and possibly several, molecular diameters, i.e. tens of nm. This would lead to a segment density distribution, and a distribution of mean square end to end distances, normal to the surface. The greater degree of freedom available at the surface will be more favourable thermodynamically to the ends of the polymer chain, than to the more restricted inner segments, with the resulting tendency for chain ends to fill the intermolecular spaces at the surface. Lastly, the rate at which such systems reach equilibrium may be slow, in practice the system being frozen into a non-equilibrium configuration dependent on the previous history of the sample, e.g. temperature used in film formation, rate of evaporation of solvent, starting concentrations and shear mechanical forces experienced. Whether at equilibrium, or not, the surface has an important role in the practical applications of polymer systems.

The techniques which are suitable for the determination of surface, or near surface, composition of polymer films have been recently reviewed by Stamm (1991). Amongst the most useful have been scattering spectroscopies using ions, neutrons or X-rays. Jones et al (1990) have used forward-recoil spectrometry to measure surface excess concentrations of deuterated poly(styrene) in poly(styrene). The same technique allowed Jones et al (1991) to detect the formation of composition fluctuations, with wave vectors normal to the surface, in poly(ethylenepropylene)/ deuterated poly(ethylenepropylene) mixtures obtained from a spinodal decomposition from an unstable quenched state.

The poly(styrene paramethylstyrene) system has been studied by small angle neutron scattering (Jung and Fischer 1988). Comparison of data from (deuterated) poly(styrene) poly(para-methylstyrene) mixtures and the corresponding copolymers showed that the interaction parameter at the spinodal is higher for the block copolymer compared to the mixture, in accordance with the occurance of a larger single phase region for the copolymer. The segregation of the paramethylstyrene block to the surface of poly(paramethylstyrene-b-deuterated styrene) copolymers has been studied by neutron reflectrometry (Götzelmann et al to be published), and evidence has been found for composition fluctuations in this system by use of nuclear reaction analysis, NRA (Stamm 1991, Giessler et al 1991). In this particular case NRA has a depth resolution better than 6nm at the surface.

This paper reports the use of static Secondary Ion Mass Spectroscopy, SIMS, to examine poly(styrene-b-paramethylstyrene) copolymers. The technique probes a depth of 0.5-1nm, enabling the extent of segregation of the paramethylstyrene block at the outermost surface to be measured

2. EXPERIMENTAL

Static SIMS spectra from polymers are characteristic of the material, as has been demonstrated extensively by Briggs (1987). A major problem arises, however, with quantification of the data. The cross section for ionisation of the molecular fragment is very dependent on matrix effects and surface condition. The relevance of the technique to the quantification of the surface composition of poly(styrene-b-paramethylstyrene) copolymers was

determined by first investigating random copolymers over a range
of composition.

Random poly(styrene paramethylstyrene) copolymers were
synthesised from the monomers with the use of azobisiso-
butyronitrile (AIBN) initiator. The monomers (Aldrich Chemical
Company) were distilled immediately before use to remove
stabiliser. A low level of conversion was used in the
polymerisation to ensure that the effects of changes in
differences in the reactivity ratios of the monomers did not lead
to a compositional drift. The polymer was precipitated from
toluene by cold methanol and the monomer ratio in the product was
obtained by H^1 nmr analysis. The molecular weights were
determined by gpc analysis, using poly(styrene) as a reference
standard, and are given in Table 1.

Table 1
Random copolymers

Copolymer	Mn	Ratio styrene/paramethylstyrene
A	30000	12:88
B	28000	32:68
C	37000	59:41
D	41000	78:22

Di-block copolymers were synthesised at Mainz by Polymer
Standards Service GmbH by anionic polymerisation, and
characterised by gpc. The block copolymer data are given below.

Table 2
Block copolymers

Copolymer	Copolymer Mn	styrene Mn (D or H)	paramethylstyrene (H) Mn	Mw/Mn
1	200000	114000 (H)	86000	1.06
2	230000	115000 (D)	115000	1.08
3	260000	52000 (D)	208000	1.16
4	272000	195000 (D)	77000	1.06

SIMS data were obtained with a Vacuum Generators 12-12 quadrupole
mass analyser, fitted with an einzel lens energy filter designed
in house. Vacuum Science Workshop mass filtered ion and electron
flood guns were used to irradiate the sample. An incident Argon
ion beam of 0.1 nA at 3 keV irradiated ca. 5 mm^2 of the sample,
and charge accumulation was compensated with 30 eV electrons from
the flood gun.

Samples were deposited from chloroform solution as films on glass cover slips, and annealed, as required, in a stream of pure nitrogen, see Table 3.

3. RESULTS and DISCUSSION

3.1. Pure Materials

The SIMS spectra of poly(styrene) and poly(paramethylstyrene) are shown in figures 1 and 2 respectively. The poly(styrene) fragmentation pattern is similar to published data (Briggs 1982), with the principal signal at mass 91 dalton, corresponding to rearrangment of the aromatic ring and adjacent backbone methyne group to the relatively stable tropyllium ion, C_7H_7+. The next most abundant ion is at 115 dalton, C_9H_7+, and involves the aromatic ring and three of the backbone carbon groups. The fragment at 77 dalton, corresponding to the aromatic ring, is relatively minor.

The signal at 73 dalton is partly contamination by siloxane impurities. It proved impossible to remove these impurities completely. In some later samples the siloxane signal was a major feature of the spectrum. The detection cross section for siloxane in SIMS is high, so the amount of impurity may still correspond to only a minor fraction of the surface. No relationship was observed between the magnitude of the siloxane signal and the ratios of signals from the copolymer components, and it was assumed that the overlying siloxane contaminent did not appreciably affect the results.

The spectrum of poly(paramethylstyrene), figure 2, shows a principal signal at 105 dalton, the C_8H_9+ ion. This species corresponds to the aromatic ring and adjacent methyne forming a seven membered ring, as with poly(styrene), but also retaining the methyl substituent. The fragments at 91 and 129 dalton are approximately second equal in intensity. The former, C_7H_7+, arises from scission of the aromatic ring-backbone bond and is relatively more abundant than the analogous C_6H_5+ ion from poly(styrene). This is not suprising as, in this case, the methylphenyl ion can rearrange to form the very stable tropyllium species. The latter, $C_{10}H_9+$, would correspond to the C_9H_7+ species from poly(styrene), but in poly(paramethylstyrene) the analogous $C_{10}H_9+$ fragment is less intense. These data are in agreement with a recent study of the mechanism of fragmentation of methyl substituted styrene homopolymers (Chilkoti et al 1991). Partial deuteration of the polymers gave evidence for the complex nature of the SIMS fragmentation process. Broadly, the spectrum of poly(para- methylstyrene) may be understood from consideration of the poly(styrene) spectrum, and the relative stabilities of

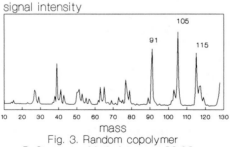

Fig. 3. Random copolymer
D-Styrene:p-Methylstyrene=32:68

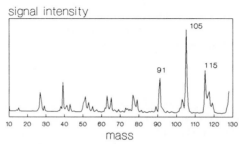

Fig. 2. SIMS spectrum of poly(p-methylstyrene)

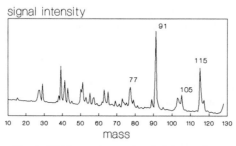

Fig. 1. SIMS spectrum of poly(styrene)

the fragment ions.

Maximum accuracy in determining the ratio of components in the copolymer, would be achieved by measuring the intensities of unique major fragments. The close chemical relationship between the components, in this study, results in all fragments of significance for poly(styrene) being observed in poly(para-methylstyrene) . A unique signal at 119 dalton was reported for poly(paramethylstyrene) by Chilkoti et al (1991). This was however only a minor feature, and taking into account also the signal to noise ratio at the low incident ion irradiation flux employed, and the expectation that the surface would be rich in paramethylstyrene, it was rejected as a measure of the para-methylstyrene concentration. The characteristic fragments for measurement of the relative amounts of each component in the copolymers were therefore chosen simply as the most abundant ion for each pure polymer, i.e. 91 and 105 daltons for poly(styrene) and poly(paramethylstyrene) respectively, to which a correction was made for the contribution from the other monomer.

3.2. Random Copolymers

A typical spectrum of a random copolymer, ratio of styrene:para-methylstyrene = 32:68, is shown in figure 3. It is immediately evident from inspection of the relative amounts of the 91 and 105 dalton signals that both monomers are present at the surface. In order to quantify the relative amounts of each component, it is necessary to allow for the cross-section for each monomer. The relative sensitivity for a polymer, in this context, depends on the fragment chosen for measurement and is not a measure of the overall sensitivity of the polymer to radiation.

An expression or the relative intensities of the 91 and 105 dalton signals at each composition of copolymer can be derived, allowing for the contribution of styrene to the 105 dalton signal and from paramethylstyrene to the 91 dalton signal, and including a parameter for the relative sensitivity.

$$\frac{I105}{I91+I105} = \frac{(X^*St^*0.21+MSt)}{(X^*St^*0.21+MSt)+(X^*St+MSt^*0.46)}$$

where I91 and I105 are the intensities of the 91 and 105 signals respectively, St and MSt are the relative concentrations of styrene and paramethylstyrene respectively, 0.21 is the contribution to I105 from pure poly(styrene), 0.46 is the contribution to I91 from pure poly(paramethylstyrene), and X is the relative sensitivity of poly(styrene) compared to poly(para-methylstyrene).

Careful control of the synthesis should have produced a
completely random copolymer. It is assumed that for compositions
in the region of 50:50, it will be unlikely that segregation of
one component can occur even at the surface. Curves were plotted
for various values of X with the constraint that the line should
coincide with the experimental data for the random copolymers in
the vicinity of compositions 50:50. Reasonable agreement between
experiment and calculation was obtained with X=1.0. This value of
X may reflect the selection of the most abundant fragment and the
close chemical relationship of the monomers. The calculated and
experimental points are shown in figure 4. The experimental data
follow the calculated curve within experimental error,
considering the expected degree of reproducibility of SIMS
spectra, and the additional errors introduced in allowing for the
contribution of both monomers to each signal.

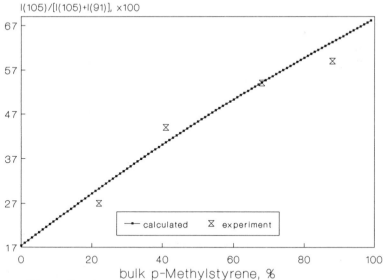

Fig. 4. Random copolymers, SIMS intensities vs bulk conc.

The above data show that SIMS provides a semi-quantitative
analysis of the poly(styrene paramethylstyrene) copolymer
surface. There is no evidence for segregation of either species
in the random copolymers, even at high relative concentrations,
though small segregation effects would not be apparent. This is
consistent with the polymer being completely random and showing
no evidence of blockiness. The reactivity ratios of the monomers
are almost equal, and theory would not predict blocking in this
situation.

3.3 Block Copolymers

The surface of a film of a poly(styrene-b-paramethylstyrene)
copolymer (table 2, copolymer 1) was examined by SIMS. In
contrast to the random copolymers, the block copolymers showed
strong surface segregation effects. The spectrum of the block
copolymer was indistinguishable from that of pure poly(paramethylstyrene) within the errors of reproducibility of data.
Inspection of figure 4 suggests that the surface content of
styrene must be less than ca. 5%.

As noted above, the coincidence of fragment masses made it
difficult to detect small amounts of one monomer in the other.
Block copolymers of paramethylstyrene and deuterated styrene were
examined, therefore, to try to increase the detection limit of
the minor surface component.

The principal signal from poly(styrene), C_7H_7 at 91 dalton,
becomes 98 dalton for the deuterated material. This region of the
mass spectrum is weakly populated in both poly(styrene) and
poly(paramethylstyrene), figures 1 and 2. Measurement of small
concentrations of the deuterated monomer at the surface of the
copolymer therefore becomes feasible.

Films of the poly(paramethylstyrene-b-deuterostyrene) copolymers
gave SIMS spectra, over the mass range 10-140 dalton, very
similar to that of pure poly(paramethylstyrene). There was
evidence, however, of a small contribution at 98 dalton, which
depended on the fraction of deuterostyrene in the copolymer. A
limited mass range and multiple scans were used to improve the
accuracy of the data. Figure 5 shows the SIMS spectra over a
selected mass range for the copolymers containing deuterostyrene
(copolymers 2-4, table 2). The spectrum for the copolymer with
the highest proportion of deuterostyrene shows a definite signal
at 98 dalton, whilst the signal intensity at this mass is reduced
for the other samples with lower proportions of deuterostyrene.

In addition to the minor contribution at 98 dalton in the spectra
of poly(paramethylstyrene-b-deuterostyrene) copolymers, a signal

signal intensity

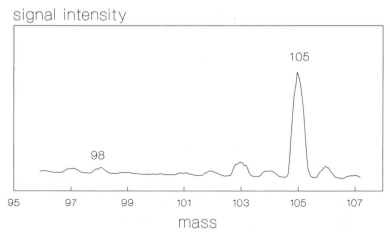

Fig. 6. Block copolymer
D-Styrene:p-Methylstyrene=195:77
Rapid evaporation of polymer solution

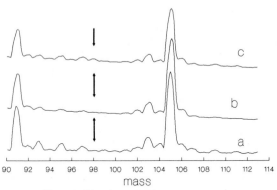

Fig. 5. Block copolymers
D-Styrene:pMethylstyrene=52:209(a);115:115(b);195:77(c)

was generally observed at 97 dalton. This signal, which was also minor, was very weak in pure poly(paramethylstyrene), but increased in intensity when deuterostyrene was used as a comonomer. It was evident, to various extents, in other samples containing deuterostyrene, from more than one synthetic source. The nature of this impurity was unknown. The 98 dalton signal did not vary in proportion to the 97 signal, so it was assumed that the 98 signal did not include a significant contribution associated with the impurity, and was a suitable measure of the deuterostyrene content.

The copolymer in the liquid will have a configuration varying from a completely random coil to, possibly, a more compact aggregated form, depending on solvent and temperature. The solid film formed by evaporation of solvent will adopt a configuration which depends on the interaction between polymer segments, and the ability of the chains to diffuse. The value of the critical molecular weight for entanglement, M_c, is expected to be close to that for polystyrene, i.e. 20000. Since the polymers have gelled, they will only be able to undergo limited chain diffusion in the time scale of the evaporation process. The above data suggest that either segment diffusion is rapid even in the time scale of evaporation of the volatile solvent employed, so the polymer has time to approach a stable configuration in the surface region as the solvent is removed, or that some form of partially phase separated structure exists in the solution phase. Two methods of influencing the surface composition were attempted. The evaporation rate of solvent from the polymer solution was increased to force the film to adopt a more metastable state. Annealing studies were also carried out to determine whether the polymer had in fact reached an equilibrium configuration.

A drop of polymer solution (copolymer 4) on a glass cover slip was exposed to a sudden vacuum. The SIMS spectrum of the resulting film is shown in figure 6. In comparison with the data for films formed by natural evaporation, figure 5, the 98 dalton signal is increased relative to the 105 dalton principal peak, when the evaporation rate is increased. Table 3 lists various conditions for evaporation, and subsequent annealing of films of copolymer 4.

Table 3

Copolymer/ run	Film deposition technique	Annealing conditions	*Ratio I98/I105 signals,%
A1	spinning	-	2.0
B1	spinning	120C, 3 days	1.3
C1	rapid evacuation	-	5.0
D1	rapid evacuation	-	4.5
D2	rapid evacuation	140C, 1 day	3.5

* The 98 dalton fragment is unique to deuterostyrene and, because of the obviously low deuterostyrene signal, the 105 dalton signal is effectively unique to paramethylstyrene. The relative sensitivity factor, X, for styrene and paramethylstyrene at these masses is one (section 3.2), assuming isotope effects can be neglected. The ratio I98/I105 is therefore equivalent to the ratio of monomer concentrations.

Rapid evacuation of the polymer solution results in an approximate twofold increase in the relative amount of deutero-styrene detected. Annealing reduces the surface concentration of deuterostyrene somewhat, but the rearrangement of the polymer chains, even in the surface region, is slow. The data suggest that, at this level of relative surface concentration, the previous history of the sample appears to be more important than annealing in determining the state of the surface. The values for Mn for these polymers is in the region of 200000, which implies that the polymers will be highly entangled, and that reptational motion will have virtually ceased in these systems.

From entropic reasons, it may be considered that some styrene should be at the surface because of the potentially greater freedom of chain ends compared to the conditions at lattice sites. Such effects would not be evident at the molecular weights of copolymer employed here. Also the bulk phases of almost pure paramethylstyrene in the styrene-paramethylstyrene system might still contain a small amount of the other component. Rather, the data from deposition and annealing (table 3) suggests that the small amount of deuterostyrene observed is trapped at the surface by cage effects during formation of the film.

4. CONCLUSIONS

The ability of static SIMS to give semi-quantitative data has been confirmed for the poly(styrene paramethylstyrene) system, using a range of random copolymers.
The static SIMS data for the poly(protonated styrene-b-para-methylstyrene) copolymer showed that the relative amount of

styrene at the surface must be small.

The use of deuterated polymers resulted in a unique signal for the minor surface component of the block copolymer, thus increasing the sensitivity. The data for the thick films, used in this study, showed that the surface of poly(paramethylstyrene-b-deuterostyrene) copolymers is almost completely dominated by the paramethylstyrene component, leading to a relative surface concentration of ca. 99% paramethylstyrene.   Even though a deuterated species has a lower surface energy than the corresponding protonated species, e.g. poly(deuterostyrene) migrates to the surface of a mixture of poly(styrene) and poly(deuterostyrene) (Jones et al 1990), it is not sufficient to compete with the affinity of paramethylstyrene for the surface.

Evidence has been obtained for the trapping of deuterostyrene at the surface, to an extent which depends on the history of formation of the film.

5. REFERENCES

Briggs D 1982 Surf. Interface Anal. 4 151

Briggs D 1987 Polymer Surfaces and Interfaces ed Feast W J and Munro H S (Chichester: John Wiley) ch 2

Chilkoti A, Castner D G and Ratner B D 1991 App. Spectroscopy 45 209

Giessler K H, Rauch F and Stamm M, to be published

Götzelmann A, Reiter G, Stamm M, Kuhn D and Rauch F, to be published

Jones R A L, Norton L J, Kramer E J, Composto R J, Stein R S, Russell T P, Mansour A, Karim A, Felcher G P, Rafailovitch M H, Sokolov J, Zhao X and Schwarz S A, 1990 Europhys. Lett., 12 41

Jones R A L, Norton L J, Kramer E J, Bates F S and Wiltzius P, 1991, Phys. Rev. Lett. 66 1326

Jung W G and Fischer E W 1988 Makromol. Chem., Macromol. Symp. 16 281

Stamm M 1991 Adv. in Polymer Sci. 100, in print

6. ACKNOWLEDGEMENT

We are grateful to the European Science Foundation and the EEC for funds to carry out this collaborative work. F H is grateful for an ECTS award.

Segregation in polymer films studied by neutron reflectometry

M. Rei Vilar[*] and M. Schott[**]

* LASIR-CNRS, 2 Rue Henri Dunant 94320 Thiais France
**Groupe de Physique des Solides, Universités Paris 7 et Paris 6, 2 Place Jussieu, 75251 Paris CEDEX 05 France

ABSTRACT: Segregation in binary polymer films with different compositions have been studied by neutron reflectometry. Incompatible polymers such as Polystyrene and PMMA segregate strongly, forming bilayers of almost pure polymers. PS isotopic blends segregate slightly with the D component at the surface. Some results on compatible PS and PVME are also given.

I- POLYMER BLENDS IN THIN FILMS

Alloying two or more atomic or molecular species is an age old method for preparing new materials with new or improved properties. Preparation and study of polymer alloys is an extremely active field of research. Determination of the phase behaviour of such alloys is a central problem in the field (Bates 1991). The conceptually simplest way for preparing a polymer alloy is by blending two types of linear homopolymers A and B, each made of a single chemical type of monomer, and each of well defined molecular weight as can be achieved by anionic polymerization techniques for instance. The theoretical study of the equilibrium phase behaviour of such blends dates back to 50 years, with the pioneering works of Flory and Huggins (see for instance Flory 1953). The salient parameters determining the phase behaviour of an A-B blend are the Flory-Huggins segment-segment (monomer-monomer) interaction parameter χ and the numbers of segments per chain N_A and N_B. It is customary to write χ as the sum of an enthalpic and an entropic term

$$(1) \qquad \chi = (\alpha/T) + \beta$$

Theory shows that there is a critical value

$$(2) \qquad \chi_c = \frac{(\sqrt{N_A} + \sqrt{N_B})^2}{2N_A N_B}$$

such that for $\chi > \chi_c$ phase separation may occur. Since $\chi_c \propto N^{-1}$, even a very small positive value of χ can generate an experimentally accessible critical temperature. For instance, the very small difference in polarizabilities between CH and CD is enough to generate phase separation in certain regions of a polymer isotopic mixture phase diagram (Buckingham and Hentschel 1980,

Bates et al. 1988). This is true for instance for the system hydrogenated polystyrene (or H-PS) and perdeuterated polystyrene (or D-PS) that we study below (Bates and Wignall 1986).

In thin films, polymer blends may behave in another way than in the bulk, since the presence of a surface and an interface introduces qualitatively new features: adsorption energies and surface energies are not negligible in the total energy balance, and in addition the system has now a special direction, perpendicular to the surface plane. Such surface or interface effects have been demonstrated in the literature recently, as shown in the following examples:

1-*Influence of the surface* Bhatia et al. (1988) have studied the surface of PS-PVME (polyvinylmethylether) miscible blends, and shown that it is enriched in PVME due to the lower surface tension of the latter. In a polymer blend thin film, a concentration gradient perpendicular to the surface is therefore expected. Similarly, surface enrichment in D-PS for an isotopic PS blend in the single phase region has been observed by Jones et al. (1989).

2- *Preferential adsorption* Leonhardt et al. (1990) and Frantz and Granick (1991) have studied adsorption of D-PS and H-PS from Cyclohexane solution at 30 °C on oxidized Silicon, and shown that D-PS is preferentially adsorbed and displaces previously adsorbed H-PS, although in these conditions Cyclohexane is a slightly better solvent for D-PS (Strazielle and Benoit 1975).

3- *Special direction* If a homogeneous blend is rapidly quenched into the two phases region of the phase diagram, spinodal decomposition sets in (Cahn 1961). In presence of a surface or an interface, to which one component is preferentially attracted, this decomposition may become one dimensional, and a regular compositional oscillation perpendicular to the surface will form and grow in amplitude and wavevector with time after quench. This has recently been observed in a polyethylenepropylene isotopic blend (Jones et al. 1991).

In all these cases, a composition gradient is observed. If one polymer is deuterated and the other is not, this corresponds to a D (or an H) surface enrichment and concentration gradient.

A few years ago, we observed by high resolution electron energy loss spectroscopy (HREELS) that the surface of a D-PS/H-PS blend thin film is enriched in D (Rei Vilar et al 1989). But HREELS is sensitive to the immediate vicinity of the surface only, down to a depth of 5 to 10 Å (see for instance Schreck et al. 1990). Neutron Reflectometry allows to study a thin film throughout its depth. In this paper, we report on preliminary experiments using the latter method, on thin films containing one of three pairs of polymers: D-PS/H-PS; D-PS/H-PMMA; and D-PS/H-PVME.

II- NEUTRON REFLECTOMETRY

This method and its application to polymers has been recently reviewed by Russell (1990), so only a few basic facts will be recalled here. A surface or an interface reflects or refracts thermal neutrons according to the same laws as for light. The refractive index n of a material for thermal neutrons is of the form

$$(3) \qquad n = 1 - (\lambda^2/2\pi)\Sigma N_i b_i = 1 - \eta\lambda^2/2\pi$$

where the summation is on the atomic *isotopic* species present in the material. N and b are the atomic number density and neutron coherent scattering length of isotopic species i respectively, and λ is the neutron wavelength, typically, in these experiments, 1 to 30 Å. b is usually positive, although it may not be, and of the order of 10^{-12} to 10^{-13} cm, so that n is smaller than one by a few 10^{-6} near $\lambda = 5$ Å. Therefore, a neutron beam incident onto a surface at an angle θ more grazing than a critical value θ_c such that $\cos\theta_c = n$, is totally reflected. To a good approximation $\theta_c = \lambda\sqrt{(\eta/\pi)}$. Different isotopes may, and often do, have different b. One of the largest differences is between Hydrogen and Deuterium for which $b(H) = -0.374 \; 10^{-12}$cm, and $b(D) = +0.667 \; 10^{-12}$cm. This isotope selectivity makes it possible to use neutron reflectometry to measure composition gradients *perpendicular to an interface* in films of polymer blends where one component is deuterated, the polymer chemical structures being otherwise identical or different. A requirement is that the film must be laterally homogeneous on a centimeter scale.

Consider then a plane parallel polymer layer, where the Deuterium concentration D varies along the direction z perpendicular to the surface only. It can be shown that the reflection coefficient $R(\lambda)$ for a neutron beam incident onto its surface is dependent on $n(z)$. Although $n(z)$ cannot be directly deduced from $R(\lambda)$, the latter can be calculated for chosen $D(z)$ and $n(z)$, and therefore a fitting procedure can be used to determine a $D(z)$ yielding the observed $R(\lambda)$ (Russell 1990). Procedures for taking into account surface or interface roughness have been developed, so they can be included in the fitting; some information about this roughness is contained in the measured $R(\lambda)$ (Névot and Croce 1980). In several recent works, the behaviour of plane interfaces, for instance the interdiffusion of polymers (Klein 1990), has been studied in this way .

III- MATERIALS AND METHODS

H-PS, D-PS and H-PMMA samples of small polydispersity $M_w/M_n < 1.05$ were purchased from Touzart et Matignon. PVME, $M_w = 72500$ and $M_w/M_n \approx 2.2$ was a purified sample kindly donated by Dr C. Strazielle; Institut Charles Sadron, Strasbourg. All polymers were atactic. Dilute polymer solutions in Toluene or CCl4 were prepared using pure solvents and mixed before casting in appropriate proportions. Concentrations were always below c*.

Films were prepared by spin-coating in air at room temperature onto Silicon wafers, several days before the neutron experiments. In all cases, the Silicon surface is covered by a ≈ 17 Å thick natural thermal oxide; so, polymer adsorption is on SiO2. Typical spinning conditions were: rotation speed 5000 t/min, acceleration 5000t/min.sec. Solvent evaporation took place within 5 to 10 seconds at most. Since, with the possible exception of some PS/PVME samples, T_g of all films is well above 295 K, and since films were not annealed, no subsequent polymer motion would occur; the polymer films are studied in the state attained within 20 seconds after deposition of the solution on the wafer surface. Therefore, one should keep in mind that the state of a film may be controlled by kinetic factors, and may not reflect a thermodynamic equilibrium.

Thickness and average refraction index for light were measured by ellipsometry for each film. Single component film values agreed with the literature for PS, PMMA, and PVME. Thickness d can also be accurately measured by neutron reflectometry (Russell 1990). Figure 1 shows the case of a D-PS film of Mw=27000 (similar results were obtained with films of the other pure polymers). The best fit is for d = 877 Å (ellipsometry yields 892 Å) and η =0.64 10^{-5} Å^{-2}, equal to the theoretical value. The interface and surface roughness are at most a few Å. This demonstrate that the films studied here meet the necessary requirements for a meaningful application of neutron reflectometry, are smooth, flat, homogeneous films, of constant thickness, with no residual solvent.

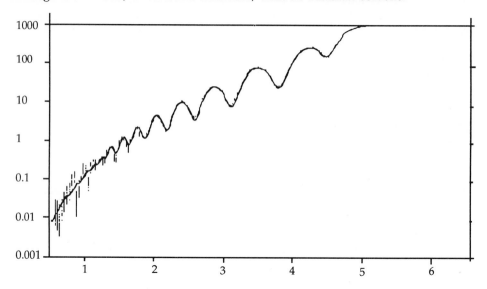

Figure 1- Reflectivity curve of a pure D-PS film. Abscissa: neutron wavelength in Å, ordinate: $R(\lambda) \times 10^3$, on this an all further figures (unless otherwise indicated). Points and vertical lines are experimental data with their standard error. The line is the fit with the parameters given in text.

The neutron reflection data were obtained on the CRISP spectrometer at the ISIS facility, Rutherford Appleton Laboratories, UK. ISIS is a pulsed neutron source, so CRISP collects data over a whole range of λ simultaneously, usually from 0.5 to 6.5 Å. This gives a high counting rate allowing to study ranges where $R(\lambda)$ is $\leq 10^{-5}$ relatively easily (Penfold et al. 1987).

IV- PS-PMMA FILMS

PS and PMMA are immiscible. But since their T_g are both around 100 °C, any solid state motion leading to segregation would be very slow at room temperature. Films were cast from solutions in Toluene, a common solvent. One may ask if segregation is fast enough to occur during solvent evaporation, and what the geometry of the segregated phases is, if they form: what is the effect of the surface?

Two examples of the results are shown on Figures 2 and 3, which show the curve $R(\lambda)$ corresponding to a single homogeneous layer of the same average composition and the best fit using a two-layer model. This best fit is such that it makes no sense to go beyond the two-layer approximation. Results at various average compositions are given in Table I (following page).

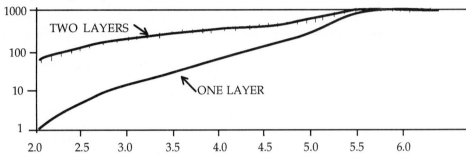

Figure 2- Reflectivity of a (50 % D-PS+50 % H-PMMA) film.

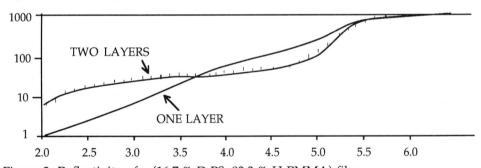

Figure 3- Reflectivity of a (16.7 % D-PS+83.3 % H-PMMA) film.

The fits are always good. The fitted total thickness is within 5 % of the ellipsometric one; the calculated average composition is within 10 % of the nominal solution composition. Clearly, the 5 to 10 seconds it takes for the solvent to evaporate are enough to allow extensive separation. In the film containing equal amounts of the two polymers, separation is practically complete. As the amount of PS is reduced, the thickness of the layer close to the interface increases, and it remains essentially pure PMMA, whereas the thickness of the surface layer decreases and its PS concentration decreases, remaining about twice the nominal concentration in solution. The computed thickness of each layer is only about 2 to 4 times the dimension of a chain in the solid, $\approx$ 60 to 70 Å for $M_W(PS) = 40000$ and $M_W(PMMA) = 46000$.

In these fits, the roughnesses of the surface and of the interfaces between the two layers and between film and substrate were neglected, since the results did not extend to small enough reflection coefficients. However, these roughnesses cannot be larger than 20 Å for the fit to be possible, and are probably smaller. The thickness of a PS-PMMA interface has been studied by Fernandez et al. (1988) - see also Anastasiadis et al. (1990)- : an homogeneous PS film was deposited on top of a thick PMMA film, and the system was extensively annealed to reach equilibrium. The interface thickness, over which there is PS-PMMA mixing, was found to be no larger than 20 Å. Thus, the interlayer interface found here seems comparable to the equilibrated one, although it is obtained by evaporation of a cosolution at 25 °C and not after annealing above T_g.

That the fits work implies films of constant thickness over $\approx$ 1 cm^2. Therefore, the segregation observed here between PS and PMMA is one dimensional over a similar area. Our results compare well with those of Jones et al. (1991). The driving force is possibly preferential adsorption of PMMA on the thin SiO$_2$ layer covering the Si surface. At larger thicknesses, and/or in other solvent evaporation conditions, lateral segregation should occur. If the lateral dimensions of the segregated areas is large enough (see Russell 1990), the R(λ) would become the (incoherent) addition of two reflectivity curves, appropriately weighted, one corresponding to a PS-rich film, the other to a PMMA-rich one. We believe we have evidence of this in one sample of another series.

V- POLYSTYRENE ISOTOPIC BLENDS

For these polymers, $\chi = 0.2T^{-1} - 2.9 \; 10^{-4}$, so that according to eq. (2) the critical temperature is 25 °C for a mixture of polymers with $N=N_A=N_B=5300$, or $M_W=5.4 \; 10^5$. However, $T_g \geq 100$ °C so that an homogeneous film prepared at room temperature will not evolve in any measurable time. Indeed, Jones et al. (1990) have prepared at room temperature films containing 15 % D-PS of $M_W=10^6$ and 85 % H-PS of $M_W=1.8 \; 10^6$, and found them homogeneus. Surface enrichment required annealing at 184 °C for 10^4 to 10^6 sec. These films are presumably relatively thick. On the other hand, HREELS shows that the surface of thin D-PS/H-PS films with $M_W=40000$ coated on Gold is always enriched in D-PS (Rei Vilar et al. 1989). Since the latter M_W is much below the critical value, the phenomenon is not associated to phase segregation, but must be due to the influence of the surface and/or the PS-Gold interface.

To study this phenomenon further, ca 600 Å thick films of D-PS/H-PS blends were prepared by spin-coating onto Silicon wafers from dilute Toluene solutions. The measured R(λ) could sometimes, but not systematically, be fitted by assuming a single homogeneous layer. An example where this seems possible is shown on Figure 4. The corresponding films contained equal weights of D-PS (M_w 27000) and H-PS (M_w 30300). The fit shown corresponds to a thickness d = 541 Å (545 Å by ellipsometry) and η = 0.368 10^{-5} (theory: 0.39). No significant enrichment in D is indeed observed by HREELS for that composition.

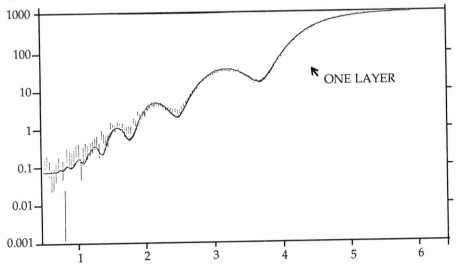

Figure 4- Reflectivity of a (50 % H-PS+50 % D-PS) film.

An example where fitting with a single layer is not possible is shown on Figure 5. The solution used to prepare this film contained 1/6 of D-PS and 5/6 of H-PS

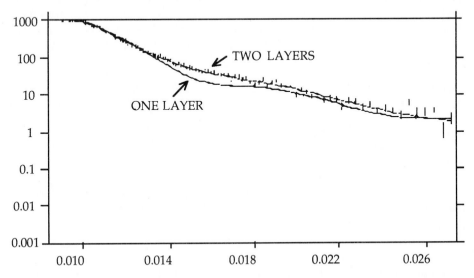

Figure 5- Reflectivity of a (83.3 % H-PS+ 16.7 % DPS). In this figure, the abscissa is in momentum transfer units q = ($2\pi \sin\theta/\lambda$) Å^{-1}

in weight. The solid line is computed using the ellipsometric thickness and theoretical refraction index. The fit shown, which reproduces very well the experimental $R(\lambda)$, is a two-layer model with $d_1 = 237Å$ and $\eta_1 = 0.257 \; 10^{-5}$ for the surface layer, and $d_2 = 343$ Å and $\eta_2 = 0.227 \; 10^{-5}$ for the second layer. The average index computed from these values is equal to the theoretical one. η_1 and η_2 correspond to D-PS concentrations of 24 % and 16 % respectively. This enrichment is similar to the 25 % surface concentration in D found in HREELS experiments, corresponding to the 10 outer Å. Here again the two layer model is certainly a crude approximation, and a continuous gradient is more likely.

These preliminary results suggest several tentative conclusions. First, concentration gradients are indeed formed in very thin films, with a surface enrichment in D. Second, this is more likely to be governed by surface energy differences than by preferential adsorption, since the same effect is observed on Gold and on SiO_2. On SiO_2, D-PS is also preferentially adsorbed (Frantz and Granick 1991), which would act to decrease the composition gradient in very thin films, and possibly to generate in thicker films two D-PS rich layers, sandwiching a H-PS rich one. Third, the effect of molecular weight, if present, is not very large. Polymer surface energies are dependent on M_w, but the effect is important for small masses only (LeGrand and Gaines 1969).

VI- D-PS/H-PVME FILMS

This is an example of chemically different yet compatible polymers, with a phase diagram showing a lower critical temperature above 100 °C, which is very sensitive to deuteration (see Shibayama et al. 1985) and references therein). The surface enrichment in PVME due to the difference in surface energies has been mentioned in §I.

In order to remove residual solvent, the films were annealed at 50 °C for 3 days under vacuum after spin-coating. PVME differs from the other polymers used in this study in that its T_g is -29 °C. T_g is at room temperature or below for all compositions containing more than 30 % PVME and $T_g \approx$ 50 °C for 15 % PVME. In this case therefore, equilibrium by polymer diffusion can occur before the neutron reflection experiment.

In our hands, preparation of PS-PVME films by spin-coating is as yet not reproducible: in some cases, apparently very homogeneous films yielding good $R(\lambda)$ curves are obtained. An example is shown on Figure 6, corresponding to a thickness d = 1930 Å, and an $\eta = 0.438 \; 10^{-5}$ Å^{-2}, close to the one expected for the proportions of polymers put in solution. In other cases however, the index of refraction is much closer to 1, as if the film was much richer in PVME than the starting solution. An example is shown on Figure 7, wher a film cast from a solution of the same composition and of similar thickness d= 2245 Å, also gives a very good fit but with η =0.32 10^{-5} Å^{-2}, close to the pure PVME value.

The homogeneity of the films may be related to the low T_g. The frequent observation of "enrichment" in PVME of the film compared to the solution may be related to strong PVME-SiO_2 interaction. More experiments would be needed to settle these points.

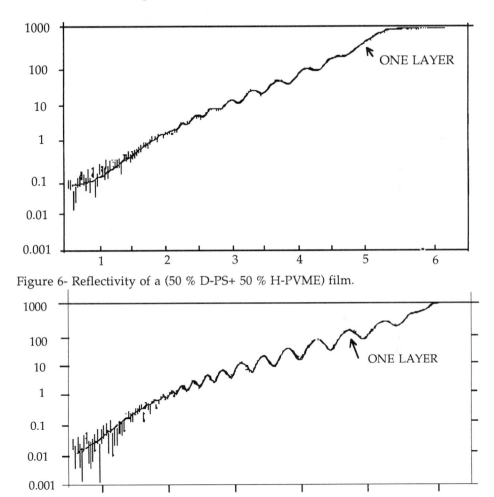

Figure 6- Reflectivity of a (50 % D-PS+ 50 % H-PVME) film.

Figure 7- Reflectivity of a D-PS/H-PVME film cast from a 50-50 solution, which has an index corresponding to a greater concentration in PVME.

ACKNOWLEDGEMENTS   This work was made possible by beam time allocation on the CRISP spectrometer at ISIS. We are grateful to Dr J. Penfold and Dr C. Shackleton for their help during the experiments and data analysis. We thank Dr C. Strazielle for the gift and characterization of pure PVME and useful discussion. We thank Mrs Eakim for help during film preparation.

REFERENCES

Anastasiadis SH Russell TP Satija SK and Majkrzak (1990) J Chem Phys 92 5677

Bates F S (1991) Science 251 898

Bates F.S. and Wignall G.D. (1986) Macromolecules 19 932

Bates F S Fetters L J and Wignall G D (1988) Macromolecules 21 1086

Bhatia Q S Pan D H and Koberstein J T (1988) Macromolecules 21 2166

Buckingham A D and Hentschel H G E (1980) J. Polym. Sci., Polym. Physics Ed. 18 853

Cahn J W (1961) Acta Metall. 9 795

Fernandez M L Higgins J S Penfold J Ward R C Shackleton C and Walsh D J (1988) Polymer 29 1923

Flory P J (1953) Principles of polymer chemistry (Cornell University Press)

Frantz P and Granick S (1991) Phys. Rev. Lett. 66 899

Jones R A L Kramer E J Rafailovich M H Sokolov J and Schwarz S A (1989) Phys. Rev. Lett. 62 280

Jones R A L Norton L J Kramer E J Composto R J Stein R S Russell T P Rafailovich M A Sokolov J Zhao X and Schwarz S A (1990) Europhys. Lett. 12 41

Jones R A L Norton L J Kramer E J Bates F S and Wiltzius P (1991) Phys. Rev. Lett. 66 1326

Klein J (1990) Science 250 640

LeGrand D G and Gaines Jr G L (1969) J. Colloid Interf. Sci. 31 162

Leonhardt D C Johnson H E and Granick S (1990) Macromolecules 23 687

Névot L and Croce P (1980) Rev. Phys. Appl. 15 761

Penfold J Ward R C and Williams W G (1987) J. Phys. E 20 1411

Rei Vilar M Schott M Pireaux J J Grégoire C Caudano R Lapp A Lopes da Silva J and Botelho do Rego A M (1989) Surf. Sci. 211/212 782

Russell T P (1990) Mater. Sci. Reports 5 171

Schreck M Abraham M Göpel W and Schier H (1990) Surf. Sci. Lett. 237 L405

Shibayama M Yang H Stein R S and Han C C (1985) Macromolecules 18 2179

Strazielle C and Benoit H (1975) Macromolecules 8 203

D-PS fraction in solution (% of total polymer)	Total thickness by ellipsometry (Å)	Surface layer			Interface layer			Calculated mean D-PS fraction in film (%)
		Thickness	index (x10⁵)	% D-PS	Thickness	index (x10⁵)	% D-PS	
50	291	145	0.602	92	148	0.15	8	49.6
25	281	112	0.42	59	189	0.16	10	27.4
16.7	285	92	0.31	38	204	0.16	10	18.7

Thicknesses in Å. Indices in units of 10^{-5} Å^{-2}.

Table 1

Parameters of three thin D-PS/H-PS films

Structure, properties and mechanism of formation of boundary (surface) polymer layers according to infrared spectroscopy data

O N Tretinnikov and R G Zhbankov

Institute of Physics, Byelorussian Academy of Sciences, 220602 Minsk, USSR

ABSTRACT: Surface phenomena in PS and PMMA films cast from solutions on glass substrates were investigated by IR spectroscopy methods (combined with contact angle measurements). It was established that, during film formation, in PS chains conformational transitions ttgg$\rightarrow$tttt type occur in the interfacial region, whilst in the case of PMMA structural perturbations proceed in the reverse direction (tttt$\rightarrow$tgtg). Glass transition temperature T_g in PS monotonically increases with decrease of the distance from the interface, whilst in PMMA, as the distance from the interface reduces, T_g decreases reaching its minimum and then it begins to increase. A mechanism of surface structure formation is proposed on the basis of new ideas about the "conformational characteristics - surface activity of polymer chain" relationship.

1. INTRODUCTION

It is well known that the surface properties of polymers, in the condensed amorphous state, significantly differ from their properties in the bulk. These differences are generally attributed to the specific character of the structure of polymeric molecules on or in the vicinity of an interface. And it is assumed, as a rule (Manson et al. 1976, Lipatov et al. 1974), that during polymeric material formation (e.g., from a polymer solution or melt) there occur processes of macromolecule adsorption at the interface "polymer solution or melt/surrounding environment (or substrate, filler, etc.)". This leads to the formation of adsorption layers of macromolecules whose conformations are perturbed by the action of the field of surface forces and molecular mobility is substantially reduced. However, there is a large amount of experimental data which

contradit the above-mentioned adsorption mechanism of surface
layers formation. Thus, experimental values of thickness of
surface layers (Lipatov 1977, Tretinnikov 1988) turn out to
be greater by a factor of 10^2-10^3 than the corresponding values predicted theoretically from the point of viev of adsorption models (Scheutjens et al. 1980, Skvortcov 1986, Silberberg 1988). Moreover, in a number of polymers the mobility of
macromolecular chains near the interface remains unchanged or
even increases (Mason 1960, Howard et al. 1981, Reid et al.
1990). Finally, at present, experimental data on the quantitative relationship between the molecular structure of surface
layers and their properties are practically lacking. It follows from the foregoing that further studies of surface phenomena in amorphous polymer solids are required.

The purpose of this work is to study the molecular structure,
relaxational properties and mechanism of surface layers formation in polystyrene (PS) and poly(methyl methacrylate) (PMMA)
films cast from solutions on glass substrates. As the tools
for surface layers characterization, infrared (IR) transmission and attenuated total reflectance (ATR) spectroscopy methods combined with contact angle measurements were used.

2. EXPERIMENTAL SECTION

Industrial predominantly syndiotactic PS (tacticity, 73% rr,
15% rm, and 12% mm triads, $M_\eta = 2.4 \cdot 10^5$) and PMMA (tacticity,
65% rr, 32% rm, and 3% mm triads, $M_\eta = 5.4 \cdot 10^5$) previously
purified by precipation from solutions were investigated.
The polymer films were cast on glass substrates from 2% (by
weight) solutions of PS in toluene and 0.5% solutions of PMMA
in chloroform. The residual solvent was removed from the
films formed at room temperature by drying films at 80° C
(PS) and 90° C (PMMA) for 10 h. The films were removed from
the substrate with distilled water and were dried over silica
gel for 5 days.

The IR spectra were recorded on a Specord-M80 spectrophotometer. The film thickness with an error of $< 2\%$ was determined
from the IR transmission spectra by the distance between the
maxima of interference fringes in the 2000-4000 cm^{-1} range.
IR-ATR spectra were obtained using a production-type attachment to Specord-M80. Reflecting elements (RE) made of KRS-5 (10
reflections at an angle of 55°) crystals were used, and the
penetration depth was 0.45 ($\nu = 3000$ cm^{-1}) - 2.4 μm ($\nu = 500$ cm^{-1}).

The glass transition temperature of the films involved was determined from the temperature dependence of the absorbance ratio of the bands at 1240 and 1270 cm^{-1} (PMMA) and at 1602 and 1030 cm^{-1} (PS) in IR transmission spectra. High-temperature IR spectra were obtained on a Specord-IR75 spectrophotometer using a production-type attachment to this instrument. The films were heated up to 140°C, held at this temperature for 1 h, and then slowly cooled down to room temperature (at a rate of 0.5 deg/min). The IR spectrum recording the temperature was held constant to an accuracy of ±1 deg, and the relative intensities of analytical bands were reproduced to an accuracy of ±1%.

IR-ATR spectra and contact angle were measured for the films separated from the substrate on the surfaces exposed to the air (air-facing side) and glass substrate (glass-facing side).

3. RESULTS AND DISCUSSION

3.1 Conformations of macromolecules in surface layers

The IR-ATR and transmission spectra of 19.6 and 1.75 μm-thick

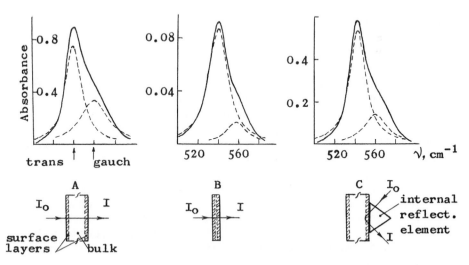

Fig. 1. Transmission (A,B) and ATR (C) IR spectra of 19,6 (A,C) and 1,75 (B) - μm-thick PS films in the region of the conformation-sensitive bands.

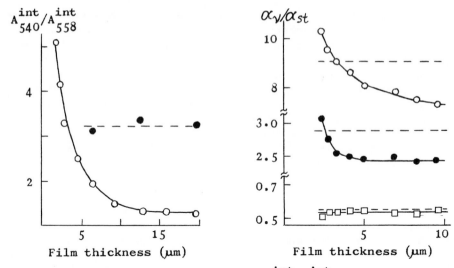

Fig. 2. Integral intensities ratio $A_{540}^{int}/A_{558}^{int}$ of the conformation-sensitive bands in transmission (o) and ATR (KRS-5-55°)
(•) IR spectra vs. PS films thickness.

Fig. 3. Absorption coefficients (normalized to the absorption
coefficient of the internal standard band at 2295 cm^{-1}) of
bands at 750 (□), 1438 (•) and 1150 cm^{-1} (o) in transmission IR spectra vs. PMMA films thickness. The broken lines
correspond to the values of α_N/α_{st} in IR-ATR (KRS-5-55°)
spectrum of 8.22 μm PMMA film.

PS films in the region of the conformation-sensitive complex band with a peak at 540 cm^{-1} are shown in Figure 1. The
spectra differ in the distribution of intensities of the elementary components at 540 and 558 cm^{-1}. As shown by Jasse
et al. (1977), these bands are due to the presence of trans
(ν_{tt} = 540-542 cm^{-1}) and gauche (ν_{gg} = 558-561 cm^{-1}) isomers in the polymer chain. The integral intensities ratio of
the bands mentioned ($A_{540}^{int}/A_{558}^{int}$) in the IR-ATR spectra does
not depend on the film thickness (Figure 2). In the transmission IR spectra, as the film thickness reduces, starting from
12-15 μm, the value of $A_{540}^{int}/A_{558}^{int}$ increases approaching the
corresponding values of this magnitude in the IR-ATR spectra.

The surface layers make an important contribution to the transmission IR spectra of thin films and to the IR-ATR spectra
(see the scheme in Figure 1). In the case of a sufficiently
thick film, the IR transmission spectrum is mainly determined

by the spectral characteristics of the polymer bulk. The more intense IR-absorption by tt-conformers (and, correspondingly, the less intense absorption by gg-conformers) in IR transmission spectra of thick PS films should therefore be attributed to the specific character of the molecular structure of surface PS layers, namely, to the higher concentration of tt-conformers of polymer chains as compared to their concentration in the bulk. According to theoretical calculations (Froelich et al. 1978) in syndiotactic PS, tttt and ttgg conformations of polymer chains are more advantageous energetically. Consequently, the increase in the tt-conformer concentration and, correspondingly, decrease in the concentration of gg-conformers in the surface layers of PS, is most probably due to the ttgg→tttt conformational transitions in macromolecular chains.

Figure 3 shows the variations of the absorption coefficients of some bands in IR transmission spectra vs. PMMA films thickness. The dotted lines show the values of these magnitudes in the IR-ATR spectrum of a "thick" (d = 8.22 μm) film. Attention is attracted by the difference between the absorption coefficients of bands at 1150 and 1438 cm^{-1} in the IR-ATR and transmission spectra of a "thick" PMMA films. Is it also seen that in the IR transmission spectra, as the film thickness is decreased, starting from d = 10 and 4 μm for bands at 1150 and 1438 cm^{-1} respectively, the absorption coefficients of the above bands increase approaching the corresponding values of these magnitudes in the IR-ATR spectrum. It was shown previously (Tretinnikov et al. 1991) that increase in the absorption coefficient for the band at 1150 cm^{-1} is associated with the tttt→tgtg conformational transition in syndiotactic PMMA chains. On this basis we can conclude that the surface layers of PMMA are characterized by the increased content of tgtg conformers of the main chain due to the decrease in the fraction of tttt conformers.

We have not found any differences in the IR-ATR spectra obtained first from one and then from the other side of the films investigated. This means that, in the process of film formation, surface layers are formed both at the polymer/glass and the polymer/air interface, the conformations of macromolecules being similar in these layers.

3.2 Relaxational properties of the polymer surface layers: relationship with the molecular structure

Figure 4 gives glass transition temperature T_g vs. the PS film thickness. The change in T_g correlates with the change in the relative content of tt and gg conformers in films (see Figure 2). According to the Gibbs-DiMarzio theory (1958), T_g depends mainly on the thermodynamic flexibility of the chain determined by the difference of energies of t and g conformers U_{gt} and on the value of intermolecular interactions E. For the vinyl-type polymers the following relation holds (Miller 1978):

$$T_g = (U_{gt} + 0.1E)/4.2R \qquad (1)$$

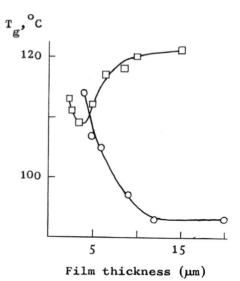

Fig. 4. Glass-transition temperature vs. PS (O) and PMMA (□) films thickness.

The increased concentration of trans conformers in the surface layers of the PS films in question means that the thermodynamic flexibility of chains in these layers is lover (U_{gt} is higher) than in the polymer bulk. With decreasing film thickness the fraction of chains with reduced flexibility increases and, accordingly, T_g increases.

In case of PMMA (Figure 4) as the film thickness is decreased, starting from d = 10 μm, T_g decreases reaching its minimum at d = 3.5 μm, and then it begins to increase. The change in T_g at d = 4-10 μm correlates with the change in the relative content of tt and tg conformers in films (see Figure 3). The change in the qualitative shape of the T_g(d) is observed jast when the absorption coefficient of the 1438 cm^{-1} band begins to increase. The concentration of gauche conformers and, accordingly, thermodynamic flexibility of chains in the PMMA surface layers are higher than in the bulk. With decreasing film thickness the fraction of chains with increased flexibility increases and T_g decreases. At d < 4 μm the content of gauche conformers still increases, but T_g stops decreasing and begins to rise. According to formula (1), this may only be due to the sharp increase in the energy of intermolecular interactions. This experimentally manifests itself as an increase in the absorption coefficient of the 1438 cm^{-1} band.

Thus, the relaxational properties of polymer chains of PS and PMMA are changed quite differently under the action of the interface. Nevertheless, in both cases these changes are in an unambiguous quantitative relation with the change in the conformational characteristics of macromolecules in the surface layers.

3.3 Mechanism of surface structure formation

As shown above, PS and PMMA macromolecules behave quite differently near the interface: in PS chains conformational transitions of the ttgg⟶tttt type occur, which lead to an increased content of trans conformers in the surface layers, whilst in the case of PMMA the conformational transformations proceed in the reverse direction (tttt⟶tgtg). This result is rather unexpected, taking into account that PS and PMMA are very similar in their physico-chemical characteristics.

At the same time (as can be seen from Figure 5) the conformations that are realised near the interface have one specific peculiarity common for both polymers: one can pass a plane through the polymer chain axis, which will spatially separate polar ($-C_6H_5$, $-COOCH_3$) and nonpolar (CH_2, $-CH_3$) functional

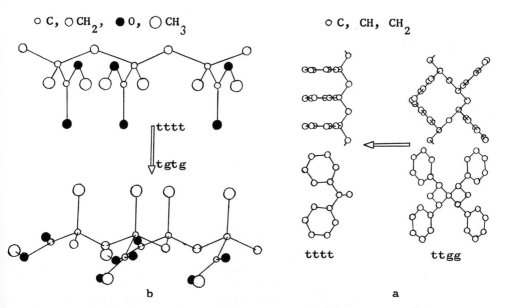

Fig. 5. Conformational transitions in PS(a) and PMMA(b) macromolecules in the vicinity of an interface.

Table I. Advancing Contact Angles (θ) of Liquids on PS and PMMA Films

Polymer	Liquid	θ^a Air-facing sides	Glass-facing sides
PS	Water	80	62
	S-tetrabromoethane	28	23
PMMA	Water	69	62
	S-tetrabromoethane	25	18

[a] In degrees, standard error of the mean $\pm 2°$ ($n > 7$).

Table II. Surface Free Energies γ and Their Dispersion γ^d and Polar γ^p Components Calculated from Contact Angle Values for PS and PMMA Films (ergs·cm^{-2})

Polymer	Air-facing sides			Polarity	Glass-facing sides			Polarity
	γ^d	γ^p		γ^p/γ	γ^d	γ^p		γ^p/γ
PS	38.2	3.8	42.0	0.09	32.1	14.4	46.5	0.31
PMMA	34.5	9.5	44.0	0.22	33.9	13.6	47.5	0.29

groups of the polymer, since they will be arranged on different sides of this plane. The conformations realised in the polymer bulk do not exibit this property. We think this fact is of fundamental importance for understanding the mechanism of surface structure formation.

The interfacial energy at the "forming film/surrounding environment" interface can be minimized by selective accumulation on the surface of polar or, vice versa (depending on environment polarity), nonpolar polymer functional group (Yasuda 1981, Zhbankov et al. 1984, Lavielle 1985). While in low-molecular compounds a simple reorientation of the whole molecule

suffices for above-mentioned surface phenomenon, in the case
of polymers a definite conformation of the polymer chain is
also required. The surface activity of the macromolecule, i.e.
its ability to dispose on the surface polar or nonpolar functional groups, is the largest when these qroups in the macromolecule are spatially separated: polar and nonpolar groups
lie at different sides of the plane passing through the chain
axis. Therefore, the content of polymer chain conformers providing this spatial separation increase in the vicinity of an
interface during the film formation. Surface layers whis a
specific molecular structure arise.

According to the proposed model the glass-facing surfaces of
the films investigated must be more polar then the air-facing
surfaces. We measured the contact angles of a few highly polar liquids on the polymer films under investigation (Table I)
and on the basis of the data obtained by means of the Girifal-
co-Good-Fowkes equation modified in the approximation of the
geometrical mean (Kaelble 1970) we calculated the surface
free energy (γ) and its dispersion (γ^d) and polar (γ^p) components (Table II). It is seen that at the glass-facing sides
the values of γ^p are greater and the values of γ^d are, conversely, smaller than at the air-facing sides. This points to
different surface concentrations of polar and nonpolar functional groups depending on the polarity of the medium which
the polymer film formed contacted.

Interestingly, when the polarity of the surrounding medium
changes, the change in the polarity of the surface in PS is
greater than in PMMA (Table II). This fact finds a fairly
simple explanation from the point of view of the proposed model of surface layer formation, which takes into account the
conformational aspect of the surface activity of macromolecules. In the case of PS the highest surface activity of the
macromolecule is attained in the plane trans-conformation
(tttt). In syndiotactic PS this conformation is energetically
most preferable. Consequently, it predominates even in the polymer bulk, and at the interface their concentration turns
out to be even greater as a result of the action of the surface forces. In the case of PMMA, the tgtg-conformation,
which provides the greatest surface activity, is a high-energy
conformation for the bulk of the polymer and is present in it
in small amountts. Near the interface the content of the sur-
face-active conformers increases, but it is clear that their
concentration will be much smaller than in the case of PS.

4. CONCLUSION

The peculiarities of the structure of thin polymer films and surface layers in thick films cast from solutions on substrates are caused by processes of adsorption of macromolecules at interfaces during the films formation. The concentration of polymer chain conformers which provide spatial separation in the macromolecule of its polar and nonpolar functional groups increases during adsorption. The surface activity of macromolecule depends on its ability to assume such conformations. Depending on surrounding environment polarity, at the surface of polymer films mainly polar or, on the contrary nonpolar functional groups are localized.

5. REFERENCES

Manson J A and Sperling L H 1976   Polymer Blends and Composites   (New York: Plenum Press)
Lipatov Y S and Sergeeva L M 1974   Adsorption of Polymers (New York: John Wiley)
Lipatov Y S 1977   Physical Chemistry of Filled Polymers (Moscow: Chimija, in Russian)
Tretinnikov O N 1988 D. Sc. Thesis   (Minsk: Inst. Phys., BSSR Acad. Sci.)
Scheutjens J M H M and Fleer G H 1980 J. Phys. Chem. 84 178
Skvortcov A M and Gorbunov A A 1986 Vysokomol. Soedin. A28 1941 (in Russian)
Silberberg A 1988 J. Colloid Interface Sci. 125 14
Mason P 1960 J. Appl. Polym. Sci. 4 212
Howard G J and Shanks P A 1981 J. Appl. Polym. Sci. 26 3099
Reid C G and Greenberg A R 1990 J. Appl. Polym. Sci. 39 995
Jasse B and Monnerie L 1977 J. Mol. Struct. 39 165
Froelich B, Jasse B, Noel K and Monnerie L 1978 J. Chem. Soc. Faraday Trans.II 74 445
Tretinnikov O N and Zhbankov R G 1991 J. Mater. Sci. Lett. (accepted for publication)
Gibbs J H and DiMarzio E A 1958 J. Chem. Phys. 28 373
Miller A A 1978 Macromolecules 11 859
Yasuda H, Sharma A K and Yasuda T 1981 J. Polym. Sci. Polym. Phys. Ed. 19 1285
Zhbankov R G and Tretinnikov O N 1984 Vysokomol. Soedin. B30 259 (in Russian)
Lavielle L and Schultz J 1985 J. Colloid Interface Sci. 106 438
Kaelble D H 1970 J. Adhesion 2 66

Structure and dynamics of grafted polymer layers: A Monte Carlo simulation

Pik-Yin Lai and Kurt Binder

Institut für Physik, Johannes-Gutenberg Universität-Mainz, Postfach 3980, D-6500 Mainz, Germany.

ABSTRACT: Polymer chains anchored with one end at a hard wall under good solvent conditions are studied by Monte Carlo simulations using the bond- fluctuation model. Both for the "quenched" case where the anchored points are fixed and the "annealed" case where anchored ends can diffuse laterally are considered. Structural properties like monomer positions and chain linear dimensions are reported and discussed in the light of current theories. The result for the relaxation time of the total chain configurations are consistent with the recent scaling prediction $\tau \sim \sigma^{2/3} N^3$ where σ is the surface coverage and N is the chain length. For the annealed case, the lateral diffusion constant is found to behave as $D_\perp \sim \sigma^{-2/3} N^{-p}$ where p crosses over from $p \approx 1$ at small σ to $p \sim 2$ at large σ.

1. INTRODUCTION

Because of their wide applications in polymer technology and their importance in understanding the fundamental problems of polymeric materials, grafted polymer layers[1, 2] have been a subject of recent interest both theoretically[3 − 26] and experimentally [27 − 34]. While there is general agreement that in the brush regime the chains are stretched out in the direction normal to the surface with the thickness of the brush being proportional to the chain length, the properties of the chains in these layers are not yet understood in full detail. Experiments usually can only provide quasi-macroscopic information such as the layer thickness, concentration profile and the force profile between two interacting brushes. Parameters such as the forces between monomers and the wall and between monomers and the solvent molecules are not known explicitly. On the contrary, in a computer simulation the system is precisely characterised and approximations in the theories are avoided, and detail information from the scale of effective monomers to the macroscopic level can be obtained. In this paper, we shall present some simulation results about the structure and dynamics of a grafted polymer layer. Two models were considered: in the "quenched" case, chains ends are fixed at randomly chosen positions at the wall, while in the second case annealing of this disorder by lateral diffusion of chain ends at the surface is permitted. More details of this work can be found in [25].

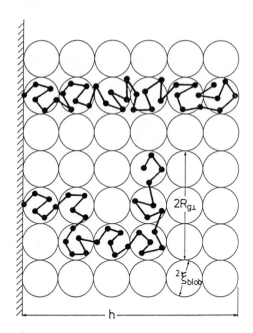

Fig. 1: Schematic illustration of the blob concept for the configurations of anchored chains at a wall.

2. BOND-FLUCTUATION MODEL AND SOME SIMULATION DETAILS

In the bond fluctuation model[35 − 37] each monomer occupies a cube of eight lattice sites in a simple cubic lattice and self-avoidance is modelled by the requirement that no two monomers can share a common site. The 108 allowed bond vectors connecting two consecutive monomers along a chain are obtainable from the set {(2,0,0), (2,1,0), (2,1,1), (2,2,0), (3,0,0), (3,1,0)} by the symmetry operations of the cubic lattice. and the bond length ranges from 2 to $\sqrt{10}$ and is characterised by the root-mean-square bond length, $\langle \ell^2 \rangle^{1/2}$. We consider a polymer brush in a good solvent and there is no monomer-monomer interaction other than self- avoidance. The Monte Carlo procedure starts by choosing a monomer at random and trying to move it one lattice spacing in one of the randomly selected directions: $\pm x, \pm y, \pm z$. The move will be accepted only if both self- avoidance is satisfied and the new bonds still belong to the allowed set. The choice of this set of bond vectors guarantees that no two bond will cross each other during the course of their motion and thus entanglement is automatically taken care of. Hence this model is especially suitable for the study of the dynamical properties of multi-chain systems[37].

A schematic picture of a grafted polymer layer is shown in Fig. 1. Monodispersed polymer chains of N monomers are placed inside a $L \times L \times 3N$ box with one end grafted to the inpenetrable $L \times L$ surface (xy plane). Periodic boundary conditions are imposed in the x and y directions. The surface coverage σ ranges from 0.025 to 0.2 and chain length from $N = 10$ to 80. As a linear dimension, we use $L = 40$ for $N \leq 40$

and $L = 20$ for $N > 40$. The number of chains in our studies ranges from 10 to 80. The system is allowed to equilibrate for a long time (typically five times the relaxation time) and statistical averages are then taken from runs extending typically over a time interval of about five times the characteristic relaxation time. Time is measured in Monte Carlo steps per monomer (MCS/monomer). One MCS/monomer means that on average each monomer has attempted to move once.

3. MONTE CARLO RESULTS AND DISCUSSIONS

We have obtained simulation results for both the quenched and annealed chains. In many cases, the results in the quenched case differ only very little from the annealed case and because of limited space, we will only present some results for the annealed case here. The equilibrium structural properties can be understood in terms of a blob picture (Fig. 1) where excluded volume interactions are not screened inside a blob of diameter $2\xi \approx \sigma^{1/2}$, containing N_{blob} monomers per blob. With increasing σ, the excluded volume interactions get more screened and the crossing over scaling behavior is verified[21]. It should be noted that though the linear dimension of the chains in the z- direction is of order $\xi(N/N_{blob})$, one must not be mistaken that the chain being stretched out strictly linearly (Fig.1 upper part); rather the chain has a random walk structure in the lateral directions (Fig. 1 lower part). The brush thickness, h can be characterised by the first moment of the concentration profile, $\langle z \rangle$ and also by $\langle R_{gz}^2 \rangle^{1/2}$ where R_{gz} is the z-component of the radius of gyration. Apart from the data for $N = 10$ and for $\sigma = 0.025$, all data in Fig. 2 fall nicely on a straight line verifying that the brush height scales as

$$h \sim \langle z \rangle \sim \langle R_{gz}^2 \rangle^{1/2} \sim \xi(N/N_{blob}) \sim N\sigma^{1/3}$$

as predicted both by the simple Alexander model[3] and the more sophisticated Self-consistent field(SCF) theory[7,9-11].

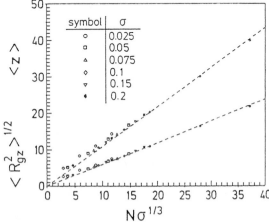

symbol	σ
○	0.025
□	0.05
△	0.075
◇	0.1
▽	0.15
✳	0.2

Fig. 2 $\langle z \rangle$ and $\langle R_{gz}^2 \rangle^{1/2}$ plotted against $N\sigma^{1/3}$. The dashed lines are just guides to the eye.

More detailed information like the "trajectory" of the grafted chain, which is the average position of the ith monomer ($\langle z_i \rangle$) along the chain from the wall, can be obtained. Fig 3 is a scaling plot with $\langle z_i \rangle / \langle z_N \rangle$ versus i/N. It is seen that $\langle z_i \rangle$ increases linearly with i for $i \leq N/2$ while for larger i $\langle z_i \rangle$ increases much slower. This feature is linked to the gradual decrease of the density profile in the outer region of the brush. If the Alexander blob picture were strictly correct, one would have $\langle z_i \rangle = \langle z_N \rangle (i/N)$ for all i. The results in Fig.3 is remarkably close to the SCF prediction (solid curve: $\sin \frac{i\pi}{2N}$), however. Using our Monte Carlo data, the structure of the grafted polymer layer has been shown[26] to be quantitatively well described by the SCF theory.

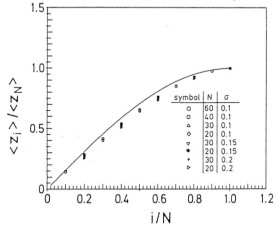

Fig. 3: Scaled position of the ith-monomer as a function of i/N. The curve is the SCF prediction.

For the dynamics, we measured the time auto-correlation functions of the chain dimensions. The relaxation time τ is estimated form the time needed for the auto-correlation function of R_{gz} to decay to $1/e$. Fig. 4 shows the dependence of τ on N and σ. The data are consistent with a behavior $\tau \sim N^a \sigma^b$ with $a = 3.0 \pm 0.1$ and $b = 0.83 \pm 0.08$.

A scaling prediction for the relaxation time was suggested[22] based on the blob picture: each blob has a relaxation time $\tau_{blob} = \xi^2 / D_{blob} = \xi^2 N_{blob} / (W \langle \ell^2 \rangle)$, where W is the monomer jump rate. Then using the anology of motion of chains restricted to tubes of radius ξ, it was concluded that the relaxation time can be estimated from the time need for density fluctuations to diffuse along the chain,

$$\tau \approx (N/N_{blob})^2 \tau_{blob} \sim N^2 \sigma^{-1/6}$$

which is in serious disagreement with the data. However a more successful scaling argument was recently suggested[20, 21] using the free draining picture. τ can be estimated from the mean- square displacement of the free chain end in the z-direction by

$$\tau \simeq \langle (\Delta z_{end})^2 \rangle / D_{Rouse} \simeq \langle (\Delta z_{end})^2 \rangle N / (W \langle \ell^2 \rangle)$$

Using then $\langle (\Delta z_{end})^2 \rangle \sim (N/N_{blob})^2 \xi^2 \sim \sigma^{2/3} N^2$, one gets

$$\tau \sim N^3 \sigma^{2/3} / W$$

which is much better consistent with the data. The discrepancy between the exponent $b \simeq 0.83$ from the data and $2/3$ is attributed to a density dependent jump rate W.

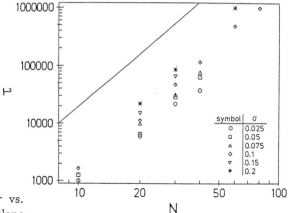

Fig. 4(a): Log-log plot of τ vs. N. Straight line indicates a slope of three.

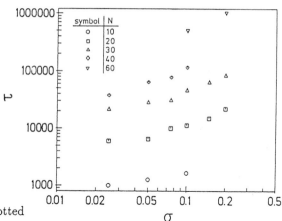

Fig 4(b): same as (a) but plotted vs. the surface coverage.

A decrease of W with increasing density is familiar from work on bulk semi-dilute solutions. Identifying $W(\sigma)$ in our case with the rate W at corresponding average densities ϕ in [37], we get the corrected effective exponent $b \simeq 0.6$. Thus we conclude that our data are reasonably well consistent with the prediction, but data for smaller σ and larger N would be needed to clearly prove this. While the Alexander picture (Fig. 1 upper part) which identifies the chain configuration with a chain in a tube of radius ξ, predicts the average stretch of chain correctly, it leads to misleading conclusions about the scale over which fluctuations occur, and its generalization to dynamics leads to completely wrong relaxation times. It has been pointed[18] out that $\tau \sim N^3$ has nothing to do with the reptation mechanism, but are due to the large distances that monomers must travel in the course of the fluctuations that renew the chain configurations.

For the case of annealed chains, we also measured the lateral diffusion constant, defined as

$$D_\perp = \lim_{t\to\infty} \langle [x_{cm}(t) - x_{cm}(0)]^2 + [y_{cm}(t) - y_{cm}(0)]^2 \rangle / 4t$$

where (x_{cm}, y_{cm}) is the center-of mass coordinate of the chain. Similar to the study of semi-dilute solutions[33], the apparent diffusion constant systematically decreases with time and high precise data data are needed to be able to exptrapolate to the long time limit in the above equation. According to the Rouse model, one expects a simple law $D_\perp \sim W/N$ while Fig. 5a suggests that such behavior is compatible with our data only for $\sigma \to 0$. For larger σ, the apparent exponent seems to increase systematically with σ and possibly this means that some suppression of lateral motions occurs due to the onset of entanglements. However a definitive statement about this question needs to await simulations with distinctly longer chains.

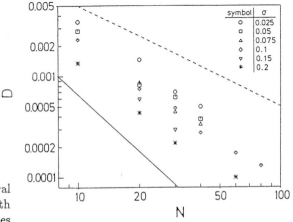

Fig. 5(a): Log-log plot of lateral diffusion constant vs. chain length N. Dashed and full straight lines indicate a slope of -1 and -2 respectively.

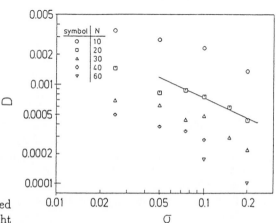

Fig. 5(b): same as (a) but plotted vs. the surface coverage. Straight line indicates a slope of -2/3.

The dependence of $D_\perp$ on σ can be obtained from the simple assumption[38] that the frictional force experienced by a monomer, ζ_F, is proportional to the averaged monomer density ϕ_{av} for a given chain length N. So one has

$$D_\perp \propto \zeta_F^{-1} \propto \phi_{av}^{-1} \propto \sigma^{-2/3}$$

Fig. 5b suggests that our data can be described by the power law above. It should be emphaise that the nature of chain motions in our quasi-two-dimensional grafted layers is very different from the diffusion in strictly two-dimensional geometry, as has been investigated in [35] where strictly Rouse-like behavior was found. The dynamics of chains in annealed grafted layers may be qualitatively similar to the dynamics of chains in diffusive adsorbed layers in which a mobility $\propto N^{-2}$ has recently been suggested[39].

To conclude, we have shown that the present model allows a convenient and economic (all simulations were done on workstations) study of all the detailed behavior of polymer brushes by computer simulations. And most of our results can be understood in term of scaling arguments.

Acknowledgement:
This research is supported by the Bundesministerium für Forschung und Technologie (BMFT) grant no. 03M4040.

REFERENCES

[1] A. Halperin, M. Tirrell and T. P. Lodge, Adv. in Polymer Science **100**, 1991

[2] S. T. Milner, Science **251**, 905 (1991)

[3] S. Alexander, J. Phys. (Paris), **38**, 983 (1977)

[4] P. G. de Gennes, C. R. Hebd. Seances Sci. **300**, 839 (1985); Macromolecules **13**, 1069 (1980)

[5] T. Cosgrove, T. Heath, B. van Lent, F. Leermakers and J. Scheutjens, Macromolecules **20**, 1692 (1987)

[6] C. Marques, J. F. Joanny and L. Leibler, Macromolecules **21**, 1051 (1988)

[7] S. Milner, T. Witten and M. Cates, Macromolecules **21**, 2610 (1988); Europhys. Lett. **5**, 413 (1988)

[8] M. R. Munch and A. P. Gast, Macromolecules **21**, 1366 (1988)

[9] A. M. Skvortsov, A. A. Gorbunov, I. V. Pavlushkov, E. B. Zhulina, O. V. Borisov and V. A. Priamitsyn, Polymer Science USSR, **30**, 1706 (1988)

[10] E. B. Zhulina, O. V. Borisov and V. A. Priamitsyn, J. Colloid Interface Sci. **137**, 495 (1990); Polymer Science USSR, **31**, 205 (1989); Polymer Science USSR **26**, 885 (1984)

[11] T. M. Birshtein and E. B. Zhulina, Polymer Science USSR **25**, 2165 (1983)

[12] A. N. Semenov, Sov. Phys. JETP **61**, 733 (1985)

[13] M. Muthukumar and S. Ho Macromolecules **22**, 965 (1989)

[14] S. Milner, T. Witten and M. Cates, Macromolecules **22**, 853 (1989)

[15] S. Milner, Z.-G. Wang and T. Witten, Macromolecules **22**, 489 (1989)

[16] T. M. Birshtein Yu V. Liatskaya and E. B. Zhulina, Polymer **31**, 2185 (1990)

[17] P.-Y. Lai and A. Halperin Macromolecules in press.
[18] J. F. Marko and T. A. Witten, Phys. Rev. Lett. **66**, 1541 (1991); Macromolecules, submitted.
[19] A. Halperin and S. Alexander, Macromolecule **22**,2403 (1989)
[20] A. Halperin and S. Alexander, Europhys. Lett. **6**, 329 (1988)
[21] L. I. Klushin and A. M. Skvortsov, Macromolecules **23**, (1991)
[22] M. Murat and G. S. Grest, Macromolecules **22**, 4054 (1989)
[23] M. Murat and G. S. Crest, Phys. Rev. Lett., **63**, 1074 (1989)
[24] A. Chakrabarti and R. Toral, Macromolecules **23**, 2016 (1990)
[25] P.-Y. Lai and K. Binder J. Chem. Phys., submitted
[26] P.-Y. Lai and E. B. Zhulina, preprint
[27] T. Cosgrove, T. Crowly and B. Vincent, in *Adsorption from Solutions*, edited by R. Otterwill, C. Rochester and A. Smith (Academic Press, New York 1982) p. 287
[28] J. Edwards, S. Lenon A., Toussaint and B. Vincent in *Polymer Adsorption and Dispersion Stability, ACS Symposium Ser. 240* Americal Chem. Soc., Washing, D.C. 1984
[29] G. Hadzioannou, S. Patel, S. Granick and M. Tirrell, J. AM. Chem. Soc. **108**, 2869 (1986)
[30] P. F. Luckham and J. Klein, J. Colloid Interface Sci., **117**, 149 (1987)
[31] A. Ansarifar and P. F. Luckham, Polymer **29**. 329 (1988)
[32] H. J. Taunton, C. Toprakcioglu, L. J. Fetters and J. Klein, Nature **332**, 712 (1988); Macromolecules **23**, 571 (1990)
[33] H. J. Taunton, C. Toprakcioglu and J. Klein, Macromolecules **21**, 3333 (1988)
[34] P. Auroy, L. Auvray and L. Leger, Phys. Rev. Lett. **66**, 719 (1991)
[35] I. Carmesin and K. Kremer, Macromolecules **21**, 2819 (1988);J. Physique (Paris), **51**, 915 (1990)
[36] H.-P. Deutsch and K. Binder, J. Chem. Phys. **94**, 2294 (1991)
[37] W. Paul, K. Binder, D. W. Heermann and K. Kremer, J. Phys. (France) II 1, **37** (1991); and J. Chem. Phys., in press.
[38] P. G. de Gennes, private communication.
[39] P. G. de Gennes, C. R. Acad. Sci. Paris II, **306**, 183 (1988)

Theoretical and experimental study of the molecular structure absorbing at 2190 cm⁻¹ in electroinitiated polyacrylonitrile

D. Mathieu, M. Defranceschi, P. Viel, G. Lécayon, and J. Delhalle[+]

CEA-Saclay, DSM-DRECAM-SRSIM, F-91191 Gif-sur-Yvette Cedex (France)

and

[+]Laboratoire de Chimie Théorique Appliquée, Facultés Universitaires N.D. de la Paix
61, rue de Bruxelles B-5000 Namur (Belgium)

ABSTRACT: A combined theoretical and experimental study is carried out to identify the molecular structure responsible for the vibrational line at 2190 cm⁻¹ observed in the infrared spectra of electroinitiated polyacrylonitrile films. The results strongly suggest that this structure is an enaminotrile resulting from the tautomerization of a β-iminonitrile defect formed during the electropolymerization. They also support the original hypothesis made on the mechanism of formation of the β-iminonitrile defect.

1. INTRODUCTION

Grafting and growing polymers directly on oxidizable metal surfaces (e.g. Fe, Ni) is possible by electropolymerization under cathodic polarization (Subramanian 1979, Bruno et al. 1075, Desbene-Monvernay et al. 1978). Films of poly(acrylonitrile) or PAN (Lécayon et al. 1990), poly(p-chlorostyrene) (Deniau et al., 1990) and poly(2-methyl-2-propenenitrile) or poly(methacrylonitrile) (Deniau et al. 1990) have recently been chemisorbed on Ni substrates in an aprotic organic solvent with monomer concentrations 10 to 10^2 times higher than the supporting electrolyte. Two mechanisms corresponding to the two classes of anionic initiators have been reported in the polymer literature (Rempp et al. 1986) and have been shown to occur with electropolymerisazation under cathodic polymerization (Boiziau et al. 1986, Lécayon et al. 1982, 1984, 1987, Deniau et al. 1990, Deniau et al.) : (a) a *direct electron transfer* from the cathode to the monomer leading to radical anions and the flow of a Faraday current in the electrochemical cell and (b) a *secondary electrode reaction* taking place simultaneously to the reduction and leading to a polymer film grafted to the metal surface. The specific conditions (electrolytic medium, electrical polarization, diffusion rate of the reactive monomers towards the electron-rich sites, etc.) in which the electroinitiated polymerization proceeds can induce side reactions that may not develop in bulk.

Infrared measurements on six electroinitiated poly(acrylonitrile) films (Viel 1990) reveal a sharp decrease in the absorbance at the nitrile stretching frequency $v(C\equiv N)$ 2246 cm^{-1} depending on the electrochemical polarization time; this decrease is accompanied by the appearance of an absorption band at approximately 2190 cm^{-1}. The observed correlation between better resistance to solubility of the films in dimethylformamide (DMF) and higher intensity of the band at 2190 cm^{-1} suggests that chemical reactions other than the main polymerization are triggered by the synthesis conditions, in particular those leading to cross-links between chains. Understanding of the chemical origin and conditions leading to the effect requires the identification of the molecular structure responsible for this absorbance at 2190 cm^{-1}.

In section 2, the conditions of the electrosynthesis are described, and infrared reflection-absorption spectroscopy (IRRAS) spectra are cross-checked with published infrared data on molecular and polymer structures absorbing in the 2190 cm^{-1} region. Quantum chemistry calculations of the nitrile stretching frequency $v(C\equiv N)$ of model molecules are reported in section 3 to provide an electronic structure based information to rationalize the $v(C\equiv N)$ band shift from 2246 to 2190 cm^{-1}. Inconsistencies in the interpretation of some IR data reported in the literature on postulated defect structures occuring in thermal treatments of PAN are also pointed out.

2. EXPERIMENTAL STUDY

In this section the details of the preparation of six electroinitiated poly(acrylonitrile) films are given, followed by an analysis of the evolution of the IRRAS line intensities at 2246 and 2190 cm^{-1}.

2.1. Film Preparation

Reactant purification and electrochemical processes are performed in an assembly of glove boxes in an argon atmosphere. Solvent and monomer purifications are carried out by storage on molecular sieves followed by fractional distillation under reduced pressure. The reaction medium is prepared by dissolution of acrylonitrile, 0.5 mol.dm^{-3}, in an acetonitrile electrolytic solution containing 5×10^{-2} mol.dm^{-3} of tetraethylammonium perchlorate (TEAP), the supporting electrolyte. Residual water content is titrated by the Karl Fischer method and found to be less than 5×10^{-4} mol.dm^{-3}.

Electrolysis in a potentiostatic mode takes place in an electrolysis cell with a standard three-electrode arrangement. Potentials are controlled with respect to an Ag/Ag$^+$ 10^{-2} mol.dm^{-3}

reference electrode; the auxiliary electrode is a platinum sheet. Electrode surfaces are prepared by cathodic sputtering of a 1 μm thick microcrystalline layer of Marz nickel (99.99%) on microscope 5 cm² glass slides.

Polymer film forms on the cathode surface during electrolysis of acrylonitrile by a three-step potential waveform selected between -0.5 V, the rest potential, and -2.8 V. Six samples of electroinitiated PAN films of increasing thickness are obtained with this procedure by raising the electrolysis (or electrochemical polarization) time from 0.2 to 10 s. Figure 1 presents the dependence on the time of electrochemical polarization of the film thickness (Viel 1990) measured by ellipsometry (SOPRA type ES2G instrument).

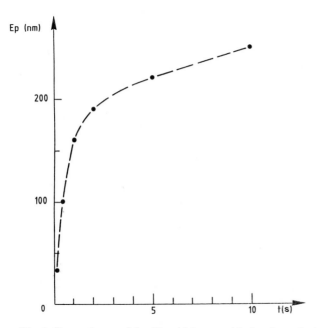

Fig. 1. Dependence of the film thickness with the electrolysis time.

2.2. IRRAS Characterization of Films

Infrared reflection-absorption spectroscopy measurements are carried out at a grazing angle of 5° on a Brucker IFS 66 instrument. The main IRRAS spectral features of the six films correspond to those recorded by IR transmission spectroscopy of a reference commercial sample (PAN Aldrich : 18,135-5 - radical polymerization), Figure 2a, and corroborate the poly(acrylonitrile) nature of the electroinitiated films. XPS and UPS measurements (Viel 1990), not reported here, confirm this correspondence. Due to space limitations, only the full spectrum of a film obtained after an electrolysis of 0.5s is shown in Figure 2b; this spectrum is fairly representative of the spectra of the other films. With respect to the commercial sample, the main differences occur in three regions : 1425 to 1455 cm⁻¹, 1580 to 1640 cm⁻¹ and 2245 to 2190 cm⁻¹. We briefly comment on the first two frequency intervals, the latter one forming the subject of the present contribution.

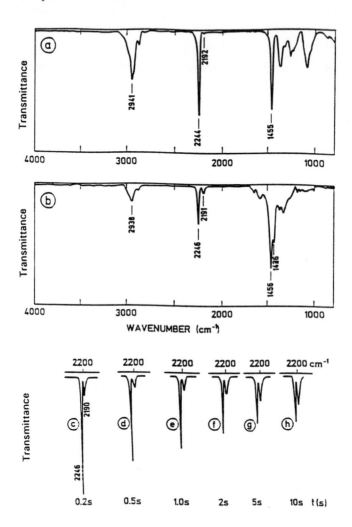

Fig.2. IRRAS spectrum of poly(acrylonitrile) films: (a) obtained by radical polymerization (Aldrich), (b) obtained by cathodic electropolymerization after 0.5 s of polarization, (c) - (h) $v(C\equiv N)$ region of the electroinitiated polymer films under 0.2, 0.5, 1.0, 5.0 and 10 s of polarization, respectively.

The band at 1455 cm⁻¹ is assigned to the in plane -CH₂- deformation in the -CH₂-CH(CN)-CH₂-CH(CN)- sequence, while the first line at 1425 cm⁻¹ probably corresponds to the same deformation in the -CH_2-CH_2-CN moiety resulting from proton addition onto the active center of the growing macroanions (Viel 1990). As in the case of the nitrile absorption band, these lines show a strong dependence on the polarization time and thus on thickness. The region extending from 1580 to 1640 cm^{-1} most likely results from perturbations in the PAN molecular structure due to cyclization reactions initiated by anions, most likely residual surface oxygen atoms and hydroxyl groups freed after reduction of the surface nickel oxides by the cathodic polarization.

Figures 2c to 2h show the appearance of the absorption band at 2190 cm^{-1}. As illustrated by these figures, increments in absorbance of the band at 2190 cm^{-1} are accompanied by a parallel decrease in the nitrile stretching absorption as the electrochemical polarization time increases. Figure 2h emphasizes the importance of the effect : after 10 seconds of electrochemical polarization the ratio of the line intensities, I_{2190}/I_{2246}, exceeds 30%. The simultaneity in the increase and decay of the lines at 2190 and 2246 cm^{-1}, respectively, and the large frequency shift (approximately 50 cm^{-1}) tend to support the idea that the line at 2190 cm^{-1} is due to a nitrile group affected by changes in its chemical environment and not by surroundings influences. This evolution signs the existence of secondary reactions developping with the polarization time at the expense of the principal polymerization reaction. This process is due to a drop in the monomer concentration in the region of reaction resulting from the competition between the transport rate of monomers from the electrochemical medium toward the reactive region and the rate of the polymerization reaction itself.The mechanism involved corresponds to a transfert of the reactive center of the growing macroanion on the carbon of a nitrile group of a neighbouring chain leading to β-iminonitrile interchain cross-links (Viel 1990). Indeed, the ν(C≡N) band in the infrared spectra of saturated nitriles lies in a relatively narrow frequency interval (Juchnovsky et al. 1983), 2230-2280 cm^{-1}, while 2215 cm^{-1} and 2220 cm^{-1} are regarded as the lower limits for ν(C≡N) of α,β-unsaturated alkyl nitriles and benzonitriles, respectively (Bellamy 1968). It is only in benzonitriles with anionic substituents that lower ν(C≡N) values, as low as 2170 cm^{-1}, are observed (Juchnovsky et al. 1983). A valuable work to interpret the 2190 cm^{-1} line is due to Baldwin (Baldwin 1961) who showed that the ν(C≡N) frequencies of a series of enaminonitrile molecules are comprised in the interval 2165-2190 cm^{-1}.

Notwithstanding its saturated nitrile character, the β-iminonitrile form (**1**) shown in Figure 3, is generally proposed (Grassie et al. 1970, 1971, Coleman et al. 1978) to account for the band observed at 2198 cm^{-1} in thermal treatments of poly(acrylonitrile). If this is consistent with the ν(C≡N) frequency observed at 2190 cm^{-1} in the spectrum of 3-imino-3-phenylpropionitrile (Pouchert) (**2**), Figure 3, it is however in contradiction with the widely accepted frequency interval for saturated nitriles (Juchnovsky et al. 1983). The 2190 cm^{-1} band has also been studied in a work on the chemical origin of thermally stimulated discharge currents in poly(acrylonitrile) (Stupp et al. 1977) where it is postulated that it could be associated with either a conjugated nitrile or a cyanide ion. As already discussed, the lower limit for $v(C \equiv N)$ of unsaturated nitriles is at 2215 cm^{-1} and as such cannot explain the line at 2190 cm^{-1}. Nitrile group frequencies of carbanions and radical anions span the 2020-2230 cm^{-1} region (Bellamy 1968) and include the line at 2190 cm^{-1}, but they ought to be disregarded in our case. Indeed, negatively charged species and their counterions would readily disperse in highly polar solvents such as DMF, which has not been observed. Besides, a strong intensity of the 2190 cm^{-1} nitrile band is on a par with an enhanced resistance to solubility of the film, which support the assumption of intermolecular cross-linking.

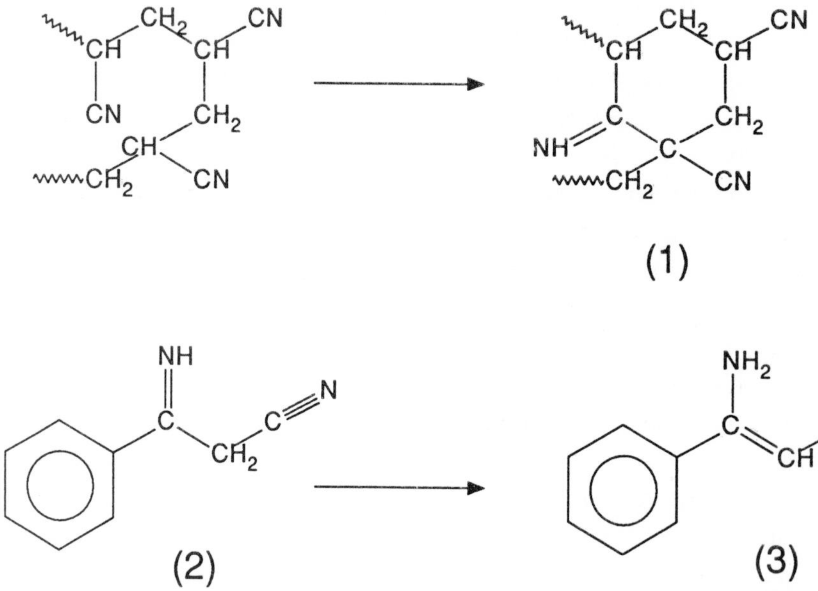

(1)

(2)   (3)

Fig.3. Postulated β-iminonitrile defect (1) occuring in thermally treated poly(acrylonitrile) samples. Structure of 3-imino-3-phenylpropionitrile (2) and its enaminonitrile tautomer, β-aminocinnamonitrile (3).

3. THEORETICAL STUDY

The reported results on thermally treated poly(acrylonitrile) do not provide consistent conclusions on the chemical origin of the line at 2190 cm^{-1} to be used with confidence in the interpretation of our own results. The assignment to the nitrile stretching frequency of the 3-imino-3-phenylpropionitrile (2) (Pouchert) band at 2190 cm^{-1} adds to the confusion.

Finally, it is surprising not to find Baldwin's results on enaminonitriles as part of recent reviews on infrared spectra of cyano groups. To help clarifying some aspects of the problem, results of theoretical calculations of the harmonic vibrational frequency $v(C\equiv N)$ for model molecules containing one nitrile group are analyzed in the next section.

3.1. Calculations and Model Molecules

Theoretical calculations have proved to be quite reliable in predicting equilibrium geometries, relative total energies (Hehre et al. 1986) and harmonic vibrational frequencies (Hehre et al. 1986, Amos et al. 1987) . Present calculations are carried out at the *ab initio* restricted Hartree-Fock level (RHF) using the Gaussian 90 series of programs (Gaussian 90) running on the CRAY II computer of the CEA-Grenoble. The requested convergence on density matrix elements is fixed to 10^{-8} and the integral cutoff to 10^{-10} a.u. (1 a.u. = 27.2 eV = 2625.50 kJ.mol^{-1}); the calculated properties are obtained for the fully optimized molecular structures. We have selected the relatively small 3-21G basis set which is suitable to calculate molecular geometries and relative stabilities of closed shell molecules (Amos et al. 1987). Similarly, frequencies calculated with this basis correlate well with the experimental values, but they are consistently larger by 10-15% (Hehre et al. 1986). As in the case of equilibrium distances, effects of electron correlation on calculated frequencies are largest for molecules with multiple $C\equiv C$ bonds (Hehre et al. 1986, Amos et al. 1987). Thus, prior to calculations directly aimed at elucidating the nature of the molecular structure vibrating at 2190 cm^{-1}, test calculations are carried out first on a series of reference molecules with the 3-21G basis and compared with experimental data to ascertain the quality of the predictions on the $v(C\equiv N)$ stretching frequency.

3.2 Results

Reference molecules are acrylonitrile (**4**), *trans*-2-butenenitrile (**5**), *cis*-2-butenenitrile (**6**), 2-methyl-2-propenenitrile (**7**), 2-iminoethanenitrile or iminoacetonitrile (**8**), ethanenitrile (**9**), propanenitrile (**10**) and 3-butenenitrile (**11**). The results on their $v(C\equiv N)$ stretching frequency are listed in Table I. As expected, the theoretical values are larger than the experimental data, but a good correlation is noted with little dispersion of the data (experimental to theoretical ratios $r_{3\text{-}21G}$ with respect to the average value $R_{3\text{-}21G}$ of 0.868). The saturated nitriles, are correctly discriminated from the unsaturated ones. The trends are also satisfactory when it comes to smaller effects. For instance, acrylonitrile (**4**) has a higher frequency than the analogs (**5**) to (**7**) differing by the position of an extra methylene group, -CH$_2$-, in the molecule. In line with experimental result, 2-iminoethanenitrile (**8**) is at the border-line between the saturated and unsaturated nitriles.

TABLE I. 3-21G theoretical v_{3-21G} and experimental $v_{exp.}$ (gas phase) C≡N vibrational frequencies of reference molecules (4) to (11) containing one nitrile group. The rightmost two columns list the individual ratios r_{3-21G} between the experimental and theoretical values, and the scaled theoretical frequencies v'_{3-21G} using the average ratio R_{3-21G} (= 0.868), respectively.

Molecule :	v_{3-21G}	$v_{exp.}$	r_{3-21G}	v'_{3-21G}
unsaturated nitriles				
(4) CH$_2$=CH-CN	2587	2239 [18]	0.865	2245
(5) *trans*-CH$_3$-CH=CH-CN	2582	2232 [19]	0.864	2241
(6) *cis*-CH$_3$-CH=CH-CN	2580	2230 [19]	0.864	2239
(7) CH$_2$=CH(CH$_3$)-CN	2580	2227 [18]	0.863	2239
(8) HN=CH-CN	2588	2250 [20]	0.869	2246
saturated nitriles				
(9) CH$_3$-CN	2595	2267 [21]	0.874	2252
(10) CH$_3$-CH$_2$-CN	2587	2265 [21]	0.875	2246
(11) CH$_2$=CH-CH$_2$-CN	2592	2263 [18]	0.873	2250

On the basis of the reported consistency of the 3-21G calculations with experimental data (equilibrium geometries (Hehre et al. 1986) and vibrational frequencies (Hehre et al. 1986, Amos et al. 1987)) as well as the good correlation obtained with our test results on v(C≡N), one may conclude that this basis is reasonably appropriate for prospective computations on the nitrile stretching frequency of structures (12) - (16) shown in Figure 4. Model β-iminonitrile are (ap)-*cis*-3-iminopropanenitrile (12), (sp)-*cis*-3-iminopropanenitrile (13) and (ap)-*trans*-3-iminopropanenitrile (14); (ap) and (sp) stand for antiperiplanar and synperiplanar, respectively. These molecules are saturated nitriles because the conjugation between HN=CH- and -C≡N groups is interrupted by a methylene spacer, -CH$_2$-. The other two molecules, *cis*-3-amino-2-propenenitrile (15) and *trans*-3-amino-2-propenenitrile (16), are enaminonitriles wherein conjugative interactions between H$_2$N- and -C≡N groups can develop through the connecting double bond. These five molecules have the same molecular formula, C$_3$H$_4$N$_2$, and differ either by their chemical structure, or configuration, or conformation. Being isomers of the type X-C≡N, the dependence of the v(C≡N) frequency on the mass of X is prevented.The relevant calculated equilibrium geometry parameters (as indicated in Figure 4), total energy and theoretical nitrile stretching frequencies scaled with R_{3-21G} = 0.868, v'_{3-21G}, are listed in Table II.

Results in Table II show a marked difference in the v'_{3-21G} vibrational frequency of the β-iminonitriles, (12) - (14), compared with those of the enaminonitriles, (15) - (16), in agreement with the early experimental results by Baldwin (Baldwin 1961). The calculated frequency shift is close to 45 cm^{-1} which also compares well with the experimental value.

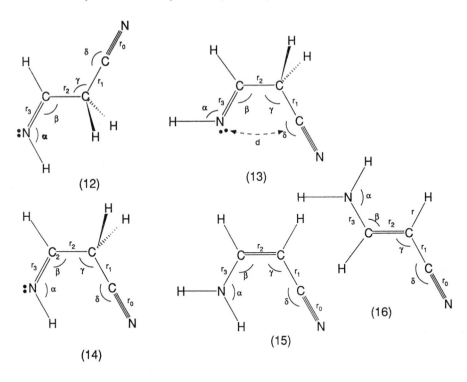

Fig.4. Molecular structure of compounds (12) - (16) and the labelling convention on their relevant geometrical parameters.

TABLE II. Bond distances (r_0 to r_3) and angles (α to δ) of the 3-21G equilibrium geometry of molecules (12) to (16) (lengths in Å and angles in degrees). Unscaled v_{3-21G} and scaled v'_{3-21G} nitrile vibrational frequencies (in cm^{-1}) and total energy E_T (in a.u.). $R_{3-21G} = 0.868$.

Molecule :	(12)	(13)	(14)	(15)	(16)
r_0	1.139	1.138	1.139	1.143	1.143
r_1	1.456	1.455	1.455	1.416	1.415
r_2	1.532	1.514	1.527	1.337	1.336
r_3	1.253	1.251	1.251	1.354	1.356
α	116.2	115.6	116.4	120.9	121.0
β	125.5	121.9	128.4	127.0	126.2
γ	112.1	113.0	112.5	121.8	121.5
δ	179.5	181.2	178.4	176.6	179.9
v_{3-21G}	2597	2605	2594	2553	2560
v'_{3-21G}	2254	2261	2252	2216	2222
E_T	-223.53309	-223.5357	-223.53732	-223.55750	-223.55449

The relative stability of enamines and β-iminonitriles is quite interesting to discuss. Compound (15) is predicted more stable than (14) by 53.0 kJ.mol^{-1}. Assuming negligible entropy variations due to molecular structure changes and approximating ΔG^0 by the 53.0 kJ.mol^{-1} energy difference, leads to a practically total displacement of the gas phase tautomeric equilibrium β-iminonitrile/enaminonitrile towards the enaminonitrile form ($K_{eq.} \approx 10^9$ at 298 K). This is in agreement with the work by Sieveking and Lüttke (Sieveking 1969) who report that, by reducing geminal dinitriles with LiAlH$_4$, the iminonitriles form first and stabilize by tautomerization to enaminonitriles (15) and (16). The initial products are *cis-trans*-mixtures, but only the more stable *cis*-enaminonitrile (15) can be isolated since the *trans*-form (16) rearranges to (15); the nitrile stretching frequency for (15) is observed (Sieveking 1969) at 2180 cm^{-1}.

The lowering of the ν(C≡N) frequency in enaminonitriles relative to β-iminonitriles can be rationalized in terms of the geometrical parameters of the molecules and their chemical bonding. In enaminonitriles the C≡N bond distance (1.143 Å) is 0.004 Å larger than the same bond in β-iminonitriles, which is significant considering the known reluctance for the C≡N bond to change upon substitution. Longer bond distances usually mean smaller force constants and frequencies. There is a substantial reduction in the C-C≡N bond length, 1.416 Å, compared to values ranging from 1.427 to 1.434 Å in conjugated nitrile compounds (4) to (7) (Hennico et al. 1991). The C=C distance (1.337 Å) is also quite large compared to that in (4) - (7) where it ranges from 1.319 to 1.329 Å. Finally the short C-NH$_2$ bond (1.354 Å) is as short as in formamide (1.353 Å) (Hehre et al. 1986) and, again as in formamide, the -NH$_2$ group is planar indicating an efficient delocalization of the nitrogen lone pair. All these geometry characteristics sign a strong conjugation that extends over the entire eneaminonitrile skeleton, H$_2$N-CH=CH-C≡N, favours charge transfer from the electron-donating H$_2$N- to the electron-withdrawing -C≡N through the -CH=CH- moiety, and leads to a net decrease of the ν(C≡N) frequency. Thus, our calculations quantify the earlier Baldwin's hypothesis (Baldwin 1961) on a strong involvement of the H$_2$N- group on the lowering of the ν(C≡N) frequency in enaminonitriles.

4. CONCLUDING REMARKS

From the results in section 3, it can be conjectured that 3-imino-3-phenylpropionitrile (2) (Pouchert), for which the ν(C≡N) frequency is observed at 2190 cm^{-1}, is not the correct molecular structure. It should rather be the β-aminocinnamonitrile (3), an enaminonitrile of which structure is shown in Figure 3. As a matter of fact, works on antiinflammatory (Lang et al. 1975) and antiarthritic (Ridge et al. 1975) properties of (3) never quote (2) as a possible

3.3 Contact angle measurements

The aim of this section was to search for modifications in polymer surface properties at fluences well below the respective thresholds for ablation. Experiments were performed at fluences below (27 mJ cm^{-2}) and above (45 mJ cm^{-2}) the threshold for the formation of gaseous products. From the measured contact angles, the polar and disperse components of the solid / air surface tension were determined according to the equations of Fowkes and Young [15].

With doped PMMA, the total surface energy remained constant within the error bar of the experiment {Fig. 4(a)}. A slight rise of the disperse part, and a corresponding decrease of the polar part, were observed at the higher fluence (45 mJ cm^{-2}). This change might be attributed to the loss of polar functional groups (i.e., C=O or –COOCH$_3$) from the polymer backbone.

For polytriazene I, the total surface energy increases with fluence {Fig. 4(b)}. Above the threshold for gaseous product formation ($\sim$ 35 mJ cm^{-2}), significant changes are occurring: The disperse part decreases, while the polar part increases with fluence. If a radical mechanism is involved in the photochemical release of N$_2$, the radical sites created could react with oxygen or atmospheric moisture, which would account for the observed increase in polarity.

Polytriazene II exhibits a distinctly different behavior {Fig. 4(c)}. The total surface energy again increases with fluence. However, with polytriazene II the disperse part increases, and the polar component of the surface energy decreases with increasing fluence, which is the opposite of what we observed with polytriazene I. This behavior of polymer II is attributed to the loss of the triazene side chain, with the polymer backbone remaining undamaged. In this situation, the radical sites apparently react (e.g. by hydrogen transfer) without the creation of polar functional groups.

3.4 XPS measurements

For the electron spectroscopic experiments, polymer films were cast on copper foils. Excimer laser irradiation was performed in air at the same fluences as used in the experiments described in Section 3.3.

Wide scan XPS spectra of the three untreated polymer surfaces are shown in Fig. 5(a). The C(1s) region has been deconvoluted into components corresponding to carbon– and

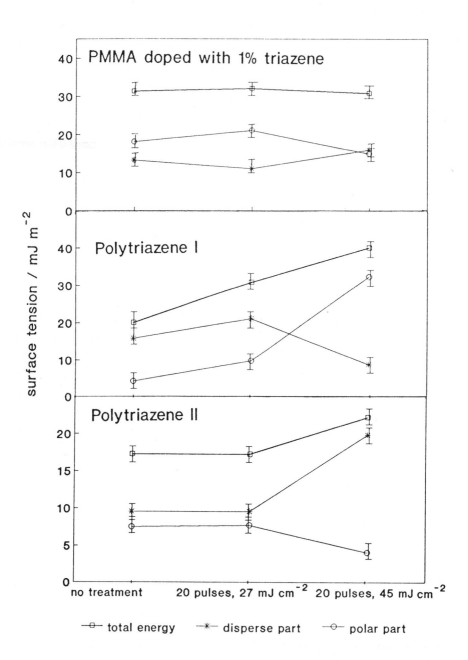

Fig. 4 Surface energy of excimer laser treated polymer surface. The total surface tension, as well as its polar and disperse components, have been determined by contact angle measurements, using ethylene glycol and water as test fluids. The untreated surface state and the effects of 3 irradiations, of increasing total fluence, are characterized.

hydrogen–bound carbon centers ('CC' and 'CH', E_b = 286.6 eV), as well as carbons attached to a singly bound (C–O, E_b = 288.5 eV) or doubly bound oxygen (C=O, E_b = 290.4 eV). A composite peak shape with 10% Lorentzian and 90% Gaussian contributions was used. Line widths were adjusted by the least squares fitting procedure; a typical value of 1.7 eV was obtained. An example of such a deconvolution is shown for doped PMMA in Fig. 5(b), before and after treatment of the sample with 5 pulses of 45 mJ cm^{-2} fluence.

With doped PMMA, no change in the overall O(1s) / C(1s) intensity ratio was observed {Fig. 6(a)}. This result confirms the conclusions of Lazare et al. [16] who found that irradiation of undoped PMMA has little net effect on the overall O / C ratio. However, a clear change in the C–O / C_{total} and C=O / C_{total} ratios is caused by the irradiation; this is evident also from the deconvolution of the signal into Gaussians in Fig. 5(b). The oxygen signal confirms the behavior observed with the carbon, i.e. the σ–bound oxgen (O_σ) increases and the π–bound oxygen decreases with irradiation {Fig. 6(a)}. A partial reversal of this behavior is observed after extensive treatment (20 pulses of 45 mJ cm^{-2} fluence).

These observations are in agreement with the contact angle measurements, from which a loss of carbonyl groups have been inferred. Apparently the gases released from the triazene dopant are capable of mechanically disrupting the PMMA chains. C=O would then be released, and radical sites on the surface would react with atmospheric moisture, to form surface hydroxyl groups.

In the XPS spectra of the polytriazenes I and II, a first effect observed upon irradiation below the threshold for gas release is the removal of carbonaceous surface impurities. At higher fluences, changes in the O / C and N / C ratios are observed. Quantification of these effects is currently in progress in our laboratory.

Positive SIMS spectra have been recorded on doped PMMA samples before and after excimer irradiation. In Table 1, the intensities of relevant fragments are indicated as observed before and after laser treatment. Irradiation results in a decrease in the intensity of fragments with masses higher than ~ 41 (which corresponds to the fragment CH_2–C=$(CH_3)^+$). Again this result points to a breaking of PMMA chains by the low fluence excimer laser pulses, accompanied by release of the –COOCH$_3$ ester functional group. It is interesting to note that these changes are occurring without topological surface changes that would be detectable in the electron microscope.

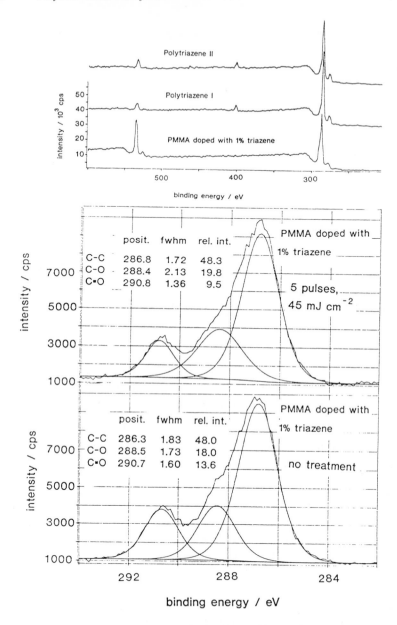

Fig. 5   XPS spectra recorded on the surface of PMMA doped with 1 % of triazene I, recorded before and after excimer laser irradiation (5 pulses of 45 mJ cm^{-2} fluence). Survey spectra are presented in the upper panel. In the lower diagrams, a deconvolution of the C(1s) peak into contributions of C–C, C–O (E_b = 288.5 eV), and C=O (E_b = 290.4 eV) sites is shown. A decrease in the signal assigned to doubly bound carbon due to irradiation, and a concomitant increase of the C–O signal are observed.

PMMA doped with 1% triazene

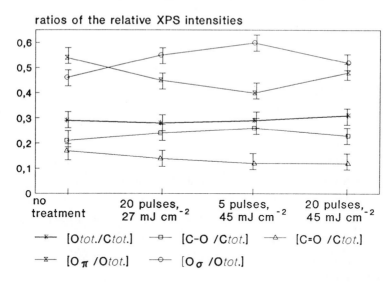

Fig. 6 Ratios of relative XPS intensities derived from the spectra shown in figure 5, as discussed in the text. Note that the total O / C ratio remains largely unchanged during irradiation, while σ–bound oxygen increases and π–bound oxygen decreases, as discussed in the text.

Table 1 SIMS intensities of untreated and laser irradiated PMMA surfaces

m / z	untreated [a] int. / counts	irradiated [b] int. / counts	assignment [17]
41	391	115	$CH_2=C(CH_3)^+$
55	179	–	$CH_3-C(CH_3)=CH^+$
59	158	41	$CH_3O-C\equiv O^+$
69	139	–	$(CH_3)CH_2-CH=C(CH_3)^+$
81	120	–	$CH_2=C(CH_3)-CH=C(CH_3)^+$
95	60	–	$CH_3(C\{CH_3\}=CH_2)_2^+$

[a] H^+ signal: 380 counts
[b] 5 pulses of 45 mJ / cm^2 fluence at 308 nm; H^+ signal: 8310 counts

4. CONCLUSIONS

The triazene functional group is well suited for sensitizing the ablation of polymer surfaces by excimer laser irradiation at 308 nm. A radical decomposition mechanism, with a quantum yield of $\sim 10^{-2}$, has been inferred for the photolysis in solution. The nitrogen released in the photolysis acts as the driving gas which results in the explosive ejection of matter from the surface.

Doping of PMMA surfaces with 1 % of monomeric 1–(3–carboxy–phenyl–)–3,3–diethyl–triazene results in efficient ablation. For a laser fluence of 8.5 J cm^{-2} at 308 nm, craters with steep edges and a depth per pulse exceeding 40 μm are created by the irradiation. This process is accompanied by a decrease in the polar part of the surface energy. XPS reveals a reduced intensity for π–bound oxygen, and an increase in the σ–bound oxygen signal. In the SIMS spectra, the intensity of higher mass fragments appears decreased after irradiation, which indicates a loss of C=O and –COOCH$_3$ fragments.

Two triazene–containing polymers have been synthesized. If the triazene functional group is placed in the main chain of the homopolymer, ablation is less efficient as a consequence of the high absorption. The observation of round crater edges indicates that melting of the material has occurred. Contact angle measurements show an increase in the polar part of the surface energy as a consequence of irradiation, as the radical sites produced during photolysis are reacting with atmospheric moisture.

As an alternative approach, the triazene functional group has been introduced as a side chain in a polyester type homopolymer. For this material the excimer laser irradiation results in steeper crater edges, and in a decrease of the polar contribution to the surface energy: the triazene side chains are detached from the polymer during the photolysis, and unpolar products are formed.

Extension of this work to the synthesis of copolymers that contain the triazene functionality at lower concentrations is in progress in our laboratories.

ACKNOWLEDGMENTS

The authors are indebted to J. Ihlemann and K. Luther for stimulating discussions, and for performing a series of ablation experiments at higher laser fluence. Sincere thanks are due to Ch. Schild and G. Sauer for the recording of XPS and SIMS spectra, and to P. Wittenbeck for discussions. Financial support of this work by grants of the Deutsche Forschungsgemeinschaft (SFB 213) is gratefully acknowledged.

REFERENCES

1. Srinivasan R and Braren B 1989 Chem. Rev. 89 1303
2. Sutcliffe E and Srinivasan R 1986 J. Appl. Phys. 60 3315
3. Srinivasan R, Braren B and Casey K G 1990 Pure Appl. Chem. 62 1581
4. Masuhara H, Hiraoka H and Domen K 1987 Macromolecules 20 452
5. Srinivasan R and Braren B 1988 Appl. Phys. A 45 289
6. Chuang T J, Hiraoka J and Mödl A 1988 Appl. Phys. A 45 277
7. Bolle M, Luther K, Troe J, Ihlemann J and Gerhardt H 1990
   Appl. Surf. Sci. 46 279
8. Ihlemann J, Bolle M, Luther K and Troe J 1990 SPIE Proceedings,
   Vol. 1361, 1011
9. Lazare S and Srinivasan R 1985 Polymer 26 1297
10. Novis Y, de Meulemeester R, Chtaib M, Pireaux J J and Caudano R
    1989 Br. Polym. J. 21 147
11. Le Blanc R J and Vaughan K 1972 Can. J. Chem. 50 2544
12. Dauth J, Lippert T, Nuyken O and Wokaun A to be published.
13. Majer J, Rehak V, Poskocil J and Cihlo L 1974 Wiss. Z.
    Tech. Hochsch. Chem. Leuna–Merseburg 16 335
14. Baro J, Dudek D, Luther K and Troe 1983 J Ber. Bunsenges. Phys. Chem.
    87 1155
15. Fowkes F M 1964 Ind. Eng. Chem. 56 40
16. Lazare S, Hoh P D, Baker J M and Srinivasan R 1984 J. Am. Chem. Soc.
    106 4288
17. Briggs D, Brown A and Vickermann J C 1989 Handbook of Static
    Secondary Ion Mass Spectrometry (New York: Wiley) pp 30–32

SECTION IV

POLYMER PLASMA SURFACE
MODIFICATIONS

Paper presented at First International Conference, Namur, Belgium
2–6 September 1991: Section IV

Dynamics of plasma treated polymer surfaces: mechanisms and effects

M. Morra, E. Occhiello, F. Garbassi

Istituto Guido Donegani S.p.A., Via G. Fauser 4, 28100 Novara, ITALY

ABSTRACT: This review deals with the aging phenomena occurring on plasma treated surfaces. After a general description of the process, discussed in the frame of polymer surface dynamics, the literature data on aging of plasma treated polymer surfaces and the effects of aging on the technological properties of polymers are reviewed.

1. INTRODUCTION

Every basic textbook of polymer physics contains at least one chapter on the dynamic properties of macromolecules, that is on the time dependent response to an externally imposed stress. While the effects of this dynamic behavior on bulk properties have been accepted and thoroughly investigated since long time, only in the last few years the dynamic approach has been extended to surface macromolecules, as described in two recent books edited by Andrade (1985, 1988) even if, in principle, there is no reason why they should not be at least as mobile as their bulk counterparts.

The appreciation of the dynamic nature of polymer surfaces is of great importance both from a fundamental and an applied point of view, since, in practical conditions, they are subjected to a kind of stress that does not have any bulk equivalent. Such stress arises from the peculiar condition of the outermost layers of a material, which are subjected to an asymmetric intermolecular field of forces, quantitatively expressed by the surface (interfacial) tension or, more precisely, by the excess specific surface (interfacial) free energy.

Many macromolecules bear both hydrophilic and hydrophobic groups (amphipatic materials): the former are expected to accumulate at the polymer-water interface, the latter at the polymer-air interface. If the interfacing medium is changed, the population of interfacial macromolecules or groups is expected to change, yielding to the stress caused by the non-minimum interfacial free energy.

An important class of amphipatic materials is constituted by surface treated polymers. In this case, in order to improve a given surface property, the polymer is subjected to some treatment which, very often, enormously increases the surface concentration of hydrophilic or hydrophobic groups (Wu 1982). This highly asymmetric surface structure causes much stress and many different stress-induced relaxation mechanisms, involving outdiffusion of untreated subsurface molecules through the modified layer, short range reorientation of macromolecular segments or side chains, chemical reactions between introduced groups and the surroundings or the chemical groups of the parent polymer.

It is not easy to study the dynamic behavior of polymer surfaces. Apart from the intrinsic difficulty of surface characterization and to define what really is a surface or an interface, another problem is to what extent the experimental observation is correlated to some intrinsic property of the polymer, usually a bulk one. For instance, the rate of detected modification process could be correlated with the T_g of the polymer, since it is a measure of the mobility of macromolecular chains. But, is the surface mobility the same as bulk mobility ? Intuitively, since surface molecules have a greater freedom than bulk molecules (as recently experimentally confirmed by Berger and Sauer (1991)) their mobility should be higher. However, it is impossible to measure a surface T_g. The same is true for short range reorientations such as the rotation of a side chain around a carbon-carbon bond (β relaxation), which could lead to a modification of the orientation of surface groups (and hence to a modification of the surface properties) and which is affected by the surface status of a macromolecule in the same, unpredictable way as the T_g.

Surfaces of treated polymers are still harder to study. In this case, the exact nature of the involved species is surely very different from the original macromolecules. Furthermore, one should expect also some heterogeneity in the surface-bulk direction, as the composition and concentration of modified molecules (and hence the properties of the material) continuously change until the bulk, unmodified material is reached.

In spite of such technical difficulties, the scientific interest and practical importance of surface modification of polymers have prompted an extensive research effort on the aging of surface treated polymers and some general features are beginning to emerge. This is particularly true in the case of plasma treated polymers surfaces, where the previously described dynamic effects often limit the usefulness of this otherwise powerful technique.

A typical example of aging of an oxygen plasma treated polypropylene (PP) sample, as described by Morra et al. (1989a), is shown in Figure 1. Both the advancing and receding angles are greatly decreased after plasma treatment, but, as time from treatment elapses, samples stored in air at room temperature loose their high wettability. In particular, the advancing angle reaches again the value of untreated PP, while the receding angle does not fully recover.

The first question on such experimental data is the following: which is the mechanism responsible for the observed increase of water contact angles ? In particular, one is concerned with contamination of ubiquitous hydrocarbons, which is a well documented source of time-dependent contact angle increase in the case of initially clean, high energy surfaces (metals and inorganic oxides) (White 1970). However, a simple experiment allows to rule out this hypothesis: the rate of aging is greatly increased if the storage temperature is increased, while aging can be stopped if the samples are stored at liquid N_2 temperature. This trend is exactly opposite to the one that would be produced by an absorption-driven aging and points clearly to a thermally activated mechanism.

While decisive with respect to the absorption mechanism, the previous experiment suggests another source of doubt. The observed aging behavior could be due to thermal decomposition of the modified layer, temperature sensitive macromolecules and oligomers could be present. However, this mechanism is ruled out by XPS analysis which, in the case of the above PP samples, shows that the surface composition does not change as the contact angle increases (Morra et al. 1989a). In the case of thermal decomposition, a decrease of the oxygen to carbon ratio is expected, as

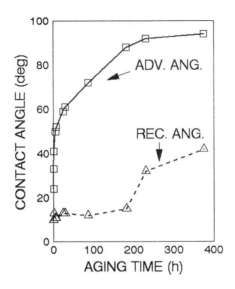

Fig.1 Room temperature aging of oxygen plasma treated PP

the modified layer volatilize and untreated surface is exposed.

An interesting comparison can be made with another surface modification technique, namely the sodium etching of fluoropolymers Recently, Rye and coworkers (1987, 1988, 1989a, 1989b) published a series of papers where XPS measurements of Na etched PTFE are coupled with thermal desorption mass spectroscopy (TDS), which allows to detect the species desorbed from the sample (not only from its surface) as a function of temperature. While XPS results substantially confirm earlier findings, TDS demonstrated that the recovery of the fluorine signal in the XPS spectrum promoted by heat exposure is not caused by the removal of a fraction of the modified surface layer, as earlier suggested by Dwight and Riggs (1974). Actually the desorption of products occurs well before the restoration of the fluorine signal, so that the heat promoted hydrophobic recovery must arise from processes in the solid phase, such as the diffusion of fluorocarbons from the subsurface region rather that through the desorption of an overlayer. In this process the etched spongy surface restructures and its surface area (Rye 1988) shows a factor 2 of loss, going from 1250 cm²/g after 40 s etching to 610 cm²/g after heating at 200°C (the surface area of untreated PTFE is below the measurable limit).

On the basis of the previous observations, it is clear that dynamic effects control the surface properties of polymers. Macromolecules can respond in several ways to external stress, either with short range motion, undetected by XPS, or with long range restructuring. Such dynamic behavior is completely unexpected if one approaches the problem in the frame of classical surface science, which, historically, is rooted on observations made on high energy surfaces, with very limited mobility and restructuring processes, such as the above described for etched PTFE, require far higher temperatures (Adamson 1990). In the case of polymers, room or slightly higher temperature is enough to allow motion.

Coming back to Figure 1 which, apart from the specific numeric values, can be considered typical of polymers hydrophilized by plasma treatment, three major points must be clarified in aging studies: 1. The immediate effect, as expressed by the contact angle or the surface composition of the just treated surface. 2. The restructuring mechanism(s), that is the kind of motion or reactions involved and the time-temperature dependence of the same. 3. The nature of the restructured surface and subsurface zone, which is a function of the recovery mechanism(s) and determines the properties of the aged surface. In the following paragraph, these points will be discussed for different classes of polymers.

2. EXPERIMENTAL STUDIES ON THE AGING OF PLASMA TREATED POLYMERS

2.1 Polyolefins

The general features of the aging behavior of oxygen plasma treated PP have been shown in Figure 1. Data refer to a 20 s treatment time, a power of 100 W, a flow rate of 8 sccm and a pressure of 2 Pa in a parallel plate, aluminium reactor, however only minor differences are observed for longer times (up to 10 mins), do not affecting the conclusions (Morra et al. 1989b).

The lack of noticeable changes of the XPS surface composition suggests that reorganization occurs in a layer thinner than (fixed angle) XPS sampling depth, that is less than 5-7 nm. Both the surface composition and receding angle (which is sensitive to high energy groups (Johnson and

Dettre 1969, Morra et al. 1990a)) of the fully recovered surface suggest that oxygen containing functionalities remain on the topmost layer after recovery. These speculations are supported by Static Secondary Ion Mass Spectroscopy (SSIMS), whose sampling depth (1 nm) makes it more surface sensitive than fixed angle XPS and very similar to contact angle measurement. In order to make the analysis more treatment-specific, Occhiello et al. (1991) used $^{18}O_2$ plasma. Aging of treated surfaces can be monitored by the ratio of peak at 13 amu (CH^-) to that at 18 amu ($^{18}O^-$). As shown in Table 1, the lower sampling depth of SSIMS with respect to XPS, allows to follow the recovery directly. In agreement with XPS and contact angle data, oxygen containing surface groups are detected also on the fully recovered surface.

The above data suggest that hydrophobic recovery of oxygen plasma treated PP goes on through rearrangement within the modified layer, that is through short-range reorientation of side groups, and that the fully recovered surface still contains polar groups. With respect to the first point, an apparent activation energy for recovery can be calculated, taking as a reference the time needed to reach the advancing angle typical of untreated PP, and using these data as input value in an Arrhenius plot (Occhiello et al. 1991a). Even if the assumptions involved in such a calculation must be taken into due account (after all, the exact relationship between the advancing angle and the surface composition of a non-ideal surface is still an open question), it is satisfactory to notice that the calculated activation energy (58.1 kJ/mole) is of the same order of magnitude of the activation energy of the bulk β relaxation of poly(vinylacetate) (PVAc) calculated by Pennings and Bosman (1979), in agreement with the suggestion that recovery involves motion of side chains.

The oxygen containing groups remaining on the treated surfaces after full recovery indicate that a pure hydrocarbon surface cannot be restored. This observation suggests that two opposite mechanisms are operating: on one side, minimization of the surface free energy tries to build up an homogeneous hydrocarbon-like surface, while, on the other side, hydrogen bonding interactions between introduced surface groups tend to avoid destruction and dippping into an hostile, apolar environment (the untreated bulk). The final reorganized surface is the best compromise between these two opposite mechanisms, and it is interesting to note

Table 1

SSIMS negative ion $CH^-/^{18}O^-$ peaks intensity ratios for
oxygen plasma treated PP samples aged in different conditions

Aging conditions	$CH^-/^{18}O^-$
16 h, 293 K	0.7
148 h, 293 K	1.3
175 h, 293 K	1.7
3 h 333 K, 44 h 293 K	2.9
2 h 393 K, 23 h 293 K	4.5

that the value of the fully recovered receding angle tends to increase as
the aging temperature increases, suggesting that, with a greater thermal
energy input, more hydrogen bonds can be broken, as discussed by Morra et
al. (1989a). Garbassi et al (1989) tested this hyphothesis by a
theoretical point of view, by modelling the surface layer with a random
copolymer of PP and PP with hydroxymethyl groups. The theoretical study
was based on macroscopic modelling of the instability of treated
surfaces,   as described by Barino and Scordamaglia (1988). Surface
rearrangement is obtained by thermally excited conformational movements
of polymeric chains driven by the lowering of the free energy of the
system. Assuming several different models of oxidized chains,   it was
shown that hydrogen bonding between hydroxil groups actually can lead to
chains more stable than PP, until a temperature (whose value of 383 K is
in full agreement with experimental data) is reached when thermal energy
overwhelms hydrogen bonding.

Contrary to PP, oxygen plasma treated high density polyethylene (HDPE)
shows a very small hydrophobic recovery and the advancing angle of the
fully recovered surface is about 24   degree, irrespective of the
temperature (at least up to 393 K) as shown by Morra et al. (1990b). A
possible explanation of this striking difference between the PP and HDPE
arises from their completely different behavior when subjected to
irradiation (Bonig 1982, McGinnis 1986): the former preferentially
undergoes chain scission, while the latter is crosslinked and extensive
crosslinking can be associated with low mobility of macromolecules and
side chains. However, literature data suggest that the behavior of PE
cannot be accounted for by cross linking alone (as discussed by Yasuda
and coworkers 1981, 1988 and Sharma 1981). In fact it is important to

note that a major difference exists between the amount of oxygen introduced by plasma treatment in PP or HDPE surfaces: in the case of a plasma treatment performed in the previously quoted conditions, the former reaches an O/C ratio from XPS analysis of 0.19, the latter of 0.31 (Morra et al. 1990b). Thus, the coupling of a high degree of crosslinking with a huge amount of oxygen containing surface groups (interacting by hydrogen bonding) appears necessary in order to greatly reduce recovery.

Since the previous results were based on commercial polymers, it is important to remind that, in these cases, it is very important to check very carefully the role played by additives, which are contained in all commercial polymers. This problem is especially important if one thinks that many additives are designed to "bloom" to the surface. Since many of them are hydrophobic, it is always important to recognize, when working with commercial polymers, if the observed hydrophobic recovery reflects a true reorganization process or an outdiffusion of additives. From an analytical point of view, the development of SSIMS greatly helped to solve this problem, since its main feature is the molecular specificity at sub-monolayer levels.

2.2 Polystyrene

The above results were obtained on commercial polymers, that is on polymers containing a wide range of different molecular weights. Polystyrene (PS) appears an interesting subject to test the effect of the molecular weight of the polymer on the hydrophobic recovery, since it is readily available as standard with a sharp distribution of molecular weights. The shake-up peak associated with the aromatic ring of the repeating unit allows an easy interpretation of the C_{1s} peak obtained by XPS analysis.

Figure 2 shows the most striking feature of the hydrophobic recovery of oxygen plasma treated PS (20 s, 100 W, 8 sccm, 2 Pa, in the same reactor used for polyolefins studies) (Occhiello et al. 1990a, 1990b, 1991b): at low aging temperature all samples reach the same limiting value of the advancing angle, irrespective of the specific molecular weight (also the kinetics is unaffected by M_w). Starting from 373 K aging temperature, however, low molecular weights exhibit a distinct behavior and the

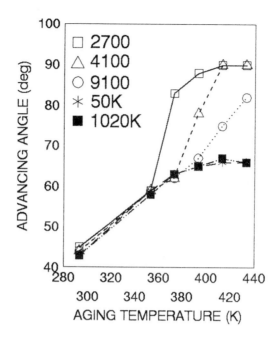

Fig.2 Limiting advancing angle as a function of aging temperature and
PS molecular weight

limiting value of the advancing angle, as well as the duration of the
process, becomes molecular weight dependent. At 433 K aging temperature,
the advancing angle of PS_{2700} and PS_{4100} is equal to the advancing angle
of untreated PS, while the receding angle (not shown) is lower. PS_{9100} is
starting do deviate from the behavior of the heavy weights, which still
follow a common path. Contact angle data are confirmed by XPS
measurements. Table 2 shows the XPS surface compositions: at low
temperature the increase of contact angles occurs without modification of
the surface composition, as in the case of PP, while, at higher
temperature, the O/C ratio starts to decrease for the lighter samples.
The deconvolution of the C_{1s} peak shows a decrease of higher binding
energy components, while the shake-up peak, which disappears after
treatment and low temperature aging, becomes again observable, in full
agreement with contact angle findings.
These results clearly point to two different recovery mechanisms: the
first one, as already observed on PP, involves short range monomeric or

Table 2

XPS surface composition (% at.) of recovered, oxygen plasma treated PS as a function of molecular weight and aging temperature

| M_w | Element | \multicolumn{6}{c}{Aging Temperature} |
		293	353	373	393	413	433
2700	C	79.5	80.3	84.8	86.0	93.8	93.6
	O	20.5	19.7	15.2	14.0	6.2	6.4
4100	C	80.2		80.5		89.0	90.4
	O	19.8		19.5		11.0	9.6
9100	C	78.7		79.8		84.9	85.4
	O	21.3		20.2		15.1	14.6
50000	C	80.8				80.5	81.1
	O	19.2				19.5	18.9
1020000	C	80.4	78.6	78.9	80.0	81.0	80.8
	O	19.6	21.4	21.1	20.0	19.0	19.2

segmental motion within the modified layer, leading to burial of polar groups away from the surface. It is expected independent of molecular weight, as actually observed. The second mechanism involves long range motion and outdiffusion of untreated molecules and is much more

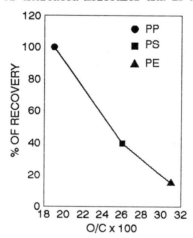

Fig.3 Degree of recovery as a function of O to C ratio

demanding in terms of activation energy, as demonstrated by the fact that it occurs only above 373 K; a strong dependence on the molecular weight is also observed.

An interesting point is the dependence on cross-linking of the two different mechanisms. By solubilization tests in tetrahydrofuran (THF) the insoluble fraction was found to increase with increasing the discharge power (ranging from 20 to 150 W) (Occhiello et al. 1990b). While minor effects were observed in the short range recovery mechanism, as detected by the measured contact angles, the onset of recovery by long range mobility was lowered, both in terms of temperature and molecular weight, the lower the crosslinking. As pointed out by Tead and coworkers (1988, 1990) crosslinked layers are effective obstacles to macromolecular diffusion.

An interesting observation involves the XPS surface concentration of oxygen and the degree of recovery (expressed as the ratio between the advancing angle after full recovery minus the advancing angle immediately after treatment and the advancing angle of the untreated sample minus the advancing angle immediately after treatment). In Figure 3, the degree of recovery is plotted as a function of the XPS surface composition, in the case of aging of PP, HDPE and PS (that is, in the case of short range reorientation recovery. Data are taken from the papers of Morra et al. 1990b and Occhiello et al. 1990a). Taken into account that the advancing angle is a crude indication of recovery (as previously discussed, it is sensitive to the most hydrophobic part of the surface and, actually, in the case of PP, the receding angle shows that recovery is not 100%), it is undeniable that a definite relationship exists between the surface concentration of oxygen and the allowed degree of overturning of introduced hydrophilic groups. These results confirm that the interactions between surface polar groups, as previously discussed in the case of the modelling of treated PP surfaces (Garbassi et al. 1989a), play a role in the control of short range reorientation at least as important as crosslinking.

2.3 Poly(dimethylsiloxane)

Silicone polymers are an interesting subject from the point of view of surface dynamic studies, since the energy of rotation about the siloxane bond in poly(dimethylsiloxane) (PDMS) elastomers is virtually zero, while the T_g is about 146 K (Oven et al. 1988a). As discussed by

Owen (1980, 1988b), the very high flexibility of siloxane backbone control the relatively high contact angle hysteresis of untreated PDMS surfaces. Oxidizing plasma or corona treatments of PDMS are known to create mainly –SiOH groups, as discussed by Owen and coworkers (1988a). According to these authors, hydrophobic recovery of corona treated PDMS goes on both through reorientation of surface groups and outdiffusion of untreated molecules.

In the case of oxygen plasma treatment, active species attack preferentially silicon (Morra et al. 1990c), increasing the average number of oxygen atoms bonded to silicon. Hydrophobic recovery is quicker than in the case of PP, as shown by comparing Figure 1 and Figure 4. Major differences are observed also in the mechanism of recovery which, in agreement with the results of Owen and coworkers (1988a) in the case of corona treatment, involves both reorientation and outdiffusion. The deconvolution of the Si_{2p} peak (Table 3, data from Morra et al. 1990c) allows to notice that immediately after treatment a high energy component is introduced, due probably to Si bonded to three oxygen atoms (Fakes et al. 1987). As aging goes on, the intensity of/ this component is reduced, suggesting a partial outdiffusion of untreated

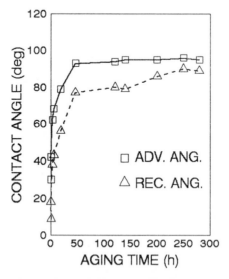

Fig.4 Room temperature aging of oxygen plasma treated PDMS

macromolecules. As shown in Figure 4, the receding angle shows a dependence on the aging time completely different from the previous

Table 3
XPS line fitting of Si_{2p} peaks for untreated and oxygen plasma treated PDMS surfaces, before and after aging

Smple	Binding energy (eV)	% of peak area
Untreated	102.0	83
	103.6	17
Aged 6 h at 293 K	101.9	21
	103.6	79
Aged 120 h at 293 K	102.0	43
	103.5	57
Aged 16 h at 293 K	102.0	50
	103.6	50

cases. Actually, advancing and receding angles show a parallel increase, with a contact angle hysteresis seldom exceeding $20°$, a very low value for plasma treated polymers. This observation suggests a further mechanism of recovery, that is silanols condensation and crosslinking. If silanols were only reoriented toward the bulk a greater contact angle hysteresis would be expected, as a consequence of water induced restructuring during contact angle measurement.

The restructured surface, even if hydrophobic, as shown in Figure 4, is rather different from pure PDMS: the advancing angle is lower, suggesting that a closely packed layer of $-CH_3$ groups can no longer be exposed, while the receding angle is higher despite the increased amount of XPS detected oxygen, a strong suggestion that the surface structure is now much more rigid and the water-hydrophilic backbone interaction cannot be further maximized when the angle is probed in the receding mode.

Thus, hydrophobic recovery on PDMS involves several mechanisms: the high bulk mobility allows outdiffusion of low surface energy chains, but irreversible modifications occurring during both treatment (oxygen attack on Si) and recovery (silanols condensation) does not allow to rebuild a pure PDMS surface.

2.4 Fluoropolymers

It is generally recognized that the small increase in hydrophilicity observed when PTFE is treated by O_2 plasma does not decrease with aging (Owen et al. 1988a, Griesser et al. 1990). This behavior is usually attributed to the low amount of oxygen introduced on the surface, rather to an intrinsic immobility of PTFE surfaces, even if crosslinking can contribute to the low mobility. Actually, reported contact angles on treated PTFE surfaces, shows a contact angle decrease that can be attributed more to a defluorination of C (according to the work of Baier et al. 1968a) than to a contribution of oxygen containing groups. A further point to take into account is that, in the case of O_2 plasma treatment, especially for long (> 1 min) treatment times, wettability is controlled by the rough morphology caused by etching, rather than interfacial energetics alone (Morra et al. 1989c). However, also PTFE surfaces can restructure, as shown by the already quoted work of Rye (1988) and by the fact that, when a high quantity of oxygen atoms is introduced on the surface, as in the case for instance of Ar plasma treatment, aging occurs (Owen et al. 1988a, Morra et al. 1990d).

On the other hand, ethylene propylene (FEP) co-polymer incorporates various oxygen containing groups after oxidative plasma treatments. In this case, changes are not permanent, and wettability decreases on storage in air for a period of three weeks, as discussed by Griesser and coworkers (1991a). FEP treated for 30 s with a 10 W power shows an advancing angle of 86° (104° in the case of untreated FEP) and recovers almost fully upon aging. On the other hand, a 120 s treatment leads to an advancing angle of 77° and only 63% is recovered on aging. The authors attribute the "permanent" part of surface modification to oxygen atoms immobilized close to crosslinks, while the displacement from the surface of mobile oxygen containing groups is responsible of the wettability loss. Interestingly, the (fixed angle) XPS surface concentration of oxygen increases with aging, contrary to the trend indicated by contact angles. As suggested by Griesser and coworkers, this unexpected results underline the extraordinary complexity of evolution processes on plasma treated surfaces: beside reorientation, monitored by the contact angle increase, sub-surface polymer oxidation chain processes, starting from remaining radicals, are going on, resulting in the oxygen increase detected by XPS.

2.5 PEEK

The aging of oxygen plasma treated PEEK was investigated by Munro and McBriar (1988) and by Brennan, Munro and coworkers (1991). Treated surfaces do not fully recover. The use of angle dependent XPS allows to observe modification in the detected elemental ratio as aging goes on. The most original feature of these works is the observation of a transient region where the aging surfaces becomes again more hydrophilic, before reaching the final plateau value. In one case this unexpected decrease of the contact angle is also reflected in a transient increase of the oxygen to carbon ratio. As a possible explanation, the authors suggest that the migration of small, polar fragments from the surface towards the bulk result in an incompatibility between these fragments and the less polar subsurface layers. This incompatibility provides the driving force for a reorganization process which enables to improve the compatibility between the bulk and the modified layers. Another possible explanation involves the delayed migration of subsurface radicals to the surface, where reaction with atmospheric oxygen occurs. As in the case of the quoted work of Griesser et al. (1991a), these results indicate the complex interplay of physical and chemical mechanisms involved in aging.

2.6 Treated polymer-water interfaces

According to the principles underlying aging, polymer surface hydrophilized by oxidizing treatments should not decay and air equilibrated surfaces should become hydrophilic again when stored in water or high energy media. The technological recognition of this phenomenon is described in some recent Japanese patents (1987a, 1987b), claiming that, to overcome aging problem, a possible solution is to interface the treated specimen with a material having a high energy surface. From an experimental point of view, this was indeed found to be the case with oxygen plasma treated PP (Occhiello et al. 1991a), PDMS (morra et al. 1990c), oxidized PTFE obtained by sputtering from a PTFE target (Ruckenstein and Gourisankar 1985), and several polymers treated with CF_4 plasma by Yasuda and coworkers (1988a, 1988b). In the latter case PET and Nylon 6 films showed an astonishing reversal of contact angles when subjected to successive cycles of hot wet and dry storage. The contact angle reversal was paralleled by a fluctuation of the XPS detected surface composition, with an accumulation of fluorine bearing groups at the solid-air interface, which disappeared at the solid-water interface.

On the other hand, it must be noted that in the quoted paper of Munro and McBriar (1988), hydrophobic recovery was observed also when oxygen plasma treated samples were stored under water, actually recovery seems greater in the case of water stored samples.

Thus, also in the case of aging against a relatively high surface energy environment, further complicated by the not completely understood details of interactions involving water molecules (Israelachvili 1985), the problem is rather complex. Again, some interesting reflection can arise considering other kinds of surface treatment. Baskzin and coworkers (1976) observed that, annealing PE surfaces oxidized by $KClO_3/H_2SO_4$ solutions, both in air and under water at 353 K, leads to an irreversible increase of the contact angle (which is lowered, due to surface oxidation, immediately after the treatment). On the other hand if PE is oxidized with $KMnO_4/H_2SO_4$ solutions, annealing in air at 353 K leads to major modifications of the surface morphology and of the XPS C_{1s} peak, in the sense of an hydrophobic recovery, but, when annealing was performed underwater, hydrophilic groups were still observed on the XPS peak (Eriksson 1984). The same was true when annealing in air was preceded by Ca^{2+} or Mn^{2+} adsorption (Eriksson 1984, Catoire 1981).

Again, the picture is extremely complex. There is no reason why an hydrophobic interface in water should be created in a purely physical process where the only variable is the interfacial free energy between the solid and the liquid phase. But, as abundantly shown by the previously quoted examples, events in plasma treated polymers are controlled by a large number of interactions, both chemical and physical, involving modified chains and the environment, surface and subsurface molecules and interactions among the treatment-introduced functional groups.

3. TECHNOLOGICAL EFFECTS OF AGING

The effects of the mobile nature of polymer surfaces are of great importance in many surface-related technological fields. In this section we will shortly review some examples.

An interesting comparison is offered by the completely different aging behavior of oxygen plasma treated polyolefins. It is well known that plasma treatment greatly improve the adhesion of both PP and PE (Wu 1982). In aging studies, it was shown that the bond strength neither of PP nor of PE is reduced after aging in shear strength measurements of polyolefins-epoxy joints, while pull tests showed a great effect of aging time in the case of PP (Morra et al. 1990b). The two tests are rather different, both in the nature of the adhesive employed and in the curing cycle, so that the different experimental results are not surprising. Actually, the most striking feature is that the dramatic increase of contact angles shown in Figure 1 is not reflected in a decrease in shear strength of PP epoxy joints. Two most likely explanations can be made: the first one is that the amount of polar groups remaining on the PP surface after aging (detected, as previously discussed, by the low value of the receding angle and by SSIMS) are

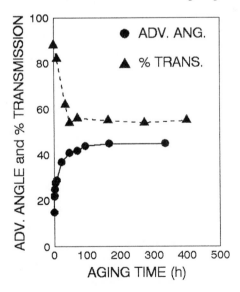

Fig.5 Advancing angle and misting behavior as a function of aging time for oxygen plasma treated PS

enough to guarantee good adhesion. The second explanation involves once again polymer surface dynamics: according to the general principles previously discussed, the aged surface is expected to restructure when the interfacing medium is changed. The surface composition of the adhesive and the curing cycle can provide the thermodynamic driving force and the kinetic mobility necessary to induce the exposure of hydrophilic groups towards the adhesive phase, as discussed by Lavielle and Schultz (1985) in the case of the adhesion between PE bulk grafted with acrylic acid and aluminum. In the present case, it must be noted that the marked hydrophobic recovery observed on treated PP surfaces is a strong evidence of great mobility of surface groups, so that the thermally activated maximization of the surface–adhesive interaction is not unlikely. Interestingly, marked decrease in bond strength with aging is always observed when adhesion on treated PP is performed either at room temperature (spray painting (Morra et al. 1989d)) or with a very quick heat exposure (evaporation of metal layers (Garbassi et al. 1990)).

Another field where the aging behavior of treated surfaces plays a major role is antimisting of transparent polymer sheets. Misting (or fogging, or clouding) is the condensation of water vapor as droplets on the solid surface, leading to loss of light transmission. This problem can be solved enhancing surface hydrophilicity, so that water condenses as a film (Ueno et al. 1990, Trotoir 1988). Plasma treatment is of course a possible solution but, as shown in Figure 5, the effectiveness of the treatment is reduced as aging goes on and the advancing water contact angle increases (Morra et. al. 1991).

The interface between synthetic materials and the biological environment is a fundamental issue in many technological problems, from antifouling to biocompatibility. In particular, the behavior of proteins at interfaces is an extremely debated and important topic, as reviewed by Norde (1986). The restructuring of complex molecules such as proteins (commonly called denaturation) is well known, however the mobility and reorganization of the solid surface can also play a role in the overall interaction. It has been demonstrated by Morra et al (1990e) that the aging routine of oxygen plasma treated PDMS surfaces employed in blood oxygenator does affect the protein absorption behavior (in particular, albumin and fibrinogen absorption were studied): the biological system (whole pig blood) is far too complex to allow definite statement on the precise nature of the overall interaction, however it is well established that the different surface structure caused by the different aging routine leads to different protein-material interactions. In this respect, Griesser (1991b) observed that aging affects the relative rate of cell growth in plasma treated cell culture dishes. Very likely they are affected by changes in protein adsorption on surfaces having different quantity and/or quality of surface groups.

4. CONCLUDING REMARKS

Plasma treated polymers surfaces respond dynamically to the stress imposed by the interfacing medium. The examples reviewed show that the aging behavior, as detected, for instance, by the increase of the contact angle of an air stored plasma-hydrophilized sample, arises from a complex interplay of several different mechanisms. Depending on the treatment, interfacing medium, parent polymer characteristics and storage temperature, aging can involve short range reorientation, long range diffusion, delayed radical-atmosphere reactions, reactions between introduced groups, subsurface reorganization. Cross-linking, often considered a way of creating aging-stable surfaces, is usually not enough to prevent short range reorientation, as shown by the work of Yasuda and coworker (1981, 1988a, 1988b) and Sharma (1981) on plasma deposited polymeric films. The stability of the topmost layers seems affected by the maximum amount of oxygen which can be introduced on the surface, probably a consequence of the trade off between lowering surface tension and loosing interaction hydrogen bonding energy, as shown by Garbassi et al (1989). This "chemical" contribution to surface stability reminds that aging is the sum of a large number of different pushes, and a guess based on the physical concept of interfacial energy alone can be misleading.

Beside short range motions, it must be remembered that for polymers even relatively low temperatures (such as dramatically shown by the quoted decrease of surface area of the rigid PTFE heated at 473 K describeb by Rye (1988)) can induce great effects on a much larger scale. A long range contribution to recovery was assessed in several of the described cases. This long range outdiffusion recovery mechanism, contrary to the short range one, is deeply affected by "bulk" properties such as the molecular weight and the T_g, at least on a qualitative basis.

From an analytical point of view, it is clear that the term "plasma treated surface" is completely useless, if one does not specify all the variables (aging time, temperature, environment) which actually affect its evolution. Incidentally, the dynamic behavior of plasma treated surfaces can explain many of the conflicting data that were reported in the early days of plasma treatments, where the effects of similar plasma exposure on wettability ranged from undetectable to very marked (Wu 1982).

From a technological point of view it was shown that the properties of plasma treated polymers are deeply affected by the aging behavior. In this respect, it is however important to note that not only the dynamic response of treated surfaces upon aging, but also the possible restructuring of the mobile surface in its working environment (as possibly shown by the unexpected constancy of PP shear strength upon aging (Morra et al. 1990b) and as discussed by Andrade in the case of protein absorption Andrade 1986)) must be taken in account in order to explain a given feature.

5. ACKNOWLEDGMENTS

We thank Dr. H. J. Griesser (CSIRO) for helpful discussions.

6. REFERENCES

Adamson A W 1990 Physical Chemistry of Surfaces (New York: Wiley)
Andrade J D ed 1985 Surface and Interfacial Aspects of Biomedical
   Polymers (New York: Plenum Press) Vol 1
Andrade J D 1986 Polymers in Medicine II eds E. Chiellini et al.
   (New York: Plenum Press) pp. 29–40
Andrade J D ed 1988 Polymer Surface Dynamics (New York: Plenum
   Press)
Baier R E, Shafrin G E and Zisman W A 1968 Science 162 1360
Barino L and Scordamaglia R 1988 Integration of Fundamental
   Polymer Science and Technology eds P J Lemstra and L A
   Kleintjens (London: Elsevier) p. 126–30
Baszkin A, Nishino M and Ter Minassian-Saraga L J. 1976
   Colloid Interface Sci. 54 317
Berger L L and Sauer B B 1991 Macromolecules 24 2096
Boenig H V 1982 Plasma Science and Technology (Ithaca: Cornell
   University Press)

Brennan W J, Feast W J, Munro H S and Walker S A 1991 Polymer 32 527
Catoire B, Bouriot P, Baszkin, L. Ter Minassian-Saraga A and
    Boissonnade M 1981 J. Colloid Interface Sci. 79 148
Dwight D W and Riggs W M 1974 J. Coll. Interf. Sci. 47 650
Eriksson J C, Gölander C G, Baszkin A and Ter Minassian
    Saraga L (1984) J. Colloid Interface Sci. 100 381
Fakes D W, Newton J M, Watts J F and Edgell M J 1987 Surf.
    Interface Anal. 10 416
Garbassi F, Occhiello E, Morra M, Barino L and Scordamaglia
    R 1989 Surf. Interf. Anal. 14 595
Garbassi F, Morra M and Occhiello E 1990 Proc. 2nd. Symp. on
    Metallized Plastics Montreal 6-11 5 1990 ed K L Mittal in press
Griesser H J, Hodgkin J H and Schmidt R 1990 Progress in
    Biomedical Polymers eds C G Gebelein and R L Dunn (New York:Plenum
    Press) pp.205-15
Griesser H J, Youxian D, Hughes A E, Gengenbach T R and Mau
    A W H 1991a Langmuir accepted for pubblication
Griesser H J 1991b personal communication
Israelachvili J N 1985 Intermolecular and Surface Forces
    (London: Academic Press)
Japan Kokay Tokkyo Koho JP 62103140 1987a to Hiraoka Shokusen KK
Japan Kokay Tokkyo Koho JP 62110973 1987b to Hiraoka Shokusen KK
Johnson Jr. R E and Dettre R H 1969 Surface and Colloid Science
    ed E Matijevic (New York: Wiley) Vol 2 pp. 85-153
Lavielle L and Schultz J 1985 J. Colloid Interface Sci. 106 438
McGinnis V D 1986 Encyclopaedia of Polymer Science and
    Technology (NEw York: Wiley) Vol 4 p. 432
Morra M, Occhiello E and Garbassi F 1989a J. Colloid
    Interface Sci. 132 504
Morra M, Occhiello E and Garbassi F, 1989b unpublished results
Morra M, Occhiello E and Garbassi F 1989c Langmuir 5 572
Morra M, Occhiello E and Garbassi F 1989d unpublished results
Morra M, Occhiello E and Garbassi F 1990a Adv. Colloid
    Interface Sci. 32 79
Morra M, Occhiello E, Gila L and Garbassi F 1990b J. Adhesion 33
    77
Morra M, Occhiello E, Marola R, Garbassi F, Humphrey P and
    Johnson D 1990c J. Colloid Interface Sci. 137 11
Morra M, Occhiello E and Garbassi F 1990d Surf. Interface Anal.
    16 412
Morra M, Occhiello E, Garbassi F, Maestri M, Bianchi R and Zonta
    A 1990e, Clinical Materials 5 147
Morra M, Occhiello E and Garbassi F 1991 Angew. Makromol.
    Chem. 189 125
Munro H S and McBriar D I 1988 J. Coatings Technol. 60 41
Norde W 1986 Adv. Colloid Interface Sci. 25 267
Occhiello E, Morra M, Cinquina P and Garbassi F 1990a Polymer
    Preprints 31 308
Occhiello E, Morra M and Garbassi F 1990b Polymers submitted
Occhiello E, Morra M, Morini G, Garbassi F and Humphrey P 1991a
    J. Appl. Polym. Sci. 42 551
Occhiello E, Morra M, Garbassi F, Johnson D and Humphrey P 1991b
    Appl. Surf. Sci. 47 235
Owen M J 1980 Ind. Eng. Chem. Prod. Res. Dev. 19 97
Owen M J, Gentle T M, Orbeck T and Williams D E 1988a Polymer
    Surface Dynamics ed J D Andrade (New York: Plenum Press) pp.
    101-10

Owen M J 1988b J. Appl. Polym. Sci. 35 895

Pennings J F M and Bosman B 1979 Colloid Polym. Sci. 257 720

Rye R R and Kelber J A 1987 Appl. Surf. Sci. 29 397

Rye R R 1988 J. Polym. Sci. Polym. Phys. Ed. 26 2133

Rye R R and Arnold G W 1989a Langmuir 5 1331

Rye R R and Martinez R J 1989b J. Appl. Polym. Sci. 37 2529

Ruckenstein E and Gourisankar S V 1985 J. Colloid Interface Sci. 107 488

Sharma A K, Millich F and Hellmuth E W 1981 J. Appl. Polym. Sci. 26 2205

Tead S F, Vanderlinde W E, Ruoff A L and Kramer E J 1988 J. Appl. Phys. Lett. 52 101

Tead S F, Vanderlinde W E, Marra G, Ruoff A L, Kramer E J and Egitto F D 1990 J. Appl. Phys. 68 2972

Trotoir J P, 1988 Modern Plastic International p. 86

Ueno M, Ugajin Y, Horie K and Nishimura T 1990 J. Appl. Polym. Sci. 39 967

White M L 1970 Clean Surfaces ed Goldfinger G (New York: Marcel Dekker) pp 361–73

Wu S 1982 Polymer Interface and Adhesion (New York: Marcel Dekker)

Yasuda H, Sharma A S and Yasuda T 1981 J. Polym. Sci. Polym. Phys. Ed. 19 1285

Yasuda T, Yoshida K, Okuno T and Yasuda H 1988a J. Polym. Sci. Polym. Phys. Ed. 26 2061

Yasuda T, Okuno T, Yoshida K and Yasuda H 1988b J. Polym. Sci. Polym. Phys. Ed. 26 1781

Plasma modification of polymer surfaces

Edward M. Liston

Branson International Plasma Corporation, P.O. Box 4136, Hayward, CA 94540 USA

ABSTRACT: The nature of RF plasma and its interaction with polymer surfaces is discussed, particularly the role of vacuum ultraviolet (VUV) radiation in the initiation of polymer surface reactions. It is shown that VUV photochemistry is the primary reaction mechanism driving plasma/polymer interactions. Plasma can be used to modify polymers to incorporate various active moieties into the surface. It is also shown that, in most cases, the "ageing" of a plasma treated surface is the result of diffusion of additives or other contaminants onto the surface.

1. INTRODUCTION

The plasma being considered here is a low pressure gas with electricity going through it. Common examples are fluorescent lights and "neon" signs. Plasma is used extensively in the semiconductor industry to etch patterns in wafers and to remove photoresist.

It has been known for at least 30 years that plasma could effect desirable changes in the surface properties of materials, particularly polymers. However, the practical application of plasma required the development of commercially available, reliable, and large plasma systems. Such systems are now available and the application of plasma to industrial manufacturing problems has been increasing rapidly for the past ten years.

In plasma treatment of polymeric materials, all significant reactions are based on free radical chemistry (Hollahan 1974). The glow discharge is efficient at creating a high density of free radicals, both in the gas phase and in the surface of organic materials, including the most stable polymers. These surface free-radicals are created by direct attack by gas-phase free-radicals or by photodecomposition of the surface by vacuum ultraviolet light generated in the primary plasma. The surface free radicals then are able to react either with each other, or with species in the plasma environment.

Free radicals created by low pressure gas plasma have four major effects on organic substrates: surface cleaning; ablation, a form of dry micro-etching; crosslinking; and surface chemistry modification. These four effects occur concurrently and, depending on processing conditions and reactor design, one or more of these effects may dominate. In all cases, these processes affect only the top few molecular layers (about 10 to 30 nanometers) so they do not change the appearance or bulk properties of the material. The net result of these effects is a major improvement (2 to 10 times) in processes that require adhesion or wettability.

Many different gases and plasma operating parameters are used to surface treat materials. Studies have been performed on the effect of these different plasmas on various polymers. It has been found that, for best results, different polymers may require different plasma treatment. In some cases it has been found that a plasma which gives excellent results on one polymer may give very poor results on a similar polymer. For example, the best process for PFA bonding results in poor bonding to FEP.

This paper will discuss the nature of plasma, the interaction between plasma and organic surfaces, the results of plasma treatment, and the ageing of plasma treated polymers.

2. NATURE OF PLASMA

By definition, plasma is an ionized gas that is spacially neutral. That is, there are an approximately equal number of positive and negative charges in a given volume. It should be stressed that these are radio frequency plasmas, not direct current plasmas. The terms "anode" and "cathode" have no meaning, on the time scale of diffusion, because the polarity of the electrodes is reversing every 37 nsec (at 13.56 MHz). The motion of the ions is almost unaffected by the RF field because there is only a few V/cm gradient in the plasma, and because of the enormous mass difference between electrons and ions.

For example, an RF field of 10V/cm will cause an ion displacement of only ±0.002 mm while the mean free path, due to thermal motion, is about 0.2 mm. Therefore, the only real effects of electrical fields are in the dark spaces over each electrode and they are symmetrical for both electrodes if they are equal area. This is a different situation than reactive ion etching or sputter etching because the pressure is much higher and there is no imposed electrical bias. Therefore, in the absence of a bias voltage on the parts, the ions have very little kinetic energy and there is almost no ion-etching of the type used in some semiconductor wafer fabrication equipment.

These plasmas are excellent electrical conductors. Therefore, large static charge does not build up on parts in an RF plasma.

The effects observed in surface treatment of polymers are caused by the chemical energy of free radicals or ions (if there is a bias voltage on the parts) or by the photochemical energy of the ultraviolet light. The plasma gas contains a few parts per million of ions (Gousset 1987), a few percent (2% to 20%) of free radicals (Egitto 1985), and a large amount of extremely energetic vacuum-ultraviolet light (VUV). The modern plasma reactor is essentially the same light producing device as those used by researchers in vacuum ultraviolet spectroscopy (Samson 1967).

The emission spectra from plasmas have been measured using an Acton Research Corporation Model VM-502 Vacuum Monochrometer. This 0.2 meter instrument is equipped with an osmium grating and mirror to maximize short wavelength performance. It is operated in the windowless mode with a turbomolecular pump. To date, measurements have been made on the emission spectra, from 20nm to 450nm, of 80 plasmas. Many cases have been found of species interactions and of significant spectral changes in mixtures of plasma gases.

It is important to stress the difference between a "pure" plasma and a "real" plasma. Essentially all of the published VUV emission data, for low pressure discharges, is from spectroscopic studies in which extreme care was taken to ensure pure gases. However, plasma reactors, by definition, are used to process materials and the plasma gas will be contaminated with the by-products of that processing. Therefore, a real plasma gas will never be pure. It will be shown later that there can be 10% to 20% contaminants in the gas. In the case of the oxidation of polymers or photoresist, these will be C,H, and O species, all of which can interact, both chemically and energetically.

In one set of experiments, the systems Ar/O_2 and He/O_2 were chosen. Electronically excited Ar^* can dissociate O_2 to give 2 ground-state O atoms (McNesby 1964). An argon ion can dissociate O_2 to give a ground state O plus an electronically excited O^* which will radiate at 130.5 nm. Excited helium (He*) will yield two ground state O's while a helium ion can give two O^*.

Figure 1 shows the intensity of the O^* radiation (at 130.5 nm) from O_2 and He/O_2 plasmas as a function of RF power. The vertical axis is in units of nanoamperes of current from the PMT. These data show a very strong relationship between 130.5 nm emission and input power. A three fold increase in power causes almost a 100 fold increase in radiant output.

These data also show that $He/5\%O_2$ plasmas have at least 5 times more radiant output than pure O_2 plasmas at the same input power. This may be due to trapping of the resonant radiation by the large amount of ground state O in the O_2 plasma. It may also be caused by an actual increase in the O^* radiation by the reaction between a He^+ and an O_2 to give two O^*.

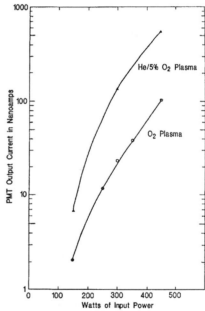

Figure 1. Intensity of 130.5 nm emission from plasma vs. RF power

Regardless of the mechanisms, these data show that the RF power level has a major effect on the intensity of the VUV reaching the surface being processed.

As was stated, in plasma reactors we are dealing with "real" plasmas. An example of this is shown in Figure 2. This was supposed to be a pure argon plasma. However, there was a small leak into the plasma reactor during this experiment. It was approximately 2 sccm. This is entirely acceptable for industrial plasma processing because it represents only about 600ppm of O_2 and 6ppm of H_2O as a background. Far more than that comes out of the polymer as dissolved gas or as reaction products during the plasma processing of polymers. However, the effects of this leak on the VUV spectra were dramatic and illustrate the effects of "real" plasma on the emission spectra.

In Figure 2 it can be seen that the intensity of the hydrogen peak at 121.5 nm is at least 4 times the intensity of the argon radiation even though the hydrogen is only present at about 6ppm. These same effects are seen in all real plasmas. The radiation spectrum in the VUV is usually very intense and complicated.

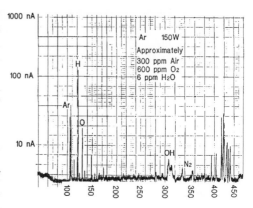

Figure 2. Emission intensity vs. wavelength nm

Figure 3 shows another example of the spectrum of a real plasma. In this case the plasma gas contains a small amount of water vapor (as seen from the OH peak at 310 nm). The spectrum shows that almost all of the radiant energy from the plasma is in the energy range below the quartz cutoff. Therefore, its presence would not be recognized and its intensity cannot be measured without using a VUV spectrometer. It is stressed that all of this radiation is photochemically active because it exceeds the bond energies in organic polymers.

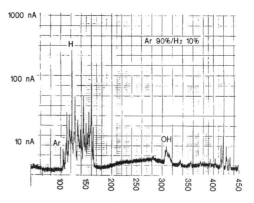

Figure 3. Emission intensity vs. wavelength nm

3. INTERACTION OF PLASMA WITH ORGANIC SURFACES

It is being proposed that, because of the large number of collisions that occur in the boundary layer, the concentration of excited free radicals from the plasma is greatly reduced during the process of diffusion to the surface being treated. As a result, these excited free radicals cannot initiate reactions on the surface. Also, in the absence of a negative surface charge on the polymer, ions cannot diffuse to the surface, either through thermal motion or by motion induced by the radio frequency field.

Therefore, the initiation reactions can only be caused by VUV photons or ions (if there is a negative surface charge). Once the initiation has occurred, the surface oxidation can progress through free-radical chain reactions with other species in the boundary layer, such as ground state O or O_2.

The interaction between a plasma and the surface of a polymer takes place in four interrelated steps:

1) The initial breaking of surface and sub-surface bonds.
2) The volatilization and reactions of organic fragments.
3) The reaction of the polymer surface with the boundary layer gas (not the plasma).
4) The reaction of the residual polymer surface after the plasma is off.

3.1 Initial breaking of surface bonds

It can be shown thermodynamically that ground state O or O_2 do not react easily with "ground state" polymers (Ranby 1975). It has also been shown experimentally by Golub (1988) that the oxidation of a polymer in an oxygen plasma, where there is VUV and ions, is about 100 times faster than it is downstream of the plasma where there are no ions and very little VUV. The polymer must either be heated, or contain accessible free-radicals or surface carbonyls for direct reaction with low energy oxygen species.

This is well known in the semiconductor industry. It is necessary to heat silicon wafers to about 230°C to get rapid oxidative removal of photoresist in "down stream" strippers. In this type of equipment there is a high concentration (10% or more) of ground state O at the wafer but there are no ions or VUV. The stripping rate follows a typical first order Arrhenius reaction.

Therefore, in the absence of high temperature, there must be some other initiation process for the oxidation of the surface. There are at least three possibilities: electronically excited species, ions, or VUV photons.

The electronically excited species (e.g. O^*) usually have too short a lifetime (10^{-9} sec) to penetrate the boundary layer directly from the plasma. However, a ground state O, deep in the boundary layer, can absorb a 130.5 nm photon and be excited. If the excited atom is close enough to the surface to reach it before radiational decay, the O^* can break organic surface bonds and initiate a chain reaction oxidation of the polymer by O or O_2. Note however that this is fundamentally a VUV photochemical process.

The second possibility is the reaction of ions with the polymer surface. This is thermodynamically favorable, if the ions can reach the surface.

The term "dark space" is used to designate the volume of plasma between the plasma that is glowing (and generating ions) and the surface of the polymer. It is important to stress that ions are formed outside the dark space, not in it (Cobine 1958). Any ions in the dark space have diffused into it with diffusional velocity plus any velocity imparted by the self-bias on the surface of the polymer. The dark space is essentially governed by electrical processes. In practice it is about 2 mm wide at a typical pressure of 40 Pa.

The term "boundary layer" is used to designate that volume of gas between the uniform gas concentration in the plasma and the surface of the polymer. It is in this layer that the first steps of oxidation occur after the organic fragments break away from the surface of the polymer. It is essentially governed by chemical/diffusional processes. In reality, the boundary layer must be thicker than the dark space because it must overlap that volume where ions and free radicals are being formed. It is probably several mm thick but, for the purposes of this discussion, it will be assumed to be the same thickness as the dark space, 2mm. The thickness and processes in the boundary layer and dark space are strongly interrelated because of the collisions that occur in the boundary layer.

If it is assumed that there are 1 ppm of ions in the plasma (Gousset 1987), the flux of ions to the interface between the plasma and the dark space will be approximately $(1.6)^{14}/cm^2$-sec. However, in the absence of a static surface-charge, these ions will have to diffuse through the dark space where no new ions are being formed. If it is assumed that the net diffusional drift is 50 cm/sec (Dushman 1962), and that the collision frequency is 10^6/sec (Benson 1960), an ion will undergo 4000 collisions during the diffusion across the dark space.

The ultimate by-products of the plasma oxidation of a polymer will be H_2O, CO, and CO_2. If it is assumed that the ablation rate is 100 nm/min and that the O_2 flow into the reactor is 200 sccm, the average atom percent concentration of by-products in the plasma will be about 15% (depending on the total polymer surface area exposed). The concentration of these by-products in the boundary layer will be much higher.

The by-product contamination, in the boundary layer, will be in the form of C,O,H compounds such as H_2O, CO, CO_2, O, OH, HO_2, COH, CH, CH_2, CH_3, etc. It is assumed that 15% of the total species in the boundary layer are by-product species and that 10% of these will be free radicals or other easily reactive species. An ion diffusing through the dark space will undergo (4000 x 0.10 x 0.15) or 60 collisions with a reactive by-product species. Each of these 60 collisions has at least a 0.5 probability of reacting with an ion.

Therefore, the net probability of an ion diffusing through the dark space (unchanged) is $(0.5)^{60}$ or 10^{-18}. The products of these collisions could be other ions. But, these would probably not be other O_2 ions because most of the species in the boundary layer have ionization potentials lower than O_2. Collisions with free radicals or electrons could destroy the ions, probably forming excited free radicals, which would then decay through other collisions. In short, in the absence of a static surface charge, an ion cannot diffuse through a contaminated boundary layer, since it undergoes too many destructive collisions.

In the presence of a negative static surface-gradient of 3 volts or more, the ions can accelerate so rapidly that they can pass through the 2 mm boundary layer in less than 10^{-6} sec. (the average collision time). Therefore, they can reach the polymer surface as ions and can initiate surface reactions.

No data have been found on the surface charge of "floating" polymer surfaces so the importance of surface-charge accelerated ions has not been established. Classical physics says that ambipolar diffusion will cause a negative charge on the polymer surface. However, three points should be made: 1) The ions will carry positive charge to the negative surface and would tend to neutralize the surface charge; 2) The VUV photons carry sufficient energy to cause photoemission of electrons from the polymer surface and create positive ions in the surface; and 3) Experimental data (Hudis 1972) show that when only VUV is present the reaction rate is still 60% to 80% as fast as when both VUV and ions are present. The importance of ions needs further study.

It should also be stressed that, even if ions can reach the surface of the polymer they will not have a great deal more energy than VUV photons. The VUV photons have sufficient energy to initiate the surface reactions, ions are not necessary from a chemical standpoint.

The calculated VUV photon flux, at the entry slits of the spectrometer, to give 100 nA of PMT current, is 10^{13} photons/cm^2-sec. If it is assumed that the absorption coefficient for O (at 130.5 nm) is 10^4/cm-atm (Okabe 1978, Mitchell 1971), the boundary layer will absorb less than 10% of the radiation from the plasma. Even if the absorption coefficient is assumed to be 10^5/cm-atm, only about 50% of the radiation will be absorbed, and some of that will be reradiated toward the polymer surface. Therefore, VUV radiation can propagate through the boundary layer with little attenuation. These photons have more than enough energy to break any organic bond (130.5 nm = 219 kcal/mole, 121.5 nm = 235 kcal/mole). When the VUV photon reaches the surface it will break surface bonds and form free radicals in the surface. It can also cause photoemission of electrons and form ions in the surface.

The essential point is that the oxidation of a polymer surface cannot be initiated by ground state atoms or molecules, because they do not have sufficient energy. Nor can electronically excited species initiate the oxidation because either their life time is too short or there will be too many collisions with reactive species in the boundary layer. Ions cannot initiate the oxidation because (unless there is a static bias voltage) they also undergo too many collisions in the boundary layer. However, the oxidation can be initiated through VUV photochemical processes, or through the diffusion of ions, if there is a negative surface charge on the polymer.

Figure 4 shows the VUV absorption spectrum for polyethylene (Painter 1980). This is typical for polymers. The absorption coefficients are extremely high so the VUV radiation is absorbed in a very shallow depth in the surface of the polymer. The absorbed photons will break surface bonds and generate surface free-radicals which can then react with the boundary layer gas, not the plasma.

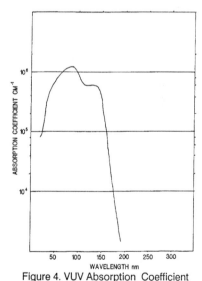

Figure 4. VUV Absorption Coefficient
vs.
wavelength for polyethylene (Painter 1980)

Table I gives some examples of the 1/e depth (the depth for 63% absorption of radiation) for various wavelengths and materials. These calculations assume a pressure of 40 Pa (0.3 Torr), and pure gases, which is not the case in real plasmas, but they do indicate that the VUV radiation can propagate through the plasma with very little attenuation.

TABLE 1

ABSORPTION DEPTHS IN MATERIALS AT VARIOUS WAVELENGTHS

Incident Radiation

Radiating Species	He	Ar	H	O
Wavelength — nm	59.	104.8	121.5	130.5
Photon Energy — eV	21.0	11.8	10.2	9.5
Photon Energy — kcal/mole	484	273	235	219

Absorbing Material

O_2 (Watanabe 1953)	—	—	11,250 cm	280 cm
Ar (Samson 1976)	3.2 cm	(NA)	(NA)	(NA)
NH_3 (Sun 1955)	6 cm	9 cm	15 cm	10 cm
Polyethylene (Painter 1980)	15.4 nm	10.5 nm	16.9 nm	16.7 nm
Polyimide (Arakawa 1981)	11.8 nm	18.2 nm	47.6 nm	66.7 nm

The data in Table 1 also show that there are differences in the absorption spectra of polymers. This may make it possible to tailor the plasma emission spectra to maximize the photochemical effect on different materials.

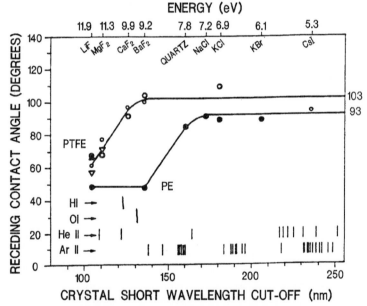

Figure 5. Effect of short wavelength cutoff
on receding water contact angle (Egitto 1990)

For example, data from Egitto (1990) are shown in Figure 5. These data illustrate why different plasma gases are required to treat different polymers. An O_2 plasma emits very strongly at 130.5 nm. This radiation is strongly absorbed by polyethylene (PE) and O_2 plasma is very efficient for treating PE. However, an O_2 plasma is not efficient for treating PTFE. One reason for this is shown in Figure 5, the 130.5 nm emission from an O_2 plasma is not strongly absorbed by PTFE and does not cause decomposition of that polymer. It is necessary to use a plasma gas that radiates at a shorter wavelength than O_2. For example, a H_2 containing plasma radiates at 121.5 nm. This radiation will be absorbed by the surface of the PTFE and will break the surface bonds. Also, the H can abstract the surface fluorine to form HF and expose carbon free-radicals to further attack.

There have been several studies that have shown, at least qualitatively, that some of the effects of plasma may be caused by ultraviolet light generated in the plasma. In these studies the parts were separated from the plasma, usually by LiF windows, and yet it was possible to get cleaning, crosslinking (Hudis 1972), fluorination (Cohen 1983), functionalization (Shard 1991), and free-radical generation (Yasuda 1976) in polymer surfaces.

3.2 Volatilization and reactions of organic fragments

During plasma treatment of polymers, bonds may be broken at both ends of small sections of polymer chains and organic fragments will be released into the boundary layer over the surface. These fragments will react, either with other free radicals in the boundary layer, or with the VUV flux. In either case, they decompose towards the ultimate by-products of CO, CO_2, and H_2O. However, in the process of this decomposition many transient species are formed that greatly complicate the chemistry of the boundary layer. It is these species that will interfere with the diffusion of active species from the plasma to the polymer surface.

The counter-current flow of the polymer decomposition products and the plasma species results in a very large number of reactions occurring within a few millimeters of the surface. This is the zone where the plasma/polymer interaction is really occurring. It is also a zone in which it is very difficult to do analytical chemistry but, analyses done outside this zone are almost meaningless.

The transients will also contain terminal or branch free-radicals that will be very reactive. If they deposit on a surface before reacting with some radical terminator they will form a crosslinked surface layer. This is the basis for plasma polymerized deposition. To maximize deposition a very low pressure and low power are used. These process variables maximize wall collisions and minimize organic fragmentation.

If deposition is not desired, as with cleaning or surface modification, higher power and pressure are used. Also, it is essential to use some plasma gas that will form permanently volatile reaction products.

For example: O_2 containing plasma will ultimately yield CO, CO_2, and H_2O; N_2 plasmas yield HN or HCN; and H_2 plasmas yield low molecular weight hydrogenated organic fragments. If a non-reactive gas is used, for example, Argon, there will be very slow cleaning or surface modification. There will also be redeposition of the molecular fragments on all exposed surfaces. This will form a very tenacious crosslinked varnish.

3.3 Reaction of the polymer surface with the boundary layer gas

Four major effects of plasma on surfaces are normally observed. Each is always present to some degree; however, one effect may be favored over another depending on substrate chemistry, reactor design, gas chemistry, and processing conditions. The effects are: removal of organic contamination from the surfaces; material removal by ablation (micro-etching) to increase surface area or to remove a weak boundary layer; crosslinking or branching to cohesively strengthen the surface; and surface chemistry modification to improve chemical and physical interactions at the bonding interphase.

Surface modification alone, or in combination with any or all the competing reactions, provides a means to dramatically improve the strength of interfacial bonds.

Cleaning — Cleaning of surfaces is one of the major reasons for improved bonding to plasma treated surfaces. Most liquid cleaning procedures leave a layer of organic contamination that interferes with adhesion processes.

For example, it is known that as little as $0.1 \, \mu g/cm^2$ (a single molecular layer) of organic contamination on a surface can interfere with bonding (Jackson 1978). This amount of contamination is the residue from $0.2 \, drops/cm^2$ of a liquid containing 10 ppm non-volatiles. It is extremely difficult to get solvents or water with less than 10 ppm non-volatiles; it is therefore safe to assume that a surface will remain contaminated after any cleaning process that finishes with a liquid rinse.

Plasma is capable of removing molecular layers from polymers and all organic contamination from inorganic surfaces, as shown by Auger analysis of the surface (Smith 1980). This results in hyperclean inorganic surfaces and polymer surfaces that are really the polymer and not the surface of some contamination on the polymer. Therefore, these surfaces give very reproducible bonds and, in many cases, make stronger bonds than normally "cleaned" surfaces.

However, it is critically important to plasma clean a polymer for a sufficient time to remove all of the contamination from the surface. Almost all polymer films, and most molded parts, contain additives or contaminants such as oligomers, anti-oxidants, mold release agents, solvents, anti-block agents, etc. which are oils or waxes. Most of these are deliberately incorporated into the polymer formulation to improve its properties or manufacturability and are designed to "bloom" to the surface of the polymer and coat that surface.

These materials often have the same, or close to the same, chemistry as the base polymer. Therefore, they are often difficult to detect with ESCA or other analytical techniques. Typically the contaminants can be present in layers 1nm to 10nm thick, even after solvent cleaning. They just continue to diffuse to the surface after solvent cleaning. Polyethylene is particularly noted for this problem.

The surface contamination will react with the plasma in the same way that the polymer will. That is, if the plasma treatment is not of sufficient duration to remove the contamination, the contaminant will become wettable and will have a modified ESCA pattern similar to that of the polymer. However, it will still be a plasma treated contaminant layer, not a plasma modified polymer surface. At normal power levels it is necessary to clean most polymers for several minutes. A treatment of a few seconds is not long enough to remove the contaminants but it is long enough to plasma treat the contaminants and give a surface that appears to be properly treated, but is not.

Ablation — Ablation is important for the cleaning of badly contaminated surfaces, for removal of weak boundary layers formed during the fabrication of a part, and for the treatment of filled or semicrystalline materials. Amorphous polymer is removed many times faster than either crystalline polymer or inorganic material. Consequently, a surface topology can be generated with the amorphous zones appearing as valleys. This change in surface can improve mechanical bonding as well as increase the area available for chemical interactions.

Crosslinking — Crosslinking occurs in polymer surfaces exposed to plasmas which are effective at creating free radicals in the polymer, yet do not provide stable moieties at the radical sites. Noble gas plasmas, such as helium and argon, are crosslinking plasmas if they are used in the total absence of oxygen or other free radical scavengers. The ions and the VUV light attack the polymer surface and break C-C and C-H bonds, leaving radicals in the surface. Once free radicals are created in this environment they can only react with other radicals and are, therefore, very stable (Yasuda 1976). If there is any flexibility in the polymer chain, or if the radical can migrate on the chain, there can be recombination, unsaturation, branching, or crosslinking. The latter effect may improve the heat resistance and cohesive strength of the surface. It may also act as a barrier layer, hindering diffusion across the interphase. The term CASING (Crosslinking via Activated Species of Inert Gas) has been applied to this treatment (Schonhorn 1966).

Surface Chemistry Modification — The most dramatic and widely used effect of plasma is the surface modification of polymers, where the outside layer of a polymer is altered to create chemical groups capable of interacting with materials deposited on the polymer. The inherently low surface-energy of untreated polymers hinders the wetting and interaction with adhesive systems (Hook 1987, Everhart 1981) or deposited metals. Typically, plasma is used to add polar functional groups which dramatically increase the surface energy of polymers.

For example, Figure 6 shows the low angle ESCA analysis of a polystyrene surface before, and after, treatment with a water vapor plasma. This demonstrates the spectacular change in chemistry of the treated surface that is typical of most polymers exposed to an oxygen

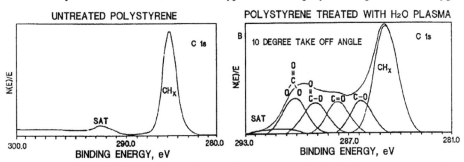

Figure 6. Effect of plasma on a polymer surface

containing plasma. This surface will be very polar, completely wettable, and receptive to reactive adhesives. It is believed that, during curing, the adhesive can react with the surface oxygen species and covalently bond to the plasma treated interphase.

Tanigawa (1991) has shown that the plasma treatment of a polymer increases the adhesion of a deposited metal layer. This may be due to the direct reaction between the depositing metal or possibly just because the polymer surface was hyperclean.

However, Mance (1989) has shown that the very acid nature of an oxygen plasma treated polymer surface inhibits electroless plating unless the surface is neutralized with hydroxide before plating.

Depth profiling has shown the plasma treatment affects only the top 10 nm to 30 nm of the polymer. Therefore, it is important that the surface be cleaned before plasma treatment and that the treatment last long enough to remove any weak boundary layer. This shallow depth of treatment is the reason why proper plasma treatment does not affect the optical, physical, or mechanical properties of the part.

It is believed the plasma treatment will leave 5% to 20% of the surface carbons with some form of organic oxygen species (Penn 1987). However, studies of the fluorination of polymers have shown almost complete replacement of the surface H with F or CFX (Corbin 1982). More work needs to be done in this area.

3.4 Reaction of the residual surface after plasma is turned off

As has been discussed, a polymer exposed to a reactive plasma will be undergoing simultaneous surface modification reactions and surface removal reactions. If the removal rates dominate or if a non-reactive plasma is used, the polymer surface will contain a large number of free radicals when the plasma is turned off. These radicals may react with themselves to form crosslinking or unsaturation or, they will react with atmospheric O_2 or water vapor to form surface oxygen moieties. These post-plasma reactions are the source of the oxygen species found in the ESCA spectra of He or Ar treated polymers.

Post-plasma reactions can also be used for grafting desired species onto the surface of polymers. For example, acrylic acid vapor will graft to surface free-radicals if the plasma treated surface is exposed to that vapor before it is exposed to air. Also, a large amount of OH will graft onto the surface of the treated polymer if the surface is exposed to water vapor.

4. AGEING OF PLASMA TREATED POLYMER SURFACES

The problem of surface contamination from additives was discussed in the section on cleaning. These contaminants, and storage materials, are the most common reason for ageing, or "hydrophobic recovery", of plasma treatment of polymers.

If a polymer contains an additive or a mobile contaminant, that material can diffuse to the surface after the plasma treatment and cover the plasma modified surface. This will appear to be a gradual reversion of the surface but in reality, it is simply a covering up of the treated surface. A truly clean surface, on a polymer with no mobile contaminants (such as a high temperature engineering polymer), can be plasma treated and the surface will remain active or wettable almost indefinitely.

Papers on the subject of "true reversion" propose that the polar surface moieties, driven by entropy, rotate about the axis of the host polymer molecule and bury themselves. This would lead to a reduction in surface energy and a "hydrophobic recovery". The author knows of no case where true reversion of the plasma treated surface has been proven, although it undoubtedly happens. The problem of such a proof is the need for proving that there are no mobile species in or on the polymer. This type of reversion could only occur on a polymer with a minimum of crosslinking. That means a "soft" polymer such as polyethylene or silicone. All commercial soft polymers contain mobile additives that will mask or modify the observation of true reversion.

Another source of apparent ageing of plasma treated surfaces is transfer of surface active contaminants from the storage medium to the plasma treated surface. Again, almost all polymers used to make plastic bags contain additives that will transfer to the plasma treated parts. One way to test storage materials is to wrap a plasma cleaned glass microscope slide in the material and see if it becomes hydrophobic. If it does, the storage material will contaminate any parts stored in it.

5. CONCLUSIONS

Low-pressure plasma can be used for the cleaning and surface modification of polymers but the details of the plasma/polymer surface interactions are extremely complex. It is unlikely that ground state oxygen atoms or molecules can initiate surface oxidation. However, they can easily react with the surface once the reaction has been initiated by VUV photons or possibly ions.

Plasma surface modification can be used to get very useful polymer surfaces if the polymer is treated for a sufficient time to remove surface contamination.

Most "ageing" of plasma treated polymers is probably due to insufficient treatment time, diffusion of contaminants from the polymer to the surface, or recontamination by the storage material.

REFERENCES

Arakawa, E.T. , M.W. Williams, J.C. Ashley and L.R. Painter 1981 J Appl Physics 52(5) 3579

Benson, S.W. 1960 "The Foundations of Chemical Kinetics", McGraw-Hill

Cobine, James, D. 1958 "Gaseous Conductors", Dover Pub.

Cohen, R.E., R.F. Baddour and G.A. Corbin Paper No. B-8-2, 6th International Symposium on Plasma Chemistry, ISPC-6, July 24-28, 1983, Montreal, Quebec, Canada

Corbin, G.A., R.E. Cohen and R.F. Baddour 1982 23 1546

Dushman, Saul and J.M. Lafferty 1962 "Scientific Foundations of Vacuum Technique", John Wiley & Sons, New York

Egitto, F,D., F. Emmi and R.S. Horwath 1985 J Vac Sci Tech B3(3) 893

Egitto, F.D. and L.J. Matienzo 1990 Polym Degrad Stab 30 293

Everhart, D.S. and C.N. Reilley 1981 Anal Chem 53 665

Golub, M.A., T. Wydeven and R.D. Cormia 1988 Polym Commun 29 285

Gousset, G., P. Panafieu, M. Touzeau and M. Vialle, 1987 Plasma Chem. and Plasma Processing, 7(4) 409

Hollahan, J.R. and A.T. Bell 1974 "Techniques and Applications of Plasma Chemistry" John Wiley & Sons, New York NY

Hook, T.J., J.A. Gardella and L. Salvati 1987 J Mater Res 2(1) 132

Hudis M. and L.E. Prescott 1972 J Polym Sci, Polym Lett B10(3) 179

Jackson, L.C. 1978 Adhesive Age Sept

Mance, A.M., R.A. Waldo and A.A. Dow 1989 J Electrochem Soc 136 1667

McNesby, James R. and Hideo Okabe 1964 "Vacuum Ultraviolet Photochemistry" in Advances in Photochemistry, Vol. 3, p. 157, Wiley-Interscience, New York

Mitchell, A.C.G. and M.W. Zemansky 1971 "Resonant Radiation and Excited Atoms", Cambridge Univ. Press

Okabe, Hideo 1978 "Photochemistry of Small Molecules", Wiley-Interscience, New York

Painter, L.R. , E.T. Arakawa, M.W. Williams, and J.C. Ashley, 1980 Rad Res 83 1

Penn, L.S., T.J. Byerley, and T.K. Liao,1987 J Adhesion 23 163

Ranby, B. and J.F. Rabek 1975 "Photodegradation, Photo-oxidation and Photostabilization of Polymers", John Wiley & Sons

Samson, J.A.R. 1976 "Techniques of Vacuum Ultraviolet Spectroscopy" John Wiley & Sons, New York

Schonhorn, H. and R.H. Hansen 1966 J Polym Sci B(4) 203

Shard, A.G. and J.P.S. Badyal 1991 Polym Commun 32 (7) 217

Smith, M.D. 1980 Insulations/Circuits May

Sun H. and G.L. Weissler 1955 J Chem Phys 23(6) 1160

Tanigawa, S., M. Ishikawa and K. Nakamae 1991 J. Adhesion Sci. Technol 5, 543

Watanabe K. , M. Zelikoff and E.C.Y. Inn, 1953 "Absorption Coefficients of Several Atmospheric Gases" AFCRL Technical Report No. 52-23 Geographical Research Papers No. 21 (AD19700)

Yasuda, H. 1976 J Macromol Sci Chem A10(3) 383

Plasma modification of polymers

J F Friedrich

Zentralinstitut für Organische Chemie, Bereich Makromolekulare
Verbindungen, Rudower Chaussee 5, D-O-1199 Berlin

ABSTRACT : This article summarizes chemical processes on polymer surfaces and in the bulk leading to the formation of
plasma-generated functional groups and radiation artifacts.
The formation of most plasma-induced groups could be explained by well-known photochemical processes indicating
the great role of plasma-uv radiation in all plasma processes. Surprisingly high penetration depths of plasma-uv radiation and plasma atoms into polymers in dimensions of
some micrometers have been measured. The equalizing effect
of each type of plasma treatment at maximum adhesion in
composites was interpreted.

1. INTRODUCTION

All processes of the plasma pretreatment of polymers and during
the formation of composites will be reviewed in this article.

Great caution is advised if the formation of plasma-generated
functional groups on polymer surfaces shall be explained by simple chemical equations. A tricky problem is the preferred destruction of aromatic rings in the plasma or on plasma-exposed
polymer surfaces. Stille (1965) explained this phenomenon by
the formation of the hexatriene diradical as intermediate followed by polymerization reactions. Plasma particle collisions
with surface atoms and the more important plasma-vacuum uv interactions lead to simple photochemical reactions. This was observed especially in the case of oxygen plasma ageing of polymers (Friedrich 1988).

A further intention of this article was to summarize the actual knowledge on the penetration ability of plasma particles into the polymer bulk during surface bombardment as well as on the depth-profiling of uv artifacts in the polymer volume.

To follow the way from the explanation of chemical processes on polymer surfaces to the interactions in the polymer composite it was necessary to investigate the rate of surface processes, their influence on adhesion, and the role of secondary reactions with oxygen from air.

The next important point was to give an answer about the type of interactions dominating in composites with plasma-pretreated polymers: chemical or physical, specific or non-specific.

Furthermore, plasma and non-plasma pretreatments in vacuum or under normal pressure are compared.

The last point which has been reviewed refers to dynamic processes along the interface. There are some indications that chemical processes along the interface do not end in static chemical bonds, complexes or other weak interactions. From the point of redox chemistry, considering the highly active nature of plasma-exposed surfaces and the heat necessary to form composites, some usual metals in composites should be able to react with the polymer (Friedrich 1985,1991). For example, copper oxide on copper foils is a strong oxidizing reagent and evaporating aluminium, magnesium or sodium are strong reducing agents being able to destroy the polymer structure near the interface thus forming a Weak Boundary Layer.

2. EXPERIMENTAL

The plasma conditions used for polymer surface modifications consisted in the application of a d.c. glow discharge in a glass bell jar at a pressure of 6 Pa, very low power input and low flow rates of plasma gases. The samples were fixed on a glass or steel cylinder rotating under floating potential through the positive column of discharge. The rotating sample holder could be heated by an ir radiator.

3. PROCESSES
2.1 Surface reactions

On surfaces of poly(ethyleneterephthalate) (PET) some oxygen-plasma generated groups could be identified by ESCA, SSIMS, acid-group titration, bromination of olefinic double bonds etc. The formation of the detected groups can be described by the following strongly simplified reactions:

- destruction of aromatic ring structures leading to polymers, polyenes, crosslinked networks and trapped radicals,
- formation of aliphatic hydroperoxides and their further reaction after the autoxidation mechanism started by the addition of molecular oxygen to radical sites,
- generation of highly oxidized C atoms (carbonates, peroxy acids)

$$\text{ester scission} \longrightarrow \text{aryl-}\overset{\nearrow O}{\underset{\cdot}{C}} \xrightarrow{\text{oxygen}} \text{aryl-}\overset{\nearrow O}{\underset{OO\cdot}{C}} \longrightarrow \text{aryl-}\overset{\nearrow O}{\underset{OOH(R)}{C}}$$

- destruction of the ester group and formation of the following intermediates: aryl·, aryl-CO·, aryl-COO·,
- stabilization of intermediates to:

$$\text{-aryl-COO·} + \text{·CH}_2\text{-CH}_2\text{-} \longrightarrow \text{-aryl-COOH} + \text{CH}_2\text{=CH}_2\text{-} ,$$

- the same as before, but after Norrish II-type reaction,
- polymerization of generated vinyl groups and separation of terephthalic units,
- hydrolytic ester scission,
- attack of hydroxy and hydroperoxy radicals to aromatic rings and formation of phenolic sites etc.

As seen at polypropylene (PP, Figure 1) the C1s deconvolution of the ESCA spectrum after oxygen plasma treatment is more of schematic nature than of clearly resolved shoulders. Therefore, little help can be given by derivatization of some plasma-generated groups.

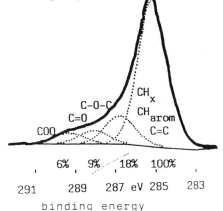

Fig. 1. Deconvolution of the C1s peak of PP after 180 s exposure to oxygen plasma

The interaction in a composite with an NCO-terminated polyurethane (PUR) can be simulated by the reaction of surface hydroxy groups with the isocyanate group of phenyl isocyanate. 1.5 % of all surface carbon atoms are involved in this reaction. Olefinic double bonds were derivatized by bromine (5 %), energetic radicals by vinyl triethoxysilane (1.5 %).

3.2 Penetration process

Two polymers with strong differences in plasma-uv absorption were examined. The PUR elastomer exhibits a browning at plasma exposure up to depths of some millimeters from the surface, as also monitored by OsO_4 derivatization of polyenes along a cross-section and by the use of a small-spot microprobe (ESMA). Poly

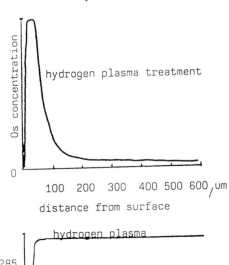

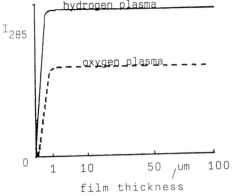

(vinylchloride) (PVC) without any additives forms a strongly absorbing layer of polyenes of 1 um thickness (Figure 2 b). This was measured by uv spectroscopy and ESCA after derivatization with bromine.

Depth profiling of PET after CF_4, CCl_4, Br_2 and I_2 plasma exposure shows surprising results. Fluorine was detected in more than 3 um depth, chlorine in traces up to 1.5 um. Bromine and iodine did not penetrate into the polymer. This was proved by AES of sample cross-sections, SIMS and the unrolling of a 12-sheet bundle of 1.5 um thick PET films welded or taped along the margins and measuring the F1s peak of each film surface. The penetration mechanism assumed is the diffusion of the neutral CF_4 molecules into the

Fig. 2. Depth profiling of Os-derivatized olefinic double bonds in PUR (a) and absorption of PVC films in dependence on film thickness at 285 nm (b)

polymer volume and their excitation (and reaction) by penetra-
ting plasma-uv radiation.

3.3 Rate of adhesion-significant processes

The plasma-surface modification is a very fast process as shown
by the maximum in adhesion after 1 s oxygen-plasma pretreatment
of PET and PP. It becomes also apparent by the rapid increase in
concentration of O-containing functional groups on the surface
of PP (Figure 3) and by contact angle measurements.

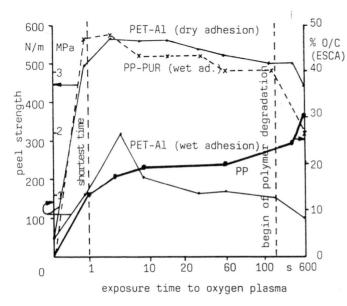

Fig. 3. Depen-
dence of adhesion
and O-concentra-
tion on the plas-
ma exposure time

exposure time to oxygen plasma

2.4 In-situ metallization

Plasma-surface activation in oxygen plasma was performed at 6 Pa
and, after lowering the pressure for 5 min to 10^{-3} Pa, the PET
film was evaporated with aluminium (100 nm). The same procedure
was done with an intermediate storage of the plasma-activated
samples in air for 12 weeks before metallization. Within the limits of the peel test no deviation from the maximum value was
detected (Figure 4). The same applied if PUR-PP composites were
used and the tensile shear strength was measured.

2.5 Interactions at the interface

At the interface between oxygen-plasma precleaned steel and phenolic resin a shift in the binding energy (1.1 eV in the $Fe2p_{3/2}$
peak) of the FeO component (Fe_2O_3) was detected indicating

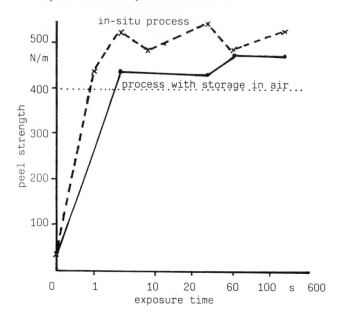

Fig. 4. Comparison in adhesion of in-situ metallized and air-stored PET

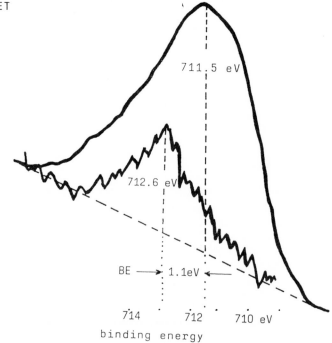

Fig. 5 Fe2p$_{3/2}$ peak of oxygen-plasma treated steel coated with a very thin layer of phenolic resin

chemical interactions (Fe··O-C type, Figure 5). For the PET-Al composite the ESCA spectra of oxygen-plasma pretreated PET were compared with the measured peel strength of the composites. At the maximum in adhesion both the C1s and the O1s peak show only one specific peak (hatched area) which is different from other peaks at lower adhesion level. This new subpeak can be ascribed to ketones or phenols (Figure 6).

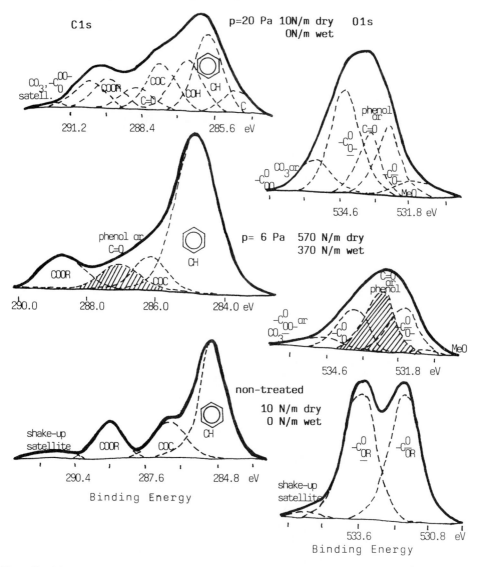

Fig. 6. C1s and O1s peaks of PET for differently adherent PET-Al composites

The vacuum deposition of Al was performed on gold substrate and nearby on oxygen-plasma pretreated PET in the same experiment. Through the thin Al layer of ca. 3 nm thickness the Au4f lines could be detected. This means that the PET-Al interface can also be measured in this way. The deconvolution strategy of both Al2p peaks in Figure 7 was the same. The FWHM's of Al^o and weakly interacting Al (Al$\cdot\cdot$O-C) were fixed to the binding energy (BE) = 0.9 eV and those of Al bound to oxygen to BE = 1.5 eV. In the case of Al on PET the Al^o subpeak at BE = 71.1 eV could not fill the broader shoulder at the low-energy range of the Al2p peak. The remarkable broadening was interpreted by the existence of a second subpeak at BE = 71.9 eV. This subpeak should be attributed to weakly bound Al^o which is interacting with the carbonyl groups of plasma-pretreated PET. As measured by peel strengths of the Al layer on differently pretreated PET no correlation between the intensity of the peak at BE = 71.9 eV and the adhesion was found.

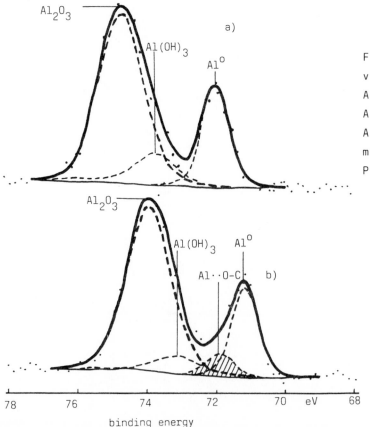

Fig. 7. Deconvolution of the Al2p peak: a) Al on gold, b) Al on O_2-plasma pretreated PET

3.6 Adhesion results

For Al and Cu films on plasma-activated PET no peeling is possible (peel limit near 400 N/m) using different types of plasma
modifications: O_2, H_2, N_2, NH_3, O_2/Br_2, Br_2, CF_4/O_2, $HCBr_3$, Ar,
$POCl_3$, PCl_3, $SOCl_2$, SO_2, vinylsilane, aminosilane etc. Furthermore, using plasma or non-plasma pretreatments, in vacuum or
in air under normal pressure high peeling forces of Al could
be measured (Figure 8). Remarkable differences were only observed in wet adhesion. Here, another type of interactions
seems to be dominant in connection with a strong influence of
substrate temperature.

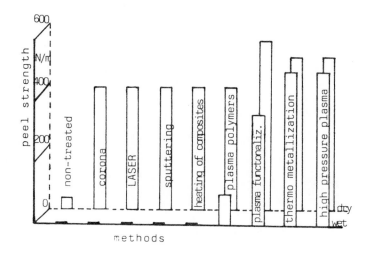

Fig. 8. Comparison in Al adhesion of different PET pretreatments (peel strength)

The methods used in Figure 8 can be divided into normal-pressure pretreatments: corona discharge in air, excimer LASER at
193 nm (Ar-F), heating of the composite by hot air ("heat gun"),
and use of a plasma jet (Ar/O_2 or N_2/O_2, about 10 000°C) as
well as into low-pressure methods: sputter technique, surface
modification by gases/vapours in the glow discharge ("plasma
functionalization") or by deposition of plasma polymer layers
acting as adhesion promoters and vacuum metallization on heated

$(160^{\circ}C)$ and plasma-activated PET.

The same result was obtained in the case of adhesion between PP and PUR. This polymer-polymer composite was manufactured from plasma-pretreated PP and PUR adhesive, cured and then aged under hot and wet conditions. Two types of PP (Stamylan and Procom) were surface-modified with normal-pressure plasmas (corona discharge on heated PP-"thermo corona", plasma jet) and different types of glow-discharge plasmas at low pressure (Figure 9).

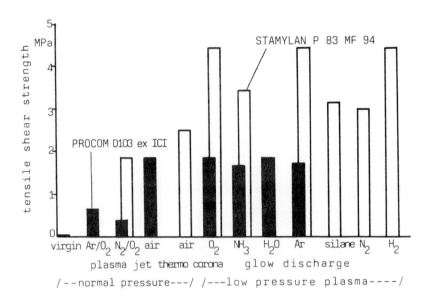

Fig. 9. Comparison of PUR adhesion of different PP pretreatments (tensile shear strength)

Within the different types of glow discharge gases and vapours no significant differences in tensile shear strength were observed. In comparison to normal-pressure plasmas a moderate increase in tensile shear strengths was detected.

3.7 Interface reactions

Different standard redox potentials between polymer and metal produce interfacial redox reactions (Friedrich 1985, 1991) initiated by additional heat required for manufacturing the composite and activating the surface by plasma.

In the process of manufacturing printed circuits from copper and polymer boards or in the cable industry redox reactions on the Cu(CuO)-polymer interface were detected much earlier.

Furthermore, Al attached on oxygen moieties on PET surface seems to be able to pick up weakly bound oxygen from the polymer under formation of Al_2O_3. This reaction is based on the high negative redox potential of aluminium:

$$Al^{+++} \quad + \quad 3 \; e^- \quad \longrightarrow \quad Al^0 \qquad E^0 = -1.51 \; V$$

Redox reactions should lead to a new and optimum arrangement of functional groups along the interface resulting in a maximum of interactions and adhesion (Friedrich 1986). The redox reaction cannot be stopped over longer periods, therefore oxidation or reduction processes penetrate into the polymer bulk and produce a Weak Boundary Layer resulting in reduced composite strengths (Friedrich 1985).

3. CONCLUSIONS

A complicated system of chemical reactions on PET surfaces was observed upon pretreatment in oxygen plasma. The majority of these processes are photochemical reactions induced by plasma radiation.

Adhesion-significant processes are completed within some seconds.
At longer exposure time to plasma a deep penetration of radiation defects and plasma particle attachment into the polymer bulk have been identified.
Chemical interactions exist at the interface but they do not dominate the adhesion. The chemical processes do not cease after the formation of bonds, however, redox reactions are likely to occur.

5. REFERENCES

Stille J K, Sung R L and van der Kooi J 1965 J. Org. Chem. 30 3116

Friedrich J F and Frommelt H 1988 Acta Chim. Hung. 125 165

Friedrich J F and Toan Le Q 1991 Adhäsion submitted

Friedrich J F et al. 1985 Acta Polymerica 36 310

Friedrich J F, Loeschcke I and Gähde J 1986 Acta Polymerica 37 687

Preparation and characterization of plasma polymerized aniline

F. FALLY, J. RIGA and J.J. VERBIST

Laboratoire Interdisciplinaire de Spectroscopie Electronique
Facultés Universitaires Notre-Dame de la Paix
Rue de Bruxelles 61, B-5000 NAMUR (Belgium)

ABSTRACT : Plasma polymers deposited from aniline in an inductively-coupled RF reactor are analyzed by XPS, IR and ^{1}H-NMR. The increase of the plasma energy leads to a greater fragmentation of the monomer. The intensity of shake-up satellites accompanying the XPS C1s primary peak and of one of the IR bands associated to aromatic rings decrease at high power, while the proportion of aliphatic and nitrile groups increases.

1. INTRODUCTION

Plasma polymerization is a technique frequently used to process solid surfaces by depositing a polymeric thin solid film. Depending on the nature of the starting monomer and on the plasma conditions, Karakelle (1989), Montalan (1989), Inagaki (1990) and Krishnamurthy (1989) synthesized polymers having different physical (adhesion, permeability, dielectric) and chemical (hydrophobic or hydrophilic) properties. However, the lack of theory to predict the disordered structure of the polymeric material obtained remains a hindrance for this technique. Recently, Yoshimura (1989) and Chilkoti (1991) characterized plasma polymers by X-ray Photoelectron Spectroscopy, Infrared and Static SIMS techniques and tried to understand the fragmentation mechanisms which need to be clarified in order to predict the polymer structure.

In this study, plasma polymerized aniline films are prepared, and their physico-chemical characteristics are determined by classical spectroscopies (NMR and IR), and also by X-ray Photoelectron Spectroscopy (XPS), a surface analysis technique. In the same way, these results are compared with those obtained on electrochemically polymerized aniline in order to characterize the disordered structures.

2. EXPERIMENT

2.1. Film preparation

Polymers are plasma deposited in an inductively coupled vacuum system (Fig. 1). After a stable argon flow rate (300 cm^3 (STP)/min) is established, the monomer placed in a reservoir kept at constant temperature (330 K) is introduced (15 cm^3 (STP)/min) by opening a needle valve and polymerized by a RF discharge at 13.56 MHz. The total pressure in the chamber during the polymerization is continuously monitored at 0.300 Torr, and if necessary automatically adjusted by varying the argon flow rate.

The effect of beam power on the polymerization process is checked by varying the plasma power level from 5 to 20 W. Substrates (glass plates for XPS and NaCl plates for IR measurements) are placed in the reactor.

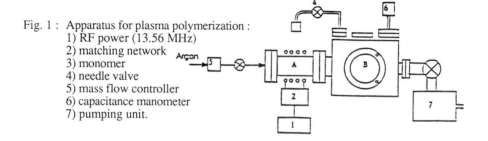

Fig. 1 :   Apparatus for plasma polymerization :
      1) RF power (13.56 MHz)
      2) matching network
      3) monomer
      4) needle valve
      5) mass flow controller
      6) capacitance manometer
      7) pumping unit.

2.2. Characterization

XPS spectra are recorded on a Hewlett-Packard 5950A spectrometer using the monochromatized AlK$_\alpha$ radiation (hv=1486.6 eV). Polymer samples are introduced as prepared in a vacuum chamber (10^{-8} Torr) to be analysed.

A Perkin Elmer 983G Infrared spectrometer is used to obtain transmission spectra of the polymer films immediately after deposition on NaCl plates.

A Fourier transform spectrometer (Brücker MSL-400) is used to perform NMR measurements. The ^{1}H-NMR spectra are obtained from samples which are soluble in deutereted acetone (plasma power = 5W). Tetramethylsilane is used as a reference (δ=0 ppm).

3. RESULTS AND DISCUSSION

3.1. X-ray Photoelectron Spectroscopy

XPS results obtained from films deposited at 10 and 5 W reveal the same atomic composition as for aniline (C/N=6.0). However, under higher power conditions (15 and 20 W), the C/N atomic ratios are lower than the stoichiometric one (C/N=5.0 and 4.5 respectively). XPS analysis also shows a low oxygen contamination. These results are not in agreement with those of Hernandez (1983) who observed a lower nitrogen content in plasma polymerized aniline films.

The evolution of the C1s and N1s spectra at various powers reveals little difference (Fig. 2). However, satellites located at 6.7 eV from the main peak are detected on the C1s spectra from films obtained at low power. These shake-up satellites associated to the π-π* transitions accompanying photoionization of unsaturated and aromatic carbon sites are absent for the polymers deposited at 15 and 20 W. This suggests that a more important fragmentation of the monomer occurs at higher plasma energy.

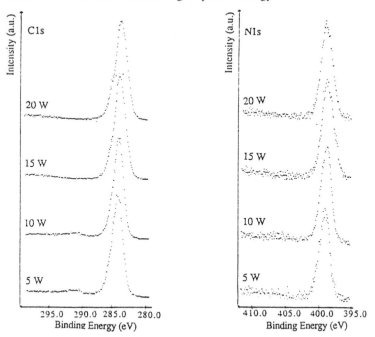

Fig. 2 : XPS C1s and N1s core-levels spectra of films obtained at different plasma powers

3.2. Infrared spectroscopy

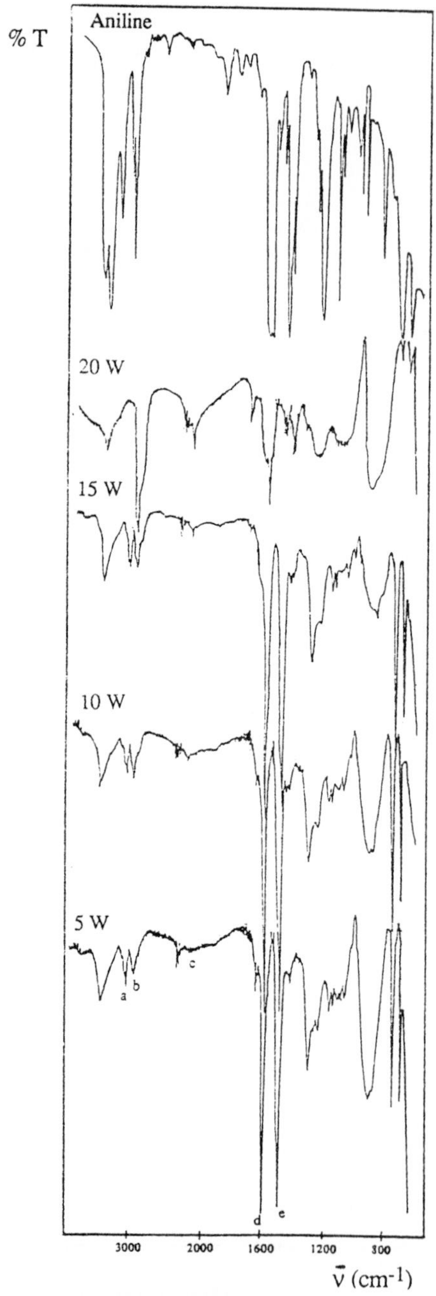

Infrared spectra of aniline and polyaniline deposited at different plasma powers are presented in figure 3. Compared to the infrared spectrum of pure monomer, differences mainly occur as broadening of bands and the appearance of some new ones : their frequences and normal vibrational mode attributions are summarized in Table 1.

The primary amine nature of the monomer is shown by two N-H stretching absorption peaks in the 3500-3300 cm^{-1} region. There is only one absorption band in this region for the plasma polymer, showing that the resultant polymer is either a secondary amine or an imine. The polymer is characterized by new bands at 2930 and 2880 cm^{-1} due to C-H stretching of aliphatic compounds. Absorption bands appear also at 2200 cm^{-1} (C≡N stretching) and at 1680 cm^{-1} (C=O, C=C stretching).

Fig. 3 :   IR spectra of aniline and of plasma polymers.

Table 1 : IR peaks (cm^{-1}) assignments for aniline and plasma polymers.

Aniline	Plasma polymer
3440⎤ N-H stretching 3360⎦ (primary amine)	3380 N-H stretching (imine, secondary amine)
3100⎤ C-H stretching 3080⎥ 3040⎥ 3020⎦	3060⎤ C-H stretching (aromatic, alkene) 3020⎦ 2930⎤ C-H stretching (aliphatic) 2880⎦ 2200 C≡N stretching
1930⎤ C-H out of plane deformation 1840⎥ (overtone and combination 1780⎥ bands) 1710⎦	1680 C=0 stretching, C=C stretching

1630⎤ Ring carbon-carbon
1610⎥ stretching
1560⎥
1530⎥
1500⎦
	1620 C=N stretching
	1600 Ring carbon-carbon stretching
	       C=N stretching
	1520 Ring carbon-carbon stretching
	1500 C=C, C=N stretching
	       C-H bending

1320 C-N stretching	1320 C-N stretching
1280⎤ C-H in plane deformation 1180⎥ 1160⎥ 1120⎥ 1060⎥ 1030⎥ 1000⎦	1260⎤ C-H in plane deformation 1180⎥ (aromatic, alkene) 1160⎥ 1080⎥ 1030⎥ 1000⎦
870⎤ C-H out-of-plane 760⎥ deformation 700⎦	880⎤ C-H out of plane deformation 760⎥ (aromatic, alkene) 700⎦

Figure 4A shows that the intensity ratios of aliphatic C-H and C≡N stretching absorptions (2930 and 2200 cm^{-1} respectively) with respect to the N-H stretching absorption (3380 cm^{-1}) increase at high power. According to these results, it is reasonable to suggest that at high energy, plasma polymerization of aniline involves a complex mechanism resulting from ring opening and crosslinking reactions, leading to a highly disordered structure.

At relatively low energy there is an increase of the relative absorption intensity in relation to the N-H absorption of some bands associated with normal modes of vibration of the aromatic moeity (Fig. 4B). For instance, two intense bands in the 1600-1400 cm^{-1} region are found on the IR spectra of polyaniline obtained at 5, 10 and 15 W. These bands are also observed for electrochemically polymerized aniline (emeraldine). For example, Sariciftci (1990) attribute the bands to benzenoid (1500 cm^{-1}) and quinoid (1600 cm^{-1}) phenyl rings. The presence of the C=N vibration mode (1600 cm^{-1}) indicates that some changes from benzenoid to quinoid forms occurs during the polymerization process.

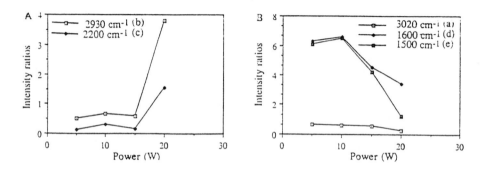

Fig. 4 :   Intensity ratios of some IR absorptions with respect to the N-H absorption (3380 cm^{-1}) at different plasma powers.

So, it is clearly shown that, in order to maintain the configuration feature of the monomer unit (aniline ring) in the plasma structure, polymerization should be performed at low power. However, even at 5 W, ring opening reactions already occur.

3.3. ^{1}H-Nuclear Magnetic Resonance

Three main regions, called A, B and C, are distinguished in the spectrum of the polymer deposited at 5 W (Fig. 5). They respectively cover a frequency range from 0 to 4 ppm ; from 4 to 6 ppm and from 6 to 10 ppm. Region A is mainly arising from ^{1}H resonance signals due to aliphatic segments but due to the complexity of the spectrum, it is difficult to give a precise interpretation of these structures.

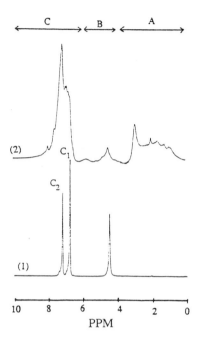

Region C is characteristic of protons resonance in olefinic and aromatic functions. By reference to the ^{1}H-NMR spectrum of pure monomer, this region in the polymer spectrum obviously contains peaks arising from ^{1}H bonded to aniline ring : the peaks C_1 at 6.7 ppm and C_2 at 7.1 ppm respectively correspond to ^{1}H in meta, para and ortho positions in the aniline ring (Fig. 5). In addition, peaks located in B region, at 4.4 ppm are attributed to hydrogen atoms bonded to nitrogen heteratom in aniline.

Fig. 5 : ^{1}H-NMR spectra of aniline (1) and of plasma polymerized aniline (5W) (2) dissolved in CD_3COCD_3.

4. CONCLUSION

XPS, IR and NMR analyses show that the plasma polymerization of aniline involves a quite complex mechanism resulting from ring opening and crosslinking reactions which are more important at high energy. At low energy, the intense satellites detected on the C1s core-levels spectra of the polymers, the presence of some bands associated with normal modes of vibration of the aromatic moeity in the IR spectra as well as the presence of ^{1}H bonded to aniline in the NMR spectrum show that the fragmentation of aniline is not the only process leading to the formation of the polymer. However, even at 5 W, ring opening reactions already occur.

5. REFERENCES

A. Chilkoti, B. D. Ratner and D. Briggs, Chem. Mater., <u>3</u>, 51 (1991).

R. Hernandez, A.F. Diaz, R. Waltman and J. Bargon, J. Phys. Chem. , <u>88</u>, 3333 (1984).

N. Inagaki, S. Tasaka, and Y. Takami, J. Appl. Polym. Sci., <u>41</u>, 965 (1990).

M. Karakelle and R.J. Zdrahala, J. Membr. Sci., <u>41</u>; 305 (1989).

V. Krishnamurthy, I. L. Kamel, J. Polym. Sci. : Part A : Polym. Chem. Ed., <u>27</u>, 1211 (1989).

D. Montalan, N. Souag, Y. Segui and C. Laurent, Thin Solid Films, <u>170</u>, 235 (1989).

N.S. Sariciftci and H. Kuzmany, J. Chem. Phys., <u>92</u>, 4530 (1990).

K. Yoshimura, T. Kitade, K. Kitamura and K. Hozumi, J. Appl. Polym. Sci., <u>38</u>, 1011 (1989).

Plasma treated polymer films: Relationship between surface composition and surface hydrophilicity

J.L. Dewez, E. Humbeek, E. Everaert, A. Doren, P.G. Rouxhet

Université Catholique de Louvain, Unité de Chimie des Interfaces,
Place Croix du Sud 2/18, 1348 Louvain-la-Neuve, Belgium

ABSTRACT : The surface composition of bis-phenol A polycarbonate (PC), poly(ethyleneterephthalate) (PET) and polypropylene (PP) treated by O_2, Ar and NH_3 RF plasma discharge, and of polypropylene treated by allyl alcohol plasma deposition has been determined by X-ray photoelectron spectroscopy and their hydrophilicity has been characterized by water contact angle measurement. The oxygen carrying a high electron density and responsible for O_{1s} peaks at or below 533.4 eV binding energy is the main factor determining the surface hydrophilicity and the part of the surface energy due to non-dispersion interactions. The water contact angle of O_2 plasma treated PC and PET reincreases during storage in air and, more slowly, during storage in a dry atmosphere. This is not related to a change of oxygen concentration as detected by XPS, and is attributed to a reorganization within the first atomic layers at the surface.

1. INTRODUCTION

The plasma discharge treatment of polymer films has been extensively studied, particularly with the aim of improving surface properties such as wettability or surface energy and adhesion (Briggs et al., 1980; De Puydt et al., 1989; Pochan et al., 1986).

It has been stressed that, after certain treatments, polymer surfaces reorient and restructure in response to their local microenvironment in order to minimize the interfacial free energy (Occhiello et al., 1991; Munro and Mc Briar, 1988; Garbassi et al., 1989). These processes, which are temperature and time dependent, may have a major influence on the properties of the interface between a polymer and a biological system (Ruckenstein and Gourisankar, 1986; Andrade et al., 1985).

The surface energy of untreated polymers and of modified bis-phenol A polycarbonate has recently been examined in relation with the surface chemical composition determined by

XPS (Dewez et al., 1991). A good correlation was obtained between the surface energy term due to non-dispersion forces, often called the surface energy polar term γ_s^p, and the surface concentration of oxygen responsible for O_{1s} peaks at or below 533.4 eV and thus carrying a high electron density. The correlation with the total concentration of oxygen was much more scattered.

This work presents further support for the dependence of surface hydrophilicity on the fraction of oxygen carrying a high electron density, as determined by XPS. Therefore several polymer films were submitted to various RF plasma treatments. The change of surface properties as a function of time after the plasma treatment was also examined in this context. This was undertaken in the frame of a research aiming to understand better how the attachment of epithelial cells can be controlled by modifying the surface properties of the substrata.

2. EXPERIMENTAL

2.1. Materials

The polymer films investigated are listed in table 1. Their surface was modified by conventional low pressure RF plasma discharges in oxygen, argon or ammonia. In addition, plasma deposition of allyl alcool was performed on PP films.

O_2 and NH_3 plasma treatments were performed with a Chemprep 130 Barrel Reactor from Chemex (capacitive coupling, 13.56 MHz). Argon plasma treatment was performed in a reactor designed and assembled at LISE, Facultés Universitaires Notre-Dame de la Paix, Namur, using a ENI POWER generator (inductive coupling, 13.56 MHz). Prior to gas admission, reactor chambers were evacuated to 8.10^{-2} mbar in the Chemex reactor and to 10^{-3} mbar in the LISE reactor.

Plasma deposition of allyl alcool (Aldrich, analytical grade) was performed in the LISE reactor. The latter was evacuated at about 10^{-3} mbar, before increasing the pressure to a stabilized value of about 0.09 mbar with a argon-allyl alcool co-vapor flow; the allyl alcool flow rate was about 30 cm^3 STP minute^{-1}. The RF generator was then turned on. The gas mixture was continuously passed over the samples for 15 to 20 minutes after the plasma reaction was terminated. For plasma deposition of allyl alcool the substrate was placed either in the quartz discharge chamber itself or in a post-discharge position. In the latter case, both a perpendicular and a parallel position with respect to the direction of the gas stream were used.

Table 1 presents the operating parameters of the various plasma treatments. Each set of parameters (gas, power, duration, and position in the case of allyl alcool deposition) defines

Table 1. Polymers investigated and main operating parameters of plasma treatments.

Polymers	Operating parameters			
Name - abbreviation	gas	pressure (mbar)	power (watt)	time (minute)
Bis-phenol A polycarbonate - PC [a]	oxygen	0.6 - 0.8	55 - 60	0.5 [d]
	argon	0.09	50	5, 10
Poly(ethylene terephthalate) - PET [b]	oxygen	0.6 - 0.8	55 - 60	0.5 [d]
	ammonia	0.6 - 0.8	55 - 60	0.5 [d]
	argon	0.09	20, 50	2, 5, 10
Polypropylene	oxygen	0.6 - 0.8	55 - 60	0.5 [d]
	argon	0.09	20, 50	2, 5, 10
- PP [c]	argon	0.09	20, 50	1, 2.5, 5
	+ allyl alcool			

a   PC : Makrofol KG (crystalline; monoaxially oriented) - Bayer.

b   PET : Mylar Type A (crystalline; biaxially oriented) - Du Pont de Nemours.

c   PP : 24 MB 200 (amorphous) - Mobil Plastics Europe.

d   The variation of the surface properties after the plasma treatment was investigated.

one sample, which was submitted to surface analysis and water contact angle measurement.

Unless specified otherwise, the specimens were placed in disposable Petri dishes and examined within a few days. The polymer films modified by O_2 or NH_3 plasma were stored at room temperature either in air or over P_2O_5 in a dessicator and a specimen was withdrawn for characterization at different time intervals up to 37 days.

2.2. XPS analysis

The surface chemical composition of the samples was analyzed by X-ray photoelectron spectrocopy (XPS) using an SSX 100 spectrometer (model 206, Surface Science Laboratories, Mountain View, California) equipped with an aluminium anode (10 kV, 11.5 mA) and a quartz monochromator. The direction of photoelectron collection made angles of 55° and 73° with the normal to the sample surface and the incident X-ray beam, respectively. The electron flood gun was set at 6 eV. A survey scan was obtained with a 1000 µm spot and a pass energy of 150 eV. A detailed scan of the C_{1s}, N_{1s} and O_{1s} lines was obtained with a 600 μm spot and a pass energy of 50 eV.

The binding energy (E_b) of the main lines (C_{1s}, N_{1s}, O_{1s}) was determined by setting the value of 284.8 eV for the C_{1s} component due to carbon involved only in C-C and C-H bonds.

The experimental peak areas were integrated by non linear background subtraction. The peaks were decomposed by using a non-linear least squares routine and assuming a Gaussian / Lorentzian (85/15) function. Intensity ratios were converted into atomic concentration ratios by using the SSI software package which uses the Scoffield cross-sections, considers a variation of the electron mean free path according to the 0.7 power of the electron kinetic energy and assumes that the transmission function is constant.

2.3. Contact angle measurement

Water contact angles, θ, were measured at room temperature with an image analyzing system, using the sessile drop technique. The water droplet volume was always kept in the range of 1-3 μl to prevent gravitational distortion of the spherical profile. The water used for measurements was HPLC grade and produced by a Milli Q plus purification system (Millipore). It was freshly taken just a few minutes before manipulations. Each determination was obtained by averaging the results of at least ten drops. A typical standard deviation was around 3°.

3. RESULTS

Table 2 presents the surface elemental composition and the water contact angle of the untreated polymer films and of representative specimens of plasma treated films. The analysis of the C_{1s} peak components and a detailed discussion of the chemical functions present on modified surfaces will not be presented here. The components obtained by decomposition of the O_{1s} peak have a full width at half maximum of 1.6 ± 0.2 eV. As indicated in Table 2, components at 531.2 and 531.6 eV may be attributed to oxygen doubly bound to carbon in ketone, ester, carbonate or amide and the components at 533.5 and 534.2 eV may be attributed to C-O-C in ester and carbonate; the component at 532.5 eV may be due to C=O of carbonate or to OH (Dewez et al., 1991).

The O_2 plasma treatment of PC provokes a substantial increase of the oxygen content. The Ar plasma treatment does not give any strong variation of the oxygen concentration, but

Table 2. Surface properties of untreated and plasma treated films of bis-phenol A polycarbonate (PC), poly(ethylene terephthalate) (PET) and polypropylene (PP) : water contact angle and surface chemical composition given by the mole fractions (omitting hydrogen) associated with the C_{1s} and N_{1s} peaks, and with the various components of the O_{1s} peak indicated by their binding energy (eV).

	C	N	total	O=C		OH, O=C	C-O-C		θ
				531.2	531.6	532.5	533.5	534.2	
PC-untreated	83.4	< 0.1	16.5			5.7		10.8	78
-oxygen [a]	71.4	< 0.1	28.6			14.3		14.3	48
-argon [b]	82.4	1.8	14.4		3.0	8.6	2.7		53
PET-untreated	69.6	< 0.1	29.2		14.5		14.7		66
-oxygen [a]	67.1	< 0.1	29.2		16.0		13.2		33
-ammonia [c]	67.9	6.0	25.7	4.3	8.9		12.5		34
- argon [b]	77.0	2.0	20.0		6.6	8.9	4.5		52
PP-untreated	100.0	< 0.1	< 0.1						95
-oxygen [c]	87.5	< 0.1	11.5			6.0	5.5		76
-argon [b]	89.2	2.1	8.7		3.5	3.6	1.5		77
-allyl alcohol [d]	80.3	1.7	18.0			18.0			10

a, stored 1 day over P_2O_5.

b, plasma treated for 5 minutes, under 50 W.

c, stored 1 day in air.

d, sample in discharge chamber, 1 minute, 20 W.

modifies profoundly the pattern of functional groups and leads to a destruction of carbonate functions.

The O_2 plasma treatment of PET modifies only slightly the oxygen surface concentration and the distribution of functional groups. The NH_3 plasma treatment leads to the formation of C-N bonds, including amide functions. The Ar plasma treatment provokes a loss of oxygen and a change of the oxygenated functions distribution.

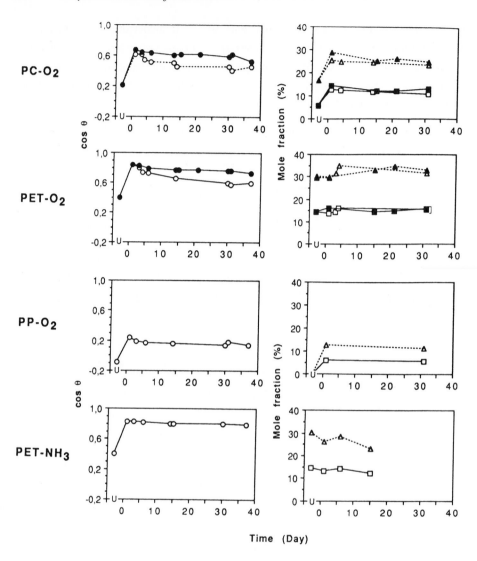

Figure 1. Plot of the cosine of the water contact angle θ (left, ○, ●) and of the mole fraction of oxygen, calculated omitting hydrogen, (right; total oxygen, Δ, ▲; oxygen responsible for O_{1s} peaks with a binding energy lower than 533.4 eV, □, ■) vs the duration of storage in air (open symbols) and over P_2O_5 (dark symbols) of O_2 plasma treated PC, PET and PP and NH_3 treated PET. U represents the untreated polymer.

Both Ar and O_2 plasma treatments of PP give rise to an incorporation of oxygen. The oxygen content of the allyl alcohol deposit varies according to the sample position in the plasma reactor and to the operating conditions; comparison between the C_{1s} and O_{1s} peak indicate that most of the oxygen is under the form of OH.

Figure 1 presents the evolution of the cosine of the water contact angle as a function of storage time for O_2 plasma treated PC, PET and PP and for NH_3 plasma treated PET. It shows also the variation of the mole fraction of total oxygen and of the mole fraction associated with the components of the O_{1s} peak at 531.2, 531.6 and 532.5 eV, i.e. the mole fraction of oxygen atoms carrying a high electron density. The mole fractions were calcultated, omitting hydrogen. The characteristics of the untreated samples are also given for comparison.

The increase of cos θ provoked by the O_2 plasma treatment of PC and PET is attenuated during storage, at a rate which is higher in normal air than over P_2O_5. This indicates that humidity accelerates surface reorganization after the modification created by the plasma. The change in surface hydrophilicity is not accompanied by a modification of the surface concentration of the oxygenated functions, over the thickness explored by XPS.

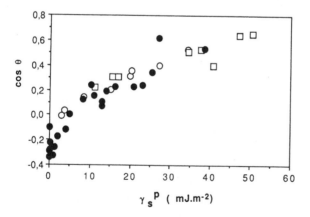

Figure 2. Plot of the cosine of the water contact angle, θ, as a function of the part of surface energy, γ_s^p, due to non-dispersion interactions, deduced from the contact angle of water-propanol mixtures and α-bromonaphtalene, using the geometric mean equation and accounting for spreading pressure. Untreated polymers (○, Dewez et al. 1991; ●, Busscher et al., 1983, Schakenraad, 1987); surface modified bis-phenol A polycarbonate (□, Dewez et al., 1991).

The hydrophilicity and the apparent surface concentration are not modified appreciably during storage of NH_3 plasma treated PET and O_2 plasma treated PP in air.

4. DISCUSSION

In recent papers (Dewez et al., 1991), the part of the polymer surface energy due to non-dispersion forces, often quoted as the polar contribution γ_s^P, has been shown to be correlated with the surface concentration of oxygen responsible for O_{1s} peaks at or below 533.4 eV, i.e. of oxygen carrying a high electron density. The obtained correlation is indeed better than the correlation with the total oxygen concentration.

Figure 2 presents the plot of cos θ vs γ_s^P obtained with model polymers and surface modified PC membranes (Dewez et al., 1991; Busscher et al., 1983; Schakenraad, 1987); γ_s^P was deduced from the contact angles of water-propanol mixtures and α-bromonaphtalene, using the geometric mean equation and accounting for spreading pressure (Busscher et al., 1983). A clear correlation would also be found using θ instead of cos θ, however the latter was prefered because it multiplies directly surface energy terms in relationships used for computing the work of adhesion and for splitting the surface energy into different contributions. The cosine of the water contact angle is much more readily determined than γ_s^P; moreover it is a strictly experimental parameter and its determination is less subject to questionable hypotheses. Therefore cos θ has been used instead of γ_s^P in the further study of the correlation between the surface properties and the oxygen surface concentration.

Figure 3 shows the plots of cos θ vs the mole fraction (omitting hydrogen) of total oxygen and of oxygen responsible for an O_{1s} peak at a binding energy smaller than or equal to 533.4 eV. Graphs a and b present the data for samples analyzed without prolonged storage. Data published (Dewez et al., 1991) for model polymers (polystyrene, poly(aryl-ether-ether ketone), bis-phenol A polycarbonate, polyetherimide, polyamide, poly(ethylene terephthalate)) and surface treated PC membranes (acid base etching, sulfatation, Corona discharge, nitration, NH_3 and O_2 plasma discharge) are also included in the graphs. Consideration of total oxygen gives a strongly scattered plot (correlation coefficient $r^2 = 0.327$) while consideration of the oxygen carrying a high electron density provides a much better correlation ($r^2 = 0.730$); the regression line passes close to 0 % oxygen and cos θ = 0, the latter corresponding to a surface energy due only to dispersion interactions. This confirms that XPS can discriminate between oxygenated functions in terms of their influence on the hydrophilicity and that oxygen carrying a high electron density plays a major role.

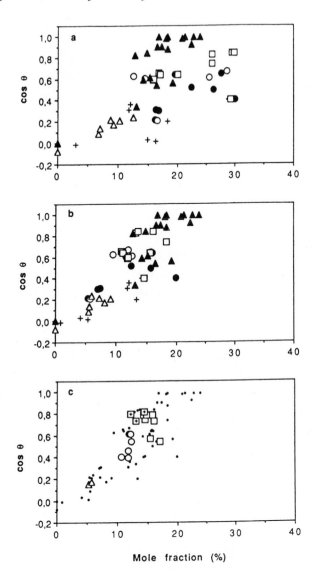

Figure 3. Plot of the cosine of the water contact angle vs. the mole fraction (calculated omitting hydrogen) of total oxygen (a) and of oxygen responsible for O_{1s} peaks with a binding energy smaller than 533.4 eV (b and c).

a and b : published data (Dewez et al., 1991) on untreated polymers (+) and surface modified PC (●); data obtained on PC (○), PET (□), PP (Δ) untreated and shortly after treatment by O_2, Ar and NH_3 plasmas; data obtained after allyl alcohol deposition on PP (▲)

c : data of part b (•); data obtained on O_2 plasma treated PC (○), PET (□) and PP (Δ) and on NH_3 plasma treated PET (□) analyzed after various times of storage in air or over P_2O_5.

The data of figure 3b are replotted in figure 3c which presents also the data obtained for the samples stored before the surface characterization.

A much lower water contact angle of O_2 plasma treated PP (24° instead of 70°) was reported by Occhiello et al. (1991) and by Garbassi et al. (1989); however it was found to increase strongly during the first 24h of storage. In the present work no significant change of water contact angle or oxygen surface concentration is observed during storage in air during 1 to 37 days (Figure 1).

The NH_3 plasma treated PET presents a hydrophilicity increase as high as the O_2 plasma treated PET (Figure 1). However, unlike the latter, it does not show any decrease of hydrophilicity during storage.

Figure 1 shows that the modification of hydrophilicity of O_2 plasma treated PC and PET during storage takes place without any appreciable change of apparent surface concentration neither of total oxygen nor of oxygen carrying a high electron density. This is also illustrated by figure 3c.

Garbassi et al. (1989), Munro and Mc Briar (1988) and Occhiello et al. (1991) have reported a decrease in polymer surface hydrophilicity at the polymer / air interface as a function of ageing time after O_2 plasma treatment without significant alteration of surface composition as determined by XPS. It has been suggested that this behavior is due to the "rotation" of surface polar groups into the bulk phase of the material in order to minimize the interfacial energy when the surface is stored in non-polar medium such as air (Occhiello et al., 1991). It has also been shown by angle resolved analysis that this reorganization process occured in a 2 nm region within the outermost 5 nm surface layer analyzed, making XPS analysis unsensitive to this change.

Another molecular process which may be responsible for a loss of surface hydrophilicity is the migration of low molecular weight polar fragments, produced by the plasma treatment, into the bulk. Recent results (Lub et al., 1989) showed that O_2 plasma treated PC surfaces carry low molecular weight polar fragments easily extractable by ethanol or water. Such surface rearrangements are generally, but not systematically, followed by a change in XPS data (Briggs et al., 1980; Munro and Mc Briar, 1988). It was however reported by Occhiello et al. (1991) that, in the case of O_2 plasma treated PP, this aging process seems forbidden, since interdiffusion of polar fragments between the surface and the bulk phase is excluded due to immiscibility between the two phases.

For O_2 plasma treated PET and PC, the decrease in surface hydrophilicity could thus be explained by a slow surface rearrangement due to molecular migration or reorientation, polar groups being buried away from the polymer-air interface. As no change of surface composition is observed by XPS, this surface reorganization would take place within the first atomic layers at the surface. It would be accelerated by the presence of humidity suggesting that water may act as a plasticizing agent. The fact that the NH_3 plasma treated PET does not change of hydrophilicity upon storage could be due to cross-linking at the polymer surface.

Figure 3 (b and c) shows that the dots of polymers of different natures are scattered in about the same may as those of samples stored for various times, when cos θ is plotted vs the mole fraction of oxygen carrying a high electron density. A typical exemple is shown by PET, the O_2 plasma treatment of which increases the hydrophilicity without any significant change of the oxygen concentration and of the oxygenated functions distribution. As explained above, such scattering may be due to a difference of chemical composition between the first atomic layer, which controls the water contact angle, and the thickness analyzed by XPS. It might also be due to the fact that polymers of similar surface elemental composition may be characterized by a different chain mobility at the polymer-water interface.

5. CONCLUSION

The data presented above show that the oxygen carrying a high electron density, detected by O_{1s} peaks at or below 533.4 eV, is the main factor determining the surface hydrophilicity. Nevertheless the hydrophilicity is also influenced by the heterogeneity of the organization within the first atomic layers at the surface, which are analyzed by XPS.

The polymer films which appear suitable for epithelial cell culture are those which present the highest surface hydrophilicity, in particular NH_3 plasma treated PET and O_2 plasma treated PC films.

ACKNOWLEDGEMENTS

The authors are members of the Research Center for Advanced Materials. They thank Coating Research Institute (Limelette) and Prof. J. Verbist, Facultés Universitaires Notre-Dame de la Paix (Namur) for access to their facilities. The support of IRSIA, FNRS and Department of Education and Scientific Research (Concerted Action Physical Chemistry of Interfaces and Biotechnology) is gratefully acknowledged.

REFERENCES

Andrade J D, Gregonis D E and Smith L M 1985 Surface and Interfacial Aspects of Biomedical Polymers - Vol 1 Surface Chemistry and Physics - Polymer Surface Dynamics ed J D Andrade (NY London : Plenum Press) pp 15-41

Briggs D, Rance D G, Kendall C R and Blythe A R 1980 Polymer 21 895

Busscher H J, Van Pelt A W, De Jong H P and Arends J 1983 Colloid Interface Science 95 23

De Puydt Y, Bertrand Y, Novis Y, Caudano R, Feyder G and Lutgen P 1989 British Polymer Journal 21 141

Dewez J L, Doren A, Schneider Y J, Legras R and Rouxhet P G 1991 Surface Interface Analysis 17 499

Dewez J L, Doren A, Schneider Y J, Legras R and Rouxhet P G 1991 Interfaces in New Materials ed P Grange (London and New York : Elsevier Applied Science) pp 84-94

Garbassi F, Morra M, Occhiello E, Barino, L and Scordamaglia R 1989 Surface Interface Analysis 14 585

Lub J, Van Vroonhoven F C, Bruninx E and Benninghoven A 1988 Polymer 30 40

Munro H S and McBriar D I 1988 Journal Coatings Technology 60 41

Occhiello E, Morra M, Morini, G, Garbassi F and Humphrey P 1991 Journal Applied Polymer Science 42 551

Pochan J M, Gerenser L J and Elman J F 1986 Polymer 17 1058

Ruckenstein E and Gourisankar S V 1986 Biomaterials 7 403

Sckakenraad J M 1987 Cell-polymer Interactions. Doctoral Thesis. Rijksuniversiteit Groningen, The Netherlands

Surface fluorination of cellulose derivatives as biomaterials

Fabienne Poncin-Epaillard[1], Gilbert Legeay[2], Jean-Claude Brosse[1]

1)Laboratoire de Chimie et Physicochimie Macromoléculaire, (URA au CNRS n°509)

Université du Maine -avenue Olivier Messiaen, 72017 LE MANS -FRANCE

2)Institut de Recherche Appliquée sur les Polymères

72 avenue Olivier Messiaen, 72017 LE MANS -FRANCE

ABSTRACT :

A film of cellulose acetate was submitted to a cold plasma of tetrafluoromethane or of sulphur hexafluoride. The interactions of these cold plasmas and cellulose acetate lead to a material whose surface has been modified by fluorination. Comparison of CF_4 or SF_6 plasma treatment shows that fluorine atoms provided by each kind of plasma induce degradation, and grafting of fluoro-carbon radicals on the surface. As a consequence, the surface energy decreases and offers the possibility of a better response of plasma modified cellulose derivatives used as biomaterials (e.g. hemodialysis membrane).

Introduction

A biomaterial, specially a blood compatible material is a synthetic or natural polymer whose surface is in direct contact with biological components. When a foreign biomaterial is dipped in blood or tissue fluids, in a few seconds adsorption of biomolecules, cells (for example macrophages cells), usually proteins, takes place. This is followed in the next minutes or few hours by cellular interactions leading to a thrombus formation. Therefore, an ideal biopolymer is a polymer which does not adsorb any proteins (Hoffman 1987, Ikada 1989), and also induce no activation of complement system (Man 1989).

The proteins adsorption depends greatly on the surface properties of the biomaterial:

- surface energy,

- surface composition and,

- surface morphology.

For example, Ikada.(1989) show that a lower protein adsorption rate is obtained either with a biomaterial "superhydrophilic" (like polyethylene grafted with acrylamide, cellulose grafted with polyvinyl alcohol), or a biomaterial "superhydrophobic" as perfluoro polymers. The surface of these biomaterials could be also modified with a simple chemical surface treatment which leads to a low protein adsorption rate.

Thus, a great deal of effort has gone into surface modifications (Legeay.1989). Surface modification could proceed by :

- a physical deposition of other compounds (surfactants...) to create a new surface chemistry in order to modify proteins adsorption) (Ueno 1990),

- a direct chemical modification (oxidation, hydrolysis, sulfonation, corona discharge or plasma treatment), to create a barrier film which reduces undesirable diffusion of small molecules from the substrate (migration of adjuvants of biomaterials(Chang 1973)), or control the rate of diffusion of drugs from the substrate (Colter 1977a-b),

- a chemical grafting of a different polymer (radiation graft copolymerization, plasma polymerization), to provide new surface chemistry for subsequent immobilization of molecules (Corretge1988, Nichols 1979).

In hemodialysis field, one of the most used membranes is a film or capillar tubes of cellulose or derivatives. During hemodialysis treatments, leukopenia is observed: a direct relation between this phenomenon and the complement activation involving the hydroxyl groups of the cellulose backbone (Law.1980, Sim 1981) has been proposed. Masking or substitution of the hydroxyl groups via plasma treatments (plasma polymerization or modification) should reduce the complement activation. Corretge (1988) proposed a grafting of poly(ethyleneglycol) on cellulose. The purpose of this publication and relatives (Man 1989) is dealing with fluorination of cellulose and derivatives surfaces through a cold plasma treatment, and its consequence on the leukopenia and the complement activation.

Cellulose and its derivatives have been treated in different plasmas for membrane permeability or textile applications (Sharma 1987, Wakida 1989, Simionescu 1976, 1977 1984, Lodesova 1987). Argon, nitrogen, air plasmas treatments lead to radical species in concentration depending on plasma conditions (Wakida 1989, Simionescu 1976, 1984), to new chemical groups or functions (ketone) and to a decrease of hydroxyl groups (Simionescu 1976, 1984). Degradation and crosslinking via hemiacetal bonds is proposed as a mechanism to these plasmas modifications (Lodesova 1987). But, although the fluorination of cellulose or its derivatives is a promising research field for its wide bioapplications, little work has been done on this subject (Yasuda 1982).

Experimental Part

plasma equipment:

The microwaves plasma equipment, has been previously described (Poncin-Epaillard 1990).

The RF reactor is a ATEA device of 300 liters. The plasma was produced by an RF (13.6MHz) capacitive excitation inside the reactor walls

Surface analyses:

The SEM pictures are made on Hitachi n° 52300 in the "Département de Génie Mécanique, Institut Universitaire de Technologie, Université du Maine" (Le Mans, France).

The FTIR spectra are recorded on FTIR Perkin-Elmer n° 1750, with a microcomputer n° 7700, transmission (20 scans), attenuated total reflexion (ATR, 200 scans, θ comprised between 30 and 60°) are used.

The electronic spectroscopy for chemical analysis (ESCA) has been developped by the "Laboratoire de Physique des Couches Minces, Université de Nantes" (Nantes, France).

Surface energies of samples have been calculated from the contact angle of distilled water, glycerol and α-bromonaphthalene (volume : 4 μl) and from Dupré relation (1969) used as described by Gleich (1989). Surface energies of different liquids are as follows :

$\gamma_{H_2O} = 72.75$ mJ.m^{-2}, $\gamma_{H_2O}^D = 21.75$ mJ.m^{-2}, $\gamma_{H_2O}^P = 51.0$ mJ.m^{-2}

$\gamma_{gly} = 63.4$ mJ.m^{-2}, $\gamma_{gly}^D = 37.0$ mJ.m^{-2}, $\gamma_{gly}^P = 26.4$ mJ.m^{-2}

$\gamma_{bro} = 44.4$ mJ.m^{-2}, $\gamma_{bro}^D = 43.6$ mJ.m^{-2}, $\gamma_{bro}^P = 0.8$ mJ.m^{-2}

films preparation:

The cellulose acetate (Aldrich, acetylation yield : 39,8 %) is dissolved in a mixture of acetone and water (23 % of cellulose acetate, 67 % acetone, 10 % H$_2$O) at room temperature during 12 hours. 1 ml of this solution is spread on a glass slide which has been previously cleaned with a sulfochromic acid then several times with H$_2$O and EtOH. The film (30μm of thickness) is maintained at 40°C during 5 mn, then dipped in ice during 1 hour, and finally dipped in hot water (65°C) during 5 mn. Then film is dried (Toprak 1979). One of the faces is mat and the other polish. The polish face seems to be more homogenous than the mat one. SEM shows a surface (mat face) with pores of 2 μm diameter.

All the samples have a fixed area : 15 cm^2.

plasma treatment:

Before applying the usual process of treatment (Poncin-Epaillard 1990), as the material is very hygroscopic, water remaining in the film must removed and the film kept in dry atmospher. Sample must be dried under vacuum at least 1 hour (Figure 1 α). When the pressure is restored back to atmospheric, the concentration of adsorbed water reaches a constant value after 16 hrs (Figure 1β).

Figure 1 : Dependence of the concentration of absorbed water in cellulose acetate:

IδmI = (initial weight of sample - final weight of sample)/initial weight of sample,

(α) on time and pressure (p = 10^{-2} mbar) conditions under vacuum

(β) on time at atmosphere pressure

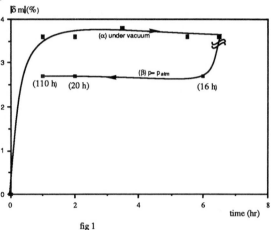

fig 1

Modification is studied in function of the different parameters of the plasma : duration (t, mn), power (Pi, W), gas flow (D, sccm : standard cm^3/mn), distance between the end of visible part of the plasma and the sample (z, cm).

All the samples, before plasma treatment, are submitted to a primary pumping during 3 hours then to a secondary pumping during 15 mn at ambient temperature.

The degradation rate (degraded layer (μm)), described as the proportion of volatil products (oligomers, CO_2, H_2O..) produced during the plasma treatment, is measured by weight difference between the untreated and treated samples.

Results and discussion

As described in previous paper (Poncin-Epaillard 1989, 1990b), the plasma modification of polymers can lead to functionalization and degradation. These different reactions will be characterized through different experiments and analyses and compared together with the idea of a potential biomedical application.

Influence of degradation

On surface topography

The CF_4 plasma treated samples, prepared from acetone solutions, show a smoother surface than the blanck one. The plasma treatment leads to an uniform erosion and also probably a fluorinated layer. No difference between treated mat or polish faces has been noticed either on surface topography or on low molecular weight products concentration probably due to the sensibility of used analyses.

On the concentration of low molecular weight products

The degradation can be characterized by the formation of oligomers, some of them are in vapor phase, the others can be blown away with the gas flux.

When one of plasma parameters (t, P or z) is increased, the degradation of cellulose acetate film also increases (Figure 2). But SF_6 plasma seems to be a more reactive plasma. The degraded layer can reach more than 5 μm thickness and fragmented macromolecules may induce a pollution in the medium (in vitro or in vivo) and toxicologic consequences when used as a biomaterial like hemodialysis membrane. Cellulose is a weak material and C-O, C-H bonds are probably dissociated via fluorine atoms substitution or addition as their strenghs (C-O: 351-389KJ.mol^{-1}, C-H: 355KJ.mol^{-1}) are low. A mild treatment needs the following conditions : t < 5 mn, Pi< 100 W, z=0 cm (substrate at the end of visible part of the plasma).

Figure 2 : Dependence of degradation rate of cellulose acetate films on plasma parameters

degraded layer = initial thickness -final thickness = (m_i / sxd) - (m_f / sxd), with s : surface of sample 15 cm^2, and d : cellulose acetate density : 1.3

(a) time (DSF_6 = 10sccm, Pi = 100W, l = 11cm, z = 0cm, p = 0.7mbar, DCF_4 = 15sccm, Pi = 40W, l = 13.5cm, z = 0 cm, p = 0.2 mbar)

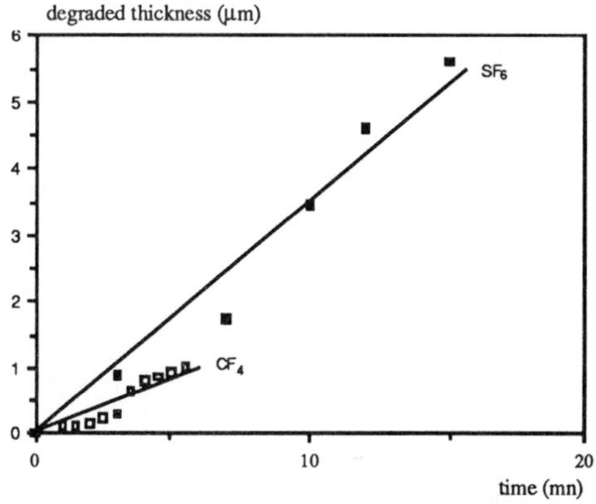

Fig 2a

(b) power (DSF_6 = 10sccm, p = 0.7mbar, t = 5mm, d = 8cm, DCF_4 = 15sccm, p = 0.2 mbar, z=0cm, t = 3mn)

(c) distance (DSF_6 = 15sccm, p = 0.2mbar, Pi = 40W, t = 3mn, l =13.5 cm, DCF_4 = sccm, p = 0.7mbar, Pi = 40W, l = 4cm, t = 3 mn)

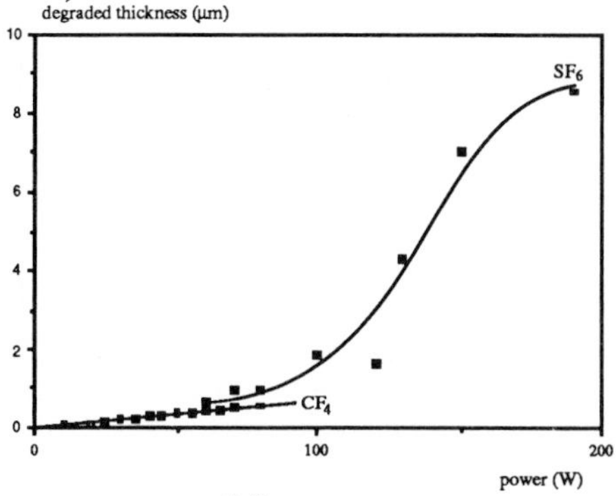

fig 2b

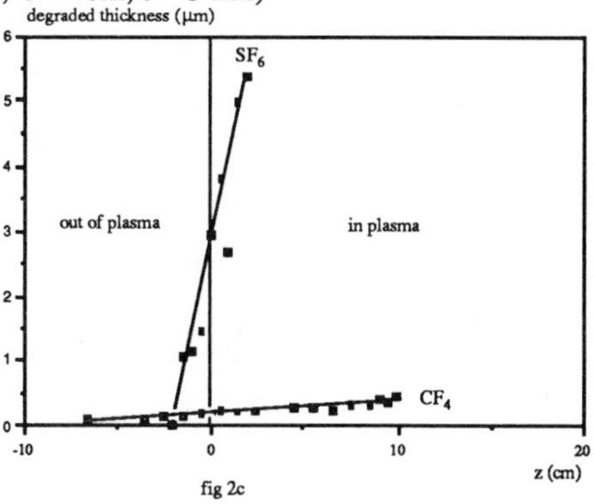

fig 2c

The difference of reactivity of CF_4 and SF_6 can be explained by the value of bond strengh of C-F (552 KJ.mol[-1]) and S-F (343 KJ.mol[-1]). The sulfur hexafluoride bonds are weaker than those CF_4 molecule, and may produce more fluorine atoms, as has been shown by Picard (1986). The silicon etch in SF_6 plasma is recognised to be about one order of magnitude greater than in CF_4 plasma in same conditions (Picard 1986). Fluorine atoms are the main etching species, produced by attachment, ionisation and dissociation of neutral species SF_6 and SF_4. As surface modified derivatives of cellulose acetate are insoluble or poorly soluble in solvents used for SEC or viscosimetry measurments, no determinatio of the molecular weight of modified product is possible. In a first step, we chose the filtration, toxicologic tests as analytical reference.

Nevertheless, with RF plasma treatment of hemodialysis membranes under mild conditions, no secondary effects (alteration of filtration properties, toxicologic consequences in vivo or in vitro tests) due to the presence of a degraded of layer of cuprophane have been detected (Man 1989). This could be explained by degradation with formation of volatile products. Work is actually done on the characterization of these volatile products.

Influence of functionalization :

Characterization of this surface functionalization will be appreciated by wettability measurements and ESCA analysis on cellulose acetate films but also on commercial product of hemodialysis membranes. An approch of ATR-FTIR analysis will also be discussed.

On water adsorption :

The CF_4 or SF_6 plasma treatments lead to a fluorinated layer. As it has a hydrophobic character, the layer should act like a barrier to hydrophilic liquid diffusion. The modified cellulose acetate film is an effective barrier water adsorption (Figure 3), the adsorption rate is divided by 3 when the sample surface is modified. As plasma modification takes place only on few layers (< 10nm), it should say that water adsorption is a surface phenomenum.

Figure 3 : Dependence of the concentration of absorbed water in cellulose acetate treated

Iδm I = (initial weight of sample - final weight of sample)/initial weight of sample,

DCF_4 = 15sccm, Pi = 40W, p = 0.2 mbar, t = 5mn, l = 13.5 cm, z = 0cm

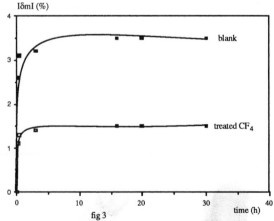

fig 3

On surface tension :

Contact angle measurement can provide considerable insight of the character and properties of polymers surface, such as physical (surface polarity, groups reorientation and mobility) and chemical (new polar or apolar functions) interactions, even if some assumptions have to be made (the solide surface must be rigid, highly smooth and homogeneous).

The surface tension and its polar and dispersive componants were calculated, but only the polar term variations with plasma conditions are represented in Figure 4 as it is the most important factor when a fluorination (creation of apolar groups on a polar surface) takes place. The other componants have also been calculated : the dispersive one is relatively constant, whereas the global energy has same evolution as the polar energy.

Figure 4 : Dependence of surface tensions of cellulose acetate films on plasma parameters
(a) time $(DSF_6 = 7sccm, p = 0.8mbar, Pi = 100W, l = 4cm, z =0cm, DCF_4 = 15 sccm, Pi = 40W, l = 13.5 cm, z=0 cm$

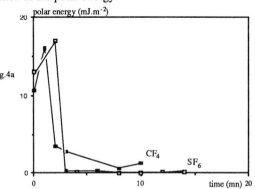

(b) power $(DSF_6=7$ sccm, p=0.6 mbar, t = 3mm, d = 3cm, $DCF_4 = 15sccm, p =0.2mbar, z = 0cm, t = 3mn)$

(c) distance $(DSF_6 = 7sccm, p = 0.6$ mbar, Pi = 100W, t = 3 mn, l = 4cm, $DCF_4 = 15sccm, p = 0.2$ mbar, Pi = 40W, l = 13.5cm, t = 3mn

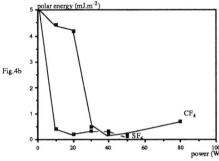

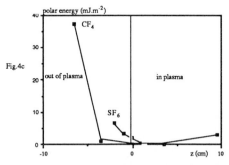

As shown with the degradation rate, the SF_6 plasma treatment leads to a more drastic effect, the polar surface tension is decreasing nearly to zero, more quickly than with CF_4 plasma when a higher duration or a higher power is applied (Figures 4a-c). When sample is in the afterglow zone or near the excitator, the polar energy is increasing again, showing a diminuation of fluorination rate (when z is negative) or a degradation effect more important than the modification (when z is positive) as the ionic bombardment is increasing (Figure 4c). A duration of 2 mn (Figure 4a) or a power of 10 W (SF_6 plasma) or 30 W (CF_4 plasma) lead to a minimum of polar surface tension.

On surface structure determined by ESCA analysis :

The ESCA analysis has been performed not only on cellulose acetate films treated in CF_4 or SF_6 plasmas but also on cuprophane films treated in CF4-RF plasma, this material being used as hemodialysis membrane (Table 1).

Table 1 : Analysis of ESCA spectra

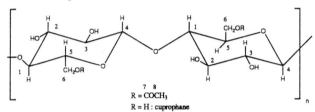

7 8
R = COCH$_3$
R = H : cuprophane

cellulose acetate				cellulose acetate treated in CF_4 plasma [a]				cellulose acetate treated in SF_6 plasma[b]			
peak	group	$E_L^{d)}$(eV)	X%	peak	group	$E_L^{d)}$(eV)	X%	peak	group	$E_L^{d)}$(eV)	X%
	8,2,3	285.3			8,2,3	285.3			8,2,3	285.3	
	6,1,5	286.8			6,1,5	286.8			6,1,5	286.8	
	4,7	289.1			4,7	289.1			4,7	289.1	
C_{1s}	C-CF$_x$		70	C_{1s}	C-CF$_x$	287.5	46	C_{1s}	C-CF$_x$	287.5	46.5
	CF				CF	288.0			CF	288.0	
	CF$_2$				CF$_2$	290.5			CF$_2$	290.0	
	CF$_3$				CF$_3$	292.2			CF$_3$	291.8	
O_{1s}		533.2	30	O_{1s}		534.8	15	O_{1s}		533.5	13
F_{1s}				F_{1s}		685.1	39	F_{1s}		689.1	40.5

cuprophane				cuprophane treated in CF_4 plasma [c]			
peak	group	$E_L^{d)}$(eV)	X%	peak	group	$E_L^{d)}$(eV)	X%
	8,2,3	285.1			8,2,3	285.1	
	6,1,5	286.8			6,1,5	286.9	
	4,7	288.5			4,7	288.5	
C_{1s}	C-CF$_x$		68.5	C_{1s}	C-CF$_x$	288.5	60
	CF				CF	to	
	CF$_2$				CF$_2$	292.3	
	CF$_3$				CF$_3$	292.3	
O_{1s}		531.2	31.5	O_{1s}		533.1	34
F_{1s}				F_{1s}		686.6	6

a) D_{CF_4} =15 sccm, Pi = 50W, p = 0.2mbar, l = 13.5cm, z = 0cm, t = 5 mn

b) D_{SF_6} = 7sccm, Pi = 100W, p = 0.8 mbar, l = 5cm, z = 0cm, t = 10mn

c) RF plasma, P = 500W, t = 15mn

d) bond energy corrected of charge effect

The microwave plasma treatment induces a high fluorination yield (CF_4 plasma : 39% (F_{1s}), SF_6 plasma : 40.5% (F_{1s})). The grafted fluorinated layer on the surface of cellulose acetate film is composed of a mixture of CF, CF_2 and CF_3 groups. As a consequence of the treatment, the oxygen concentration is decreased (15% instead of 30 %).

The cuprophane film treated in RF plasma in such conditions leading to a good hemodialysis test (Man 1989), has a fluorinated surface with a minor fluorine concentration (6 %), but sufficient for a biomedical application. When changing the X-ray beam position from 0° to 50°, the fluorine concentration is nearly constant (6-8%), the subsurface is also modified. This treatment induces also an oxidation mechanism since the oxygen concentration is increasing, probably due to a post-reaction or impurities interactions in the plasma phase.

Concerning the balance between concurrent reactions : degradation and fluorination , ESCA analysis could give the thickness of fluorinated layer (f) (Picard 1986) :

$$f = \lambda \, Ln \, \frac{I_{C_{1s}} \, blanck}{I_{C_{1s}} \, treated}$$

with λ : attenuation length of C_{1s} peak (λ = 1.6 nm using Mg Kα radiations),

I : intensity of C_{1s} peak.

The fluorinated layer thicknesses were found to be of the same order of 1 nm (0.88 nm in SF_6 cellulose acetate, 0.60 nm in CF_4 cellulose acetate, 0.63 nm in CF_4 cuprophane). These results show that fluorinated layer deposition is a minor phenomena in respect of degradation (0.7 μm in CF_4 atmosphere, 3.3 μm in SF_6 atmosphere).

On surface structure determined by ATR-FTIR :

The ATR-FTIR analysis is sometimes a simple and usefull technics for surface modification characterization[25] (Poncin-Epaillard.1989, 1990a-b). However, the complexity of the cellulose acetate film does not allow the direct application of such a spectrometric analysis. A derivation of modified and unmodified sample spectra emphasizes the apparition of new peaks which are not yet attributed, futher investigations are in progress not only on modified cellulos acetate and on use of FTIR spectroscopy but also on model molecules and use of ESCA, SIMS spectroscopies and chromatographies (SFC..).

Conclusion

Modification of cellulose acetate films with CF_4 or SF_6 plasmas lead to a fluorinated layer. The modified materials can be used as a biomaterial (hemodialysis membrane), but this study emphasizes the importance of reactions concurrent to fluorination such as degradation, as for a biomedical application, by-products must be avoided. CF_4 plasma treatement seems to be well adapted for biomedical applications, rather than SF_6 plasma treatement as degradation is a minor reaction. The plasma- material interactions mechanisms are not really cleared up and much work has to be carried out on plasma modification of model compounds of cellulose acetate.

Acknowledgements:

The authors would like to thank Mr. B.Chevet, Mr. R.Sarih (Laboratoire de Chimie et Physicochimie Macromoléculaire) and Dr N.K. Man (Hopital Necker, service de nephrologie, Paris) for their contribution to this work.

Références

Chang F.Y., Shen M., Bell A.T.1973 J.Appl.Polym.Sci. 17 2915

Colter K.D., Bell A.T., Shen M.1977a Biomat.Med.Dev.Art.Org. 5(1) 1

Colter K.D., Bell A.T., Shen M.1977b, Biomat.Med.Dev.Art.Org. 5(1) 13

Corretge E., Kishida A., Konishi H., Ikada Y.1988 "Polymers in Medicine, III" (Amsterdam: Elsevier Science Publishers) 61

Dupré A 1969 "Théorie Mécanique de la Chaleur" (Paris: Gauthier Villars) 369

Gleich H., Criens R.M., Mosle H.G., Leute H.1989 Int.J. Adhesion and Adhesives 9 88

Hoffman A.S.1987 Polymeric Materials Sci.Eng.56 699

Ikada Y., Suzuki M., Tamada Y.1989 Polym.Preprints 24(1) 19

Law S.H., Lichtenberg N.A., Levine R.P.1980 Proc.Natl.Acad.Sci. USA 77(12) 7194

Legeay G., Poncin-Epaillard F., Brosse J.C.1989 Proc.Adhecom. 89, 52

Lodesova D., Tran Xuan M., Sura S., Blecha J. 1987 Text.Chem. 17(1), 49 Chem.Abstr. 108 : 152 355 f

Man N.K., Legeay G., Jehenne G., Tiberghein D., De la Faye D. 1989 Proc.ISAO

Nichols M.F., Hahn A.W., Easley J.R., Mayhan H.L.1979 J.Biomed-Mat.Res. 13 299

Picard A., Turban G., Grolleau B. 1986 J.Phys.D. Appl.Phys. 19 991

Poncin-Epaillard F., Chevet B., Brosse J.C.1989 Le Vide, Les Couches Minces supl. n° 246 207

Poncin-Epaillard F., Brosse J.C. 1990a Makromol.Chem. 191 691

Poncin-Epaillard F., Chevet B., Brosse J.C. 1990b Eur.Polym.J. 26(3) 333

Toprak C., Ager J.N., Falk M. 1979 J.Chem.Soc.Faraday I 75 8036

Sharma C., Jubaira Y.1987 Polymeric Materials Sci.Eng. 56, 342

Sim R.B., Twose T.M., Paterson D.S., Sim E.1981 Biochem.J. 193 115

Simionescu C.I., Macoveanu M.M., Olara N.1976Cellulose Chem.Techn. 10 197

Simionescu C.I., Macoveanu M.M.1977 Cellulose Chem.Techn. 11 87

Simionescu C.I., Denes F., Macoveanu M.M., Negalescu I. 1984 Makromol.Chem.suppl.8 17

Ueno M., Ugajin Y., Horie K., Nishimura T.1990 J.Appl.Polym.Sci. 39 967

Wakida T., Takeda K., Tanaka I., Takagishi T.1989 Textile Res.J. 59(1), 49

Yasuda T. 1982 Proc.Org.Coat.Appl.Polym.Sci. 47 313

Modification of surfaces designed for cell growth studies

T. G. Vargo and J.A. Gardella, Jr.

Department of Chemistry, State University of New York at Buffalo, Buffalo NY, 14214.

ABSTRACT:The intention of this review is to: 1) briefly summarize methods for modifying surfaces of polymeric materials, 2) review these methods with respect to their past and current applications for cell culture and growth, 3) discuss current investigations which elucidate the application and importance of small chain polypeptides with respect to their ability to promote specific cellular responses, and 4) present the current use of surface modification methods and their applications towards designing new substrata which incorporate the attachment of specific growth factors.

1. INTRODUCTION

The need to understand the biology which occurs at and on manufactured materials has obvious importance in the medical field where such materials are utilized in constructing artificial devices (e.g., vascular grafts, prothesis, etc.) (Andrade et. al. 1985). The biology and chemistry at surfaces and interfaces which governs cellular adhesion and growth involves a *series* of complicated events. It is now understood that *all* cellular adhesion and growth is influenced by the initial adsorption of a glycoproteinacious "conditioning film" (Baier et. al. 1984), (Dankert et. al. 1986). In defining cellular events at surfaces, theories of bioadhesion (Baier and Meyer 1991) are stressed and refer to the initial event of adsorption of proteinacious material which precedes the subsequent cellular response to a particular surface. Historically, studies of cell biology on polymeric articles have depended almost entirely on measured surface energies and/or surface morphologies. From a microstructural viewpoint, studies of cellular response to specific chemical functionality and their geometries at surfaces represents a critical "next step" towards understanding better the biological responses at interfaces. Investigations from this microstructural perspective appear to be within sight due to the advent of highly specialized analytical instrumentation.

It is clear that influence and/or control over protein adsorption and the subsequent molecular orientation of a protein can lead to a better understanding of cellular response mechanisms. One possible method for controlling and understanding protein adsorption is to use well defined model surfaces. In this paper a review of various surface modification and construction schemes designed for investigating protein adsorption and cellular response is presented.

Also important towards achieving a better understanding of cellular response is the use of serum controlled or serum free culture media. Typically, cells are cultured in media containing a variety of different proteins and other constituents. This makes isolation of specific cellular response mechanisms difficult. In order to address this difficulty, researchers have begun to use well defined serum controlled culture media in conjunction with well defined model surface chemistry. Further, the preparation of surfaces with immobilized small chain polypeptides (Massia and Hubbell, 1990, 1991) has the potential to allow studies of cellular response to specific minimal peptide sequences (Pierschbacher and Ruoslahti, 1984) in serum free media. Review and discussion of these last two topics are included.

2. MODIFICATION OF POLYMERIC SURFACES

The synthesis of copolymers or the manufacture of polymer blends afford a wide range of versatility towards providing various surface properties e.g., surface energy or surface functionality/chemistry (Patel et. al.1988). These techniques however, lack the ability to preserve the physical and chemical properties of a bulk homopolymer whose bulk properties may have been chosen for a particular use. Thus, the development of other modification methods which could selectively modify a polymer surface with retention of original bulk properties were forthcoming. Some of these methods included various physical treatments e.g., corona discharge or radio frequency glow discharge (RFGD) plasma (Cho and Yasuda 1986) and plasma polymerization (Yasuda 1985), as well as chemical methods such as surface grafting (Ikada et. al. 1990) and interpenetrating networks (IPN) (Hourston 1986). The original difficulty with these methods was the inability to precisely control the final properties of the modified surfaces. Typically, one could change the surface energy (e.g., create a hydrophilic surface on a hydrophobic bulk or vice versa), however, control of surface residing functionality and orientation of the functionality is not trivial. For example, plasma treatments as well as plasma polymerization methods result in highly modified surfaces made up of a variety of crosslinked functionality such that the control over what type of functionality and its concentration is minimal. On the other hand, chemical modifications such as grafting or IPN, although more specific, usually result in only low levels or concentrations of surface modification.

The goal to generate specific surface chemistry through the addition of selected functionality with control over the orientation of such functionality has led to technological advances in the preexisting techniques as well as to the development of new ones - all tuned to provide better control of the surface chemistry (Ward and McCarthy 1989). For example, McCarthy has published a series of papers (e.g., Bening and McCarthy 1990) dealing with chemical modifications at surfaces resulting in addition at high surface concentrations of specific reactive functionality. The techniques to be reviewed here with respect to constructing model interfaces for biological studies will be limited to current state of the art plasma polymerization (Ratner et. al.1990), plasma treatment (Vargo et. al. 1991a), grafting (Massia and Hubbell 1991a, 1991b), (Shoichet and McCarthy 1991), surface chemical modifications (Dulcey et. al. 1991), and finally self-assembled films (Prime and Whitesides 1991).

3. MODIFIED SURFACES AND THEIR USES IN CELL CULTURE TECHNOLOGY:

Interest in addressing the cellular response to specific functionality has led to studies where surfaces have been created using techniques to provide specific control of functionality, its concentration, and (in some cases) orientation. Plasma polymerized films, which have been comprehensively characterized with respect to their surface energies and surface residing functionality, have shown various cellular responses which can be controlled depending on the deposited film (Ratner et. al. 1990). For example, in comparing various plasma polymerized films created from mixtures of oxygen gas and various liquid vapors (e.g.,acetone) Ratner et. al. (1990) observed that endothelial cell growth tends to increase as a function of increased oxygen content in the plasma mixture. Also, it was observed that poly(ethylene terepthalate) (PET), while not supporting the growth of bovine aortic endothelial cells, could be induced to support good attachment and growth of these cells after the deposition of a methanol based plasma polymer.

Investigations of plasma treatment processes have also shown control and effect on cellular response (Dewez et. al. 1991), (Vargo et. al. 1991b).   For example, we have recently utilized a two step process where step one required a plasma treatment of a poly(tetrafluoroethylene-co-hexafluoropropylene) (FEP) film (resulting in a hydroxylated surface) (Vargo et. al. 1991b) and step two, covalently attached controlled mono and multilayers of amine containing siloxanes to the modified surface.   These surfaces were prepared such that 200 to 300 micron linewidths of -OH, mono, and di-amine functionality with defined orientation were incorporated onto the FEP.   Mouse neuroblastoma (subclone Nb2a) cellular response was then monitored and shown to be restricted to the modified lines such that each of the different regions supported different cell adhesion and subsequent outgrowth.

Cellular studies were performed using Dulbecco's modified Eagle's medium (DMEM) (GIBCO) containing 10% fetal calf serum (FCS) (GIBCO), penicillin G (100ug/ml) and streptomycin (100ug/ml).   Preferential adhesion of the neuroblastoma to the 200um modified regions (separated by 300um linewidths of unmodified FEP) was observed for the FEP-OH and FEP-aminopropyl siloxane (FEP-APS).   Cell adhesion, growth, and definition between unmodified FEP and the two different surfaces (i.e., FEP-OH and FEP-APS) showed much greater cellular adhesion and growth on the FEP-APS materials as compared to the FEP-OH materials.   In the case of the FEP-ethylenediamine siloxane (FEP-EDAS) cell adhesion and growth was minimal.   Currently we feel this was due to the greater extent of silane polymerization of the di-amine siloxane.   Unlike the APS, where mono to a few multilayers of siloxane films could be controllably bonded to the interface, Electron Spectroscopy for Chemical Analysis revealed thick overlayers of the EDAS.   Because of this the interfacial concentration and amine orientation was probably different than that of the well controlled monolayer APS films and ultimately leading to a different cellular response.   To check this hypothesis the APS monoamine material was reacted under basic conditions which facilitates the growth of a thick APS film similar to the thickness measured for the EDAS films. Results showed that neuroblastoma adhesion and growth were greatly reduced on the thick FEP-APS versus the thin FEP-APS films.

The attachment of well defined lines of amino containing siloxanes has also been shown where chemical methods were utilized to attach the siloxane lines onto silicon substrates (Kleinfeld et.al. 1988).   Interestingly, in this work cell adhesion and growth were found to be best on EDAS versus APS which is opposite the results observed in our experiments. The reasons are not clear however, another result reported by Kleinfeld et. al. (1988) was that striping of the cells was only posssible in serum containing culture. Later in this review we report results where cell alignment and striping have been shown using serum free media with albumin as the only present protein.

Attachment of amine containing siloxanes in patterned lines has also been done using deep ultra-violet photochemistry (Dulcey et. al. 1991). Here, siloxanes were chemisorbed as monolayer films onto fused silica, Pt films on Si substrates, and Si wafers. Using masks, patterns were then created by desorbing the siloxane molecule through exposure to ultra-violet radiation. In both these cases cell adhesion and growth could be constrained to patterned regions.

The ability to pattern substrata with varying surface chemistries is important for a couple of reasons. First, the ability to control and direct cell growth may be important. For example, the ability to regenerate nerve fibers towards target structures (Goldberg and Burmeister 1989) would play an important role in the fields of neural physiology.

Secondly, it allows one to study cellular response as well as protein adsorption under competing conditions. This is important in cases where it is difficult to comparitively quantitate concentrations of cells or protein adsorption to different materials. Also this becomes valuable in cases where the presence of a preferred chemistry inhibits the adsorption of protein or cell growth normally observed on another present chemistry. Currently we have observed varying degrees of albumin adsorption to FEP, FEP-OH, and FEP-amino propyl siloxane (FEP-APS). However, when patterned materials were used, autoradiographs of radioactive iodine labeled albumin showed preferential adsorption to FEP versus FEP-OH. More striking were the results which showed that albumin adsorption to FEP was negligible with albumin adsorption strictly confined to FEP-APS regions. Subsequent studies of Nb2a mouse neuroblastoma showed adhesion and growth only on the albumin adsorbed to the FEP-APS. These studies were performed in serum controlled cell culture media using Hank's electrolyte (GIBCO) solution with 100 ug of albumin in 200 ml of Hank's solution.

Another recent example of a modified surface used for cell biological studies, utilized poly(ethylene oxide) (PEO) which was incorporated onto a PET substrate through grafting (Desai and Hubbell 1991). A series of varying molecular weight PEO's were used and they observed that the molecular weight of the surface grafted PEO influenced both degree of albumin (protein) adsorption and subsequent cell growth.

4. CELLULAR RESPONSE CONTROLLED BY SMALL POLYPEPTIDE FRAGMENTS:

Only recently (with the advent of advanced molecular and electron microscopic technology) has the elucidation of the components which make up cell-cell adhesion been shown (Campbell and Terranova 1988). The ability to define the ultrastructure and the components comprising, for instance, a basement membrane has resulted in isolation of molecular components which have been identified as key players in particular biological functions (Campbell and Terranova 1988). For example, laminin (a basement membrane glycoprotein) is known to have multiple structural and functional roles that depend on the stage of development and biologic state of the organism. Further, laminin has been broken down into subunits each made up of different polypeptide and amino acid macromolecules. Studies have shown specific biological functions to be associated with particular subunits.

Pierschbacher and Ruoslahti (1984) have shown in one of the original studies in this field that one could duplicate cell attachment activity observed in the presence of fibronectin (another glycoprotein) by using a synthetically prepared small chain fragment which is contained in the fibronectin molecule - tripeptide Arg-Gly-Asp (RGD). Others have followed with similar studies which investigate the effect short chain polypeptide sequences have on specific cellular responses. Gunderson and Barret (1979) discuss the use of minimal peptide sequences to locally influence the direction taken by neurites. As one further example, a pentapeptide growth factor Tyr-Ile-Gly-Ser-Arg (YIGSR); which can be isolated from laminin, has been implicated in inhibiting metastasis formation (Iwamoto et. al. 1987) as well as promoting neurite outgrowth (Campbell and Terranova 1988).

5. SURFACE IMMOBILIZATION OF MINIMAL PEPTIDE SEQUENCES

Discussion so far has dealt with reviewing first, surface modification techniques and the use of surface modification in the field of cell biology and secondly, the isolation of minimal peptide sequences and subsequent studies of their effect on biological function. One might envision the use of surface modification schemes for 1) placing molecules which mimic functionality and orientation of a peptide sequence onto a surface, 2) immobilizing peptide sequences directly onto polymeric substrates and 3) applying surface reactive functionality which can be used as docking sites for covalent attachment of peptide sequences. Through these methods, cell biology as related to well defined model systems might be accessed. Studies of this nature are currently beginning and represent a significant step towards better understanding of cellular adhesion, growth, and function at artificial biomaterial surfaces.

Immobilization of macromolecules, in this case small chain peptide sequences, can be as simple as physisorption i.e., adsorption with no covalent bond formation. In fact, until recently this was the typical method of immobilizing various growth inducers and inhibitors onto substrates. As an example Snow et. al. (1990) showed that two glycosaminoglycans, keratan sulfate and chondroitin sulfate, could be immobilized in patterns onto nitro-cellulose coated petri dishes. Results showed that these two proteoglycans were effective in inhibiting neurite outgrowth in vivo and in vitro. The difficulty with this type of immobilization is that the interface still remains undefined to a large degree. Taylor and Osapay (1990) discuss the difficulties in determining functional group conformations of biologically active peptides at model interfaces. Thus, the impetus to provide model interfaces comprised of chemisorbed or covalently bonded immobilized peptide sequences. Through attachment of a small chain peptide sequence to specific reactive sites it can then become easier to characterize the actual molecular conformations (which induce specific cellular response mechanisms).

Prime and Whitesides (1991) discuss the importance of using well defined interfaces for studying protein adsorption to interfaces. In this paper they support the use of self-assembled monolayers (SAM's) as a controlled adsorption interface and report on the adsorption of various proteins to SAM's intended to mimic three materials which normally resist protein adsorption. For each model system they prepared a series of mixed SAM's from hydrophobic alkanethiols to hydrophilic alkanethiols. Conclusions include results that protein adsorption is inhibited on hydrophilic SAM's whereas protein adsorption occurs more readily to hydrophobic SAM's.

Using light-directed peptide synthesis Fodor et. al. (1991) describe the capability to construct surfaces on glass comprised of arrays of short chain peptide sequences (in this paper up to

1024). These interfaces can then have the potential for elucidating a variety of biological response mechanisms including ligand-biological receptor interactions as well as principles which govern molecular interactions.

Massia and Hubbell (1990) describe the attachment of cell adhesion peptides Arg-Gly-Asp (RGD) and Tyr-Ile-Gly-Ser-Arg (YIGSR) to a glass substrate as well as to PET-OH and PTFE-OH (Massia and Hubbell 1991) (the PTFE-OH modification is described by Costello and McCarthy (1987). It is important to note that immobilization was predicted to occur via the N-terminus of the peptide chains. To support the attachment mechanism i.e. via a tresyl chloride reaction described by Nillson and Mossbach (1981) the N-terminus was concluded to be covalently bonded to the substrata and the C-terminus was radioiodinated to determine the presence of the peptides. Autoradiographs indicated significantly less peptide incorporation onto the polymer substrata. This is probably resultant of only a low percentage of primary alcohol sites to which the tresyl chloride reaction is specific.

With respect to cell growth (human umbilical vein endothelial cells) (Massia and Hubbell 1991) they observed adhesion to not occur to any of these surfaces in the presence of serum controlled (only albumin in an electrolyte) culture media. However, after attachment of RGD and YIGSR, cell adhesion and spreading were observed on all surfaces.

Recent work in our laboratory has shown that using the tresyl chloride reaction, as described, only low concentrations of YIGSR were incorporated onto the FEP-OH material These results were obtained via Electron Spectroscopy for Chemical Analysis (ESCA) analyses. Further, via light microscopy the surface morphology of these materials was greatly altered indicating severe morphological damage. From this we concluded that the YIGSR observed by ESCA was probably physisorbed having no particular orientation. These results were not unexpected in that the tresyl chloride reaction is extremely specific to primary alcohol sites and from the structure of our modified FEP (Vargo et. al. 1991) the alcohols predicted are all of secondary or tertiary nature. However, from previous experiments the -OH sites incorporated to the FEP material were found to be extremely reactive proton donators. Thus a simple reaction using micro-molar solution of YIGSR in an aprotic solvent (dimethyl sulfoxide (DMSO)) and a 2-3 molar excess (with respect to the YIGSR) of potassium carbonate, covalent attachment of YIGSR could be facilitated at the

C-terminus. ESCA results showed a proportionate amount of signal from the nitrogen 1s region indicating a concentration commensurate with what one would expect for a monolayer coverage. Also ToF-SIMS analysis revealed only little signal from mass ions associated with FEP. It remains for us to properly identify the mass fragments which we associate with the pentapeptide YIGSR as well as to begin cell culture experiments.

6. CONCLUSION

To quote Prime and Whitesides (1991) "no system is available that permits the structure and properties of the interface to be controlled in detail sufficient for the investigation of hypotheses concerning protein adsorption at the molecular level". This sentiment is the cornerstone of this review as well as the driving force behind much of the work contained herein.

With the present and rapid development of new surface analytical techniques and instrumentation (Gardella and Pireaux, 1990) however, the ability to control, characterize and construct well defined interfaces under complicated conditions seems to be within reach. The development of well defined surfaces which can illicit desired cellular response in simple and well defined culture media is presently a challenging area of research. Within this review many of the ongoing investigations represent significant advances with respect to understanding molecular biology at interfaces and represent the current state-of-the-art accomplishments.

REFERENCES:

Andrade J D, Gregonis D E, and Smith L M 1985 Surface and Interfacial Aspects of Biomedical Polymers ed J. D. Andrade (New York: Plenum Press) 1 1.

Baier R E and Meyer A E 1991 Fundamentals of Adhesion ed L. H. Lee (New York: Plenum Publishing Corporation) 407.

Baier R E, Meyer A E, Natiella J R, Natiella R R, and Carter J M 1984 J. Biomed. Mater. Res. 18 337.

Bening R C and McCarthy T J 1990 Macromolecules 23 2648.

Campbell J H and Terranova V P 1988 J. Oral. Pathol. 17 309.

Cho D L and Yasuda H J. 1986 Vac. Sci. Technol. A 4(5) 2307.

Dankert J, Hogt A H and Feijen J 1986 CRC Crit. Rev. Biocompat. 2 219.

Desai N P and Hubbell J A 1991 J. Biomed. Mat. Res. 25 829.

Dewez J L, Doren A, Schneider Y J, Legras R, and Rouxhet P G 1991 Interfaces in New Materials, Proceedings of the Workshop ed P. Grange (London: Elsevier).

Dulcey C S et. al. 1991 Science 252 551.

Fodor S P et. al. 1991 Science 251 767.

Gardella J A Jr. and Pireaux J J 1990 Anal. Chem. 62(11) 645A.

Hourston D J, Huson M G, and McCluskey J A 1986 J. Appl. Polym. Sci. 31 709.

Hook D J, Vargo T G, Gardella J A Jr., Litwiler K S and Bright F V 1991 Langmuir 7 142.

Ikada et. al. 1990 J. Appl. Polym. Sci. 41 677.

Iwamoto Y et. al. 1987 Science 238 1132.

Kessel C R and Granick 1991 Langmuir 7 532.

Kleinfeld D, Kahler K H, and Hockberger P E 1988 J. Neurosci. 8(11) 4098.

Massia S P and Hubbell J A 1990 Anal. Biochem. 187 292.

Massia S P and Hubbell J A 1991 J. Biomed. Mat. Res. 25 223.

Nillson K and Mosbach K 1981 Biochem. Biophys. Res. Commun. 102(1) 449.

Patel N M, Dwight D W, Hedrick J L, Webster D C, and McGrath J E 1988 Macromolecules, 21 2689.

Pierschbacher M D and Ruoslahti E 1984 Nature 309 30.

Prime K L and Whitesides G M 1991 Science 252 1164.

Ratner B D, Mateo N B, Ertel S I, and Horbett T A 1990 Proceedings of the 199th Meeting of the American Chemical Society: Division of Polymeric Materials: Science and Engineering 62 250.

Shoichet M S and McCarthy T J 1991 Macromolecules 24(6) 1441.

Snow D M et. al. 1990 Exp. Neurol. 109 111.

Vargo T G, Gardella J A Jr., Meyer A E, and Baier R E 1991a J. Polym. Sci. Polym. Chem. 29 555.

Vargo T G, Thompson P M, Gerenser L J, Valentini R F, Aebischer P, Hook D J, and Gardella J A Jr. 1991b Langmuir in press.

Ward W J and McCarthy T J 1989 Encyclopedia of Polymer Science and Engineering 2nd edition eds H. F. Mark, N. M. Bikales, C. G. Overberger, G. Menges, J. I. Kroschwitz (New York: John Wiley and Sons) suppl vol 674.

Yasuda H Plasma Polymerization 1985 ed H. Yasuda (Orlando: Academic Press, Inc.)

Author Index

Keyword Index